PRAISE FOR *ECOSOCIAL THEORY, EMBODIED TRUTHS, AND THE PEOPLE'S HEALTH*

"This book provides a clear, accessible entry into one of the author's main contributions to public health literature—the ecosocial theory of disease distribution—with helpful examples for the application and importance of this theory for a general audience. Many communities of color will resonate with her explanation of the interdependence of societal and environmental situations on unequal and disparate health outcomes."

—RANDALL AKEE,
Department of Public Policy and American Indian Studies,
University of California, Los Angeles

"A *tour de force*, Nancy Krieger's latest book weaves together decades of her own pioneering work integrating ecosocial theory, empirical research, and transformative policy and politics. This clear-headed *cri de coeur* is guided by Krieger's dazzling intellect, deep historical and contextual understanding, methodological knowhow, and above all is motivated by her lifetime of commitment to social and health justice."

—ANNE-EMANUELLE BIRN,
Global Development Studies,
University of Toronto

"Nancy Krieger's groundbreaking concept of ecosocial theory has influenced a generation of environmental and public health scholars. Her expanded framework on discovering truths explores how diverse points of pollution, social stratification, and poverty intersect through the human body. This is a must-read for anyone seeking to advance environmental and racial justice in the public health field."

—MICHAEL MÉNDEZ,
Department of Urban Planning and Public Policy,
University of California, Irvine, and author of *Climate Change from the Streets: How Conflict and Collaboration Strengthen the Environmental Justice Movement*

"In this landmark book, Nancy Krieger makes a compelling case for not simply working to address health inequities but grounding that work firmly in ecosocial theory and a deep understanding of the 'embodied truths our bodies tell.' A masterpiece, from one of the most important public health scholars of the last half century."

—MEREDITH MINKLER,
School of Public Health, University of California, Berkeley, and co-editor, *Community Organizing and Community Building for Health and Social Equity*

"This book connects all the dots—structural racism, class, power, gender, white supremacist culture, policy, ableism, and more—providing the most elegant and accessible explanation of how they all interact, connect, and shape not only embodied health but our environment and public policy. The stories, the data, and the analysis are deftly on point. This book is an absolute game changer."

—MAKANI THEMBA,
Higher Ground Change Strategies

"Nancy Krieger's conceptual thinking has been pushing the boundaries of epidemiological theory for decades now. This 'small book' will rapidly become essential reading for all those who use epidemiology to tackle the multiple dimensions of inequality affecting our societies."

—CESAR VICTORA,
International Center for Equity in Health,
Universidade Federal de Pelotas, Brazil

"Building on decades of research, Nancy Krieger's eloquent writing takes us on a journey through history, science, and sociology to peel back surface explanations and reveal what truly shapes our health. This exposé of how our bodies reflect the embodied truths of society should be required reading for anyone seeking to understand health disparities."

—STEVEN WOOLF,
Center on Society and Health,
Virginia Commonwealth University

Small Books, Big Ideas in Population Health
Nancy Krieger, Series Editor

1. J. Beckfield. *Political Sociology and the People's Health*
2. S. Friel. *Climate Change and the People's Health*
3. J. Breilh. *Critical Epidemiology and the People's Health*
4. N. Krieger. *Ecosocial Theory, Embodied Truths, and the People's Health*

ECOSOCIAL THEORY, EMBODIED TRUTHS, AND THE PEOPLE'S HEALTH

Nancy Krieger
PROFESSOR OF SOCIAL EPIDEMIOLOGY AND AMERICAN CANCER SOCIETY CLINICAL RESEARCH PROFESSOR AT THE HARVARD T.H. CHAN SCHOOL OF PUBLIC HEALTH

Oxford University Press is a department of the University of Oxford. It furthers the University's objective of excellence in research, scholarship, and education by publishing worldwide. Oxford is a registered trade mark of Oxford University Press in the UK and certain other countries.

Published in the United States of America by Oxford University Press
198 Madison Avenue, New York, NY 10016, United States of America.

Library of Congress Cataloging-in-Publication Data
Names: Krieger, Nancy, author.
Title: Ecosocial theory, embodied truths, and the people's health / by Nancy Krieger.
Other titles: Small books with big ideas ; 4.
Description: New York, NY : Oxford University Press, [2021] | Series: Small books, big ideas in population health ; 4 | Includes bibliographical references and index.
Identifiers: LCCN 2021024150 (print) | LCCN 2021024151 (ebook) |
ISBN 9780197510728 (hardback) | ISBN 9780197510742 (epub) |
ISBN 9780197510759 (ebook)
Subjects: MESH: Health Status Disparities | Socioeconomic Factors | Social Justice | Health Equity | Social Medicine | Epidemiologic Methods
Classification: LCC RA563.M56 (print) | LCC RA563.M56 (ebook) |
NLM WA 300.1 | DDC 362.1089—dc23
LC record available at https://lccn.loc.gov/2021024150
LC ebook record available at https://lccn.loc.gov/2021024151

DOI: 10.1093/oso/9780197510728.001.0001

9 8 7 6 5 4 3 2

Printed by Integrated Books International, United States of America

I dedicate this book to my parents, Dr. Dorothy T. Krieger (1927–1985) and Dr. Howard P. Krieger (1918–1992), who taught me to value knowledge for the good we can do with it in the world, and to Mrs. Montez Davis (1913–1997), who helped raise me and further opened my eyes to injustice and to living a loving life.

EPIGRAPHS

. . . the crucial distinction for me is not the difference between fact and fiction, but the distinction between fact and truth. Because facts can exist without human intelligence, but truth cannot.

TONI MORRISON (2019)[1]

. . . true stories are worth telling, and worth getting right, and we have to behave honestly towards them and to the process of doing science in the first place. It's only through honesty and courage that science can work at all. . . . The more we discover, the more wondrous the universe seems to be, and if we are here to observe it and wonder at it, then we are very much part of what it is. . . . The story continues, and the rest is up to us.

PHILIP PULLMAN (2017)[2]

Discourse is not just ideas and language. Discourse is bodily. It's not embodied, as if it were stuck in a body. It's bodily and it's bodying, it's worlding. This is the opposite of post-truth. This is about getting a grip on how strong knowledge claims are not just possible but *necessary*—worth living and dying for.

DONNA HARAWAY (2019)[3]

CONTENTS

PREFACE

Health, illness, birth, and death: they comprise the embodied truths of existence on our planet Earth for every single biological being. It is an elementary truth that to live is to live embodied. People and all other living beings are constantly engaging with—and depending upon and shaping—the social, biophysical, and ecological contexts in which life transpires. This translates to literally incorporating, bringing into the corps, the body—that is, embodying biologically—the dynamic contexts in which our lives are enmeshed. Patterns of population health—including, in the case of people, health inequities—constitute the living record of how each population and our planet are faring.

Yet elementary truths are never as simple as they appear to be. This book aims to provide a systematic rendering of the ideas and causal claims entwined with the notion of *embodying (in)justice* and its implications for public health and social justice in its many interlinked forms—including but not limited to racial justice, economic justice, reproductive justice, environmental justice, climate justice, Indigenous justice, queer justice, disability justice, and more. The basis of this construct is the ecosocial theory of disease distribution, which I explain in Chapter 1, and which I first articulated in 1994 and have been elaborating since.[1] Theories and their causal constructs matter because they can be a source of power, for good and for ill, especially when deployed and contested in systems of governance, economics, and politics that set the terms by which people and this planet can either thrive or be treated as entities to be exploited for the private gains of a few. Using concrete examples to illustrate critical concepts, the goal of this book is to use the *ecosocial theory of disease distribution* to promote clear thinking about the distinct but connected realities of *embodying (in)justice* and *embodied truths*. The intent is to inform critical and practical research, actions, and alliances to advance health equity in a deeply troubled world on a threatened planet.

Throughout, the focus on embodying (in)justice and health equity is central—albeit with no claims that health is the sole or most important consideration, since there are so many facets and features of social justice that warrant deep analysis and concerned and concerted action. To me, however, concerns about health are compelling, complex, and multifaceted. So too is the critical work of critical science, done by real people in real societies, in ways that can contribute publicly testable and tested ideas and evidence. I offer this brief book to share insights I have gained through my 35+ years of professional work as a social epidemiologist and as an advocate and activist linking issues of social justice and public health. The intent is to provide an invitation and opportunity to reflect on ideas that can lead to deeper, more rigorous, and actionable analysis, not a comprehensive review of the literature.

This book is one of several volumes for a series I initiated with Oxford University Press, on *Small Books, Big Ideas in Population Health*.[2] I conceived of this series because in my view the practice and science of the intermingled fields of public health and population health sciences could benefit from strengthening the critical conceptual tools of our trade—that is, the ideas we use to guide our research and practice in the real world. Reaching out beyond my own particular expertise, I wanted to bring a sharp focus to a diversity of key debates and insights in public health and kindred disciplines that could help hone these tools for thinking—and I recognized that this required a format longer than the standard brief scientific article and shorter than a full-fledged tome. Hence small books with big ideas!

The first two books in this series engage with big ideas under the rubrics of (1) *Political Sociology and the People's Health*, by Jason Beckfield, published in 2018,[3] and (2) *Climate Change and the People's Health*, by Sharon Friel, published in 2019.[4] The former cogently explicates how to understand, research, and reveal the "rules of the game" that structure population health and health inequities; the latter incisively offers one of the first in-depth analyses of links between the climate crisis and health inequities, their common roots in consumptogenic systems, and possibilities for progressive policy systems change. The third book, published in January 2021, is by Jaime Breilh; titled *Critical Epidemiology & the People's Health*,[5] it deftly presents critical Latin American perspectives that play an influential role in the epidemiology and public health (and collective health) of the Global South but that are less familiar in the Global North. Two other books in preparation are on epigenetics and the people's health, and on causal inference and the people's health, and others are in discussion, including on Indigenous well-being, settler-colonialism, and the people's health, with more to come. I am grateful to my colleagues who have stepped forward to be included in this series.

My own writing of this book comes at a time of heightened awareness of the urgent need for critical analyses of the structural drivers of population health and health inequities—and the power of people to change conditions for the better, including for the people's health. I began working on the text in mid-January 2020, when only whispers of a possible new infectious disease with pandemic potential began circulating into global awareness. I continued writing until mid-March and then had to halt, when the demands of dealing with COVID-19 and health inequities in the United States, on top of all the other COVID-19 disruptions to work and life, became my priority. Added to this was the horrific public murder of George Floyd on May 25, 2020, by the Minneapolis police, one that, building on far too many before, sparked subsequent weeks and months of mass protest throughout the United States and globally—against not only police violence but also structural racism more broadly.[6] Addressing racial injustice and health, including in relation to other forms of social injustice, has been central to my work as a social epidemiologist since the start of my professional career—and indeed part of why I entered this field.

In March 2020, I thus put aside work on this book, given the urgent need to put my epidemiologic skills swiftly to use to the critical work at hand, regarding the data, debate, and action around COVID-19, police violence, and structural racism more generally.[7] I was able to revisit the writing in brief moments during the summer of 2020, then put it aside again for much of the fall, given the demands of school, the pandemic, and the presidential and other elections.[8] Nevertheless, in late December, as contest over the elections began to recede and vaccines started to become available, even as COVID-19 in the United States continued to spin out of control, I could return to my writing. I completed a first draft just days before the anti-democratic violent assault on the US Capitol on January 6, 2020, and in February, with comments from colleagues in hand, revised while the impeachment trial was underway.[9] This is writing in the real world, in context.

Living through 2020 and the myriad efforts to mobilize for health justice served only to strengthen my resolve to prepare this book. My intent is to provide clarity about the nature of the embodied truths revealed by population patterns of health—so as to hold accountable the systems, institutions, and individuals that promote the degradation and plunder of people and this planet for the benefit of a few. During the past decade, there has been a global rise in reactionary attacks on the ideas, policies, scientific evidence, and movements needed for people and life on this planet to thrive equitably.[10] The urgency of repelling and exposing these repellent attacks is more important than ever.

Yet when it comes to the people's health, we need not only critical political and economic analysis but also a deep engagement with biology, in societal, ecological, and historical context. Such knowledge is necessary on its own terms, for effective action, and also to counter dominant narratives that continue to attribute primary causal agency to people's allegedly innate biology (aka genomes) and their allegedly individual (and decontextualized) health behaviors.

Hence this book: to illuminate critical *ecosocial theorizing* about *embodying (in)justice* and *embodied truths*, so as to strengthen work for the people's health. To its pages, I bring my own sense of living in history, both to explicate how I have developed my ideas and why, and, beyond this, to look forward: to who and what come next. May the arguments presented in this book better equip you, the reader, to take on the work of your generation in context—whatever it is, and whenever and wherever you may be. The collective challenge is for we who value health equity and human rights to contribute our part to the global project of ensuring the terms whereby all can thrive, alive to the sensuous possibilities of living engaged, generative, and loving lives on this planet, for generations to come.

ACKNOWLEDGMENTS

I would like first to thank my editors at Oxford University Press who have worked closely with me on this book series: first Chad Zimmerman, and presently Sarah Humphreville. I would also like to thank Lisa Dorothy Moore, George Davey Smith, and Jason Beckfield, as well as Sarah Humphreville, for taking the time to read an early draft and offer their helpful comments (and suggested references!). I am grateful to staff throughout the Harvard library system, who have responded to my many requests for books and articles over the years—and then, as COVID closed the libraries, continued to assist by scanning materials I sought. I would also like to thank Pam Waterman and Jarvis Chen for their many years of creative collaboration, helping translate my ways of using ecosocial theory into real-world research. Support for my work on this project derives from my 2019 sabbatical funding (Harvard T.H. Chan School of Public Health, to develop the book prospectus) and from my American Cancer Society Clinical Research Professor Award (2015–2020) and its renewal (2020–2025).

I also offer my thanks for support in the many ways it showed up from my brother Jim Krieger and from the many others in my family of choice (you know who you are!). I close with thanks to Emma (1981–1996), Samudra (1996–2014) and Bhu (1996–2010), and Amber (b. 2014) and Sky (b. 2014), for reasons only cats can know.

1 FROM EMBODYING INJUSTICE TO EMBODYING EQUITY

EMBODIED TRUTHS AND THE ECOSOCIAL THEORY OF DISEASE DISTRIBUTION

Is it a mystery that people subjected to economic deprivation, discrimination, and hazardous working and living conditions, compounded by histories of enslavement and colonization, typically have worse health, have worse health care, and die younger than people with economic, social, and legal privileges?[1] It shouldn't be. Observations about associations between societal power, position, and health status, that is, the societal patterning of population health, appear in the earliest known medical writings, dating back several millennia—for example, in texts from the ancient Egyptian, Greek, Indian, and Chinese civilizations, to name a few.[2] Systematic documentation of such associations was also central to many of the founding reports, in the mid-19th century CE, of the field of public health in Europe and the Americas.[3]

However, it is one thing to observe an association. It is another to explain it. This is why theory, causal assumptions, and frameworks are key, not just the observable "facts."

Ecosocial Theory, Embodied Truths, and the People's Health. Nancy Krieger, Oxford University Press.
 DOI: 10.1093/oso/9780197510728.003.0001

Ecosocial Theory, Embodied Truths, and Health Justice: Parsing Population Health Patterns

The crux of the argument, as conventionally posed, is who bears responsibility for the observed social patterning of population health: individuals or their societal context, past and present?[4]

The stakes in this debate are high since they concern accountability—and these debates repeatedly founder on the ubiquitous individual/population divide that permeates both individualistic and social analyses of health. The standard poles of the argument are as follows:

- If the fault for poor health, or the privilege and praise for good health, lies within individuals, their innate biology, and the social groups with whom they individually and independently choose to affiliate, then, per the dominant status quo framing, social group differences in health are simply a reflection of innate biology, values, and choices.[5] Health, from this standpoint, is an individual resource.
- If, however, responsibility lies within societal systems in which some groups have power at another group's expense, in ways that affect options for living a healthy life, then the social group differences in health constitute health inequities, that is, differences in health status that are unfair, avoidable, and in principle preventable.[6] Health, from this standpoint, is socially contingent, and improving health equity becomes a collective resource.

The rub is that the computation of population rates—be they of birth, health, illness, or death—requires counting individuals in both in the numerator (i.e., the "cases") and denominator (i.e., the population in which the "cases" emerge). Does this mean that population health simply reflects aggregated individual health status? No. But it takes a new way of theorizing health—as *emergent embodied phenotypes*—to understand, explain, and act to change the *embodied truths* of population health.

This book enters these debates with three premises:

- First, the familiar framing of individual versus society is dangerously wrong, especially in relation to health. On our planet Earth, no individual (of any species) ever lives—or ails or dies—separate from this world. Rather, we inhabit a planet in which every living being necessarily (1) is simultaneously an individual and part of a population shaped by its history, and (2) engages dynamically with members of their own and other species in their broader ecological context.[7]
- Second, every living being's body tells stories of its experiences[8]—what I here newly term *embodied truths*—which both reflect and shape its engagement with other organisms and the rest of the biophysical world. Stated another way, all organisms live their phenotype(s), not genotype—and this phenotype is not fixed.[9] What we live is our *emergent embodied phenotype*,[10] one that emerges through engagement with the dynamic social and biophysical features of the dynamic changing world we inhabit and alter. A corollary is that the *embodied truths of individuals' lives are inseparable from the embodied*

truths revealed by analysis of distribution and causes of population rates of health, disease, and well-being.

- Third, the reason to analyze health inequities is not to prove that injustice is wrong, since injustice is wrong by definition.[11] Rather, the point is to illuminate how both injustice and equity can respectively shape people's health and the health of our planet for bad and for good, so as to guide action and allocation of resources for prevention, redress, accountability, and change.[12]

In this first chapter, I accordingly introduce key concepts and arguments concerning embodiment and people's health, as grounded in the ecosocial theory of disease distribution.[13] In Chapter 2, I provide a range of supporting empirical examples. In Chapter 3, I step back and consider the critical challenges and contributions the *ecosocial* constructs of *embodiment*, *embodying (in)justice*, *emergent embodied phenotypes*, and *embodied truths* can offer for sparking new questions and producing new scientific knowledge that can help advance health justice in its myriad forms.

Debating "Individual" Versus "Social" Causes of Health and "Gene-Environment Interaction": "Déjà Vu All Over Again"[14]

One can be forgiven a deep sense of fatigue when jumping into current controversies over causes and patterns of population health. However, some background and context is necessary. Given the stakes, it should be no surprise that current debates still follow contours of contention—individual versus societal responsibility for population health—traced out over two centuries ago in the founding documents of the field of public health.[15] They likewise echo the worldwide arguments over eugenics in the 1920s–1940s spurred both by US Jim Crow and anti-immigration politics and, related, by Nazi and other fascist regimes.[16] They are once again rehearsed in contemporary clashes over whether racial/ethnic health divides reflect "cultures of poverty" and "Black pathology" versus structural racism.[17]

Nor are these debates unique to public health. Again, not surprisingly, given the stakes, parallel arguments pitting individuals versus society—as causal agents, as units of analysis—are littered across kindred fields, including sociology, anthropology, economics, medicine, medical and health geography, psychology, philosophy, and science and technology studies, to name a few.[18] The ubiquity and persistence of these debates, endlessly updated with the latest evidence afforded by whatever the newest technology permits, attests to relationships between the causal frameworks public health scientists and other scholars use—and contest—and the political and societal systems and issues at stake.[19]

In the case of public health, what specifically is at issue is whether, as noted earlier, social group differences in health status are (1) "natural" and fair, versus (2) societal in origin and unfair. Framed in terms of "bodies," the core causal question is whether causal agency and explanations for population health patterns

1. reside in individuals, by virtue of their innate biology—aka the "body natural"—and their individually chosen or possibly genetically determined behaviors, values, and social group or cultural affiliations,[20] versus
2. reside in the "body politic"—aka the priorities, policies, and practices of the political and economic systems governing the conditions in which individuals live.[21]

In the first case, population patterns of health simply arise from the aggregation of individuals, and the corresponding interventions, whether biomedical or behavioral, are focused on individuals. In the second case, population patterns of health reflect societally structured ways of living, thus requiring interventions focused on equitable societal changes to enable healthier living. While both accounts can (and should) recognize that inherently stochastic random events can affect both individual risk and population rates of disease, they differ in whether they frame these chance occurrences (for good or for bad) as being a matter of private individual luck versus socially structured chance.[22]

Of course, the posing of an "either/or" argument is stark—and can be viewed as a simplistic polemic.[23] The past half-century's conventional "solution" has been to proffer "gene × environment interaction" (GEI) as an alternative.[24] But this "solution" remains vexed by problems it cannot solve. First, contention continues over what and who counts as "the environment"—since entities comprising "not genes" can variously extend anywhere from non-DNA molecules within cells to macroeconomic systems.[25] Second, in the case of living beings, "genes" don't interact with environments: organisms do.[26] While analysis of literally disembodied genes (as well as genes inserted from one type of organism into another) can be designed and executed in laboratories, that is not the same as the lived experience of genomes becoming expressed as emergent embodied phenotypes.[27] The seeming "concreteness" of seemingly apolitical "genes" versus the "fuzziness" and perhaps more readily politicized "environments," and the greater possibilities for the "manipulation" of the former versus the latter by empirically oriented health scientists, means that "genes" consistently get first seat for funding and causal attention.[28]

Beyond this, GEI founders on the terms of debate set by the first explicit partitioning of "nature versus nurture" as propounded in the late 19th century CE by Sir Francis Galton (1822–1911)—an English Victorian elite investigator who came down squarely on the "nature" side and, related, coined the term "eugenics."[29] Since then, endless debates, in and outside of public health, have vigorously disputed which matters more.[30] One repeated and profoundly erroneous exercise has been to try to apportion the respective causal contributions so that the sum adds up to 100%—for example, 10% genetic, 90% environmental, or 70% genetic, 30% environmental.[31] These exercises, however, profoundly and wrongly ignore what interaction entails. Specifically: interaction—whether between "genes" and "the environment," or between multiple genes, or between different components of "the environment"—by definition means the sums must add up to more than 100%.[32]

It is not a new insight that taking interaction seriously requires understanding that "nature" and "nurture" cannot be neatly partitioned. In the 1930s, Lancelot Hogben (1895–1975), a prominent medical statistician, experimental zoologist, and population geneticist, first formally introduced the fundamental concept of the "interdependence of nature and nurture."[33] He demonstrated mathematically, and with real-world data, that the very question of "which matters more" is at its core fallacious. If, say, a plant and its clones on average grow only 3 inches tall in soil type A, but its numerous clones on average grow 6 inches tall in soil type B, then there is not one answer to how tall a plant will grow, given its genome, because it depends on context. By implication, if two independent causes contribute to an outcome, and their interaction also contributes, then the causal contribution of the two causes necessarily adds up to more than 100%.[34] This interdependence of nature and nurture, moreover, is built into the very essence of life on Earth, because an organism's gene expression and phenotypic development across the lifecourse necessarily depends on its dynamic interactions with the complex changing biophysical and social world in which it originates (by asexual or sexual reproduction), lives, and dies.[35]

Further complicating the picture, new evidence indicates that an individual organism's biological development (e.g., from zygote to adult) can require both "external cues" and the literal incorporation of other beings (e.g., the microbiome), leading new scholarship to posit that the conventional human construct of the "inside/outside" divide—of individual bodies versus context—may be deeply artificial.[36] The larger point is that simply saying both "genes" and "environment"—or "individuals" and "society"—matter for population health affords little clarity for critically analyzing causes of observed population patterns of health within and across diverse societies, over time.

Embodying (In)justice and Embodied Truths in Context: Critical Ideas in Contentious Times

Still another set of arguments point to why critical engagement with debates over "individual" versus "societal" causes of population health patterns is necessary—and why grappling with embodying (in)justice may be helpful. All involve intensifying conflicts regarding "truth" and science.[37]

I write this book at a time when "dark money" and overt political donations are increasingly funding efforts to seed doubt and deception about scientific findings—especially regarding climate change and COVID-19—that inconveniently challenge current structures of power, both secular and religious.[38] Monetary support and political clout are provided by ultra-rich families and corporations who believe the sole purpose of government is to protect their private property, especially those whose wealth depends on the fossil fuel and petrochemical industries, along with those invested in Big Food and Big Pharma, tobacco, alcohol, weapons, and financial speculation.[39] Meanwhile, religious fundamentalists in the major denominations worldwide—especially Judeo-Christian and Islamic—are ramping up their fight against what they refer to as "gender ideology," with their polemics dismissing if not denouncing any scientific evidence that supports laws and policies that promote gender equality, reproductive rights, or civil and political rights for lesbian, gay, bisexual, and transgender people.[40] Within scientific communities, yet another set of longstanding debates are again heating up over the nature of scientific objectivity and whether or not ideological beliefs can bias who becomes scientists and the science they produce, especially in relation to race/ethnicity and gender.[41]

Tellingly, despite their distinctions, all three sets of controversies strikingly revolve around what I would term *embodied truths*—that is, evidence regarding the people's health.

Why? Because when the impacts of ideas, policies, and laws become measured by metrics of health, especially human health, the evidence crosses over from being a matter of opinion to a matter of life and death. Such evidence has standing not only in the court of public opinion but also in courts of law (at least in countries not subjected to authoritarian and/or corrupt rule).[42] It is bodily evidence that links the "body natural" to the "body politic" and illuminates the impacts of governments' priorities on both the people's health and planetary health.

Even more bluntly: the evidence afforded by "embodied truths" raises the stakes. Once scientific evidence exists to show that the actions of some are harming the health of others—whether in relation to pollution, climate change, commercial products, infectious disease, second-hand tobacco smoke, or social and economic policies—the grounds shift vis-a-vis issues of liability, prevention, and even reparations.[43] The terrain of debate and action likewise shifts once evidence exists to show how human action can improve population health, planetary health, and health equity.[44]

The fundamental importance of "embodied truths" explains why it is no accident that the emergence of the professional field of public health in the mid-1800s was heralded by a series of pathbreaking landmark governmental reports—and nongovernmental critiques—regarding the health of the population in relation to social and economic conditions. The trigger was the rapid rise in England and other European countries of a new coal- and steam-powered industrial factory system joined with intensified land enclosures and imperial expansions of global commerce and investments.[45] Together these created two enmeshed and co-defining groups: a dominating new political class of self-proclaimed capitalists, coupled with vast new precariously employed and housed sickly urban populations whose terrible health and political unrest posed a challenge to governance, commerce, and investments.[46] Responding to these conditions, the iconic

1842 *Report on the Sanitary Condition of the Labouring Population in Great Britain*, presented by Edwin Chadwick (1800–1890) to both Houses of Parliament, "by Command of Her Majesty,"[47] was the first massive government report of its kind and set the basis for the world's first modern public health laws, agencies, and action. Yet, whereas the official government reports focused principally on the need for better sanitation and better morals, numerous influential nongovernmental exposés laid bare the class politics responsible for these "embodied truths," including Flora Tristan's 1842 *Promenades dans Londres: L'Aristocracie et les Prolétaires Anglais* and Friedrich Engel's 1845 *The Condition of the Working Class in England*.[48]

Such "embodied truths" played a similar role in the United States, albeit in relation to not only class but also racialized patterns of health. In 1845, in a critical work akin to Engel's, John Griscom published *The Sanitary Condition of the Laboring Population of New York with Suggestions for Its Improvement*.[49] In 1847, the American Statistical Association, founded in Boston, Massachusetts, in 1839, issued its first publication, which notably included critical data on the health of American Indians and "Negroes," both enslaved and free, and concluded with a petition to the state of Massachusetts asking for a report on the health of the population.[50] In response, in 1850 the landmark state-commissioned review, the *Report of the Sanitary Commission of Massachusetts*, was published; its authors, led by Lemuel Shattuck (1793–1859), one of the founders of the American Statistical Association, had been "appointed under a resolve of the legislature of Massachusetts."[51] Like the Chadwick report that inspired it, the Shattuck Report shaped the formation of both US state and national public health agencies and laws.[52] Moreover, challenging government policies, during the 1850s and 1860s the first wave of credentialed US Black physicians—including Dr. James McCune Smith (1813–1865), Dr. John S. Rock (1825–1866), and Dr. Rebecca L. Crumpler (1831–1895)—published powerful critiques, informed by health data, to challenge both slavery and dominant ideologies of scientific racism, white supremacy, and racial inferiority.[53] The history of anti-racist science is long, having long been in contestation with racist science.[54]

Jumping to the late 20th and early 21st century CE, the stakes embroiled in "embodied truths" remain high. In 1977, reflecting growing concerns that universal health care, as epitomized by the United Kingdom's National Health Service (NHS), could not alone secure everyone the right to health, the UK Labor Government commissioned the 1980 *Black Report*, named after its chairman, Sir Douglas Black (1913–2002), and coauthored with Peter Townsend, Jerry Morris, and Cyril Smith.[55] The report's documentation and analysis of profound class gradients in health across the lifecourse sparked new rounds of government-issued reports on social class inequities in health worldwide, setting a model followed by myriad global, state, and local health agencies to this day. Globally, the 1978 Declaration of Alma Ata, cosponsored by the World Health Organization (WHO) and the United Nations Children's Fund (UNICEF), declared: "The existing gross inequality in the health status of the people particularly between developed and developing countries as well as within countries is politically, socially and economically unacceptable and is, therefore, of common concern to all countries."[56] Together, these influential documents posited that people's health status reflected and revealed political and economic priorities and conditions, and prescribed societal changes needed to promote equity and well-being.

Their arguments, however, were not heeded—and, rather, were resolutely rejected—by the post-1980 global neoliberal economic policies that imposed austerity budgets, deregulation (including rollbacks of public health regulations), and defunding of the public sector (including public health agencies), while enabling ever greater private concentrations of wealth.[57] Thirty years later, as global evidence of growing health inequities began to mount, the World Health Organization (WHO) issued its landmark 2008 report *Closing the Gap in a Generation: Health Equity Through Action on the Social Determinants of Health*, concerned chiefly with socioeconomic health inequities.[58] In the United Kingdom, the subsequent UK 2010 Marmot Review on *Fair Society, Healthy Lives* and the 2020 update *Health Equity in England: The Marmot Review 10 Years On* documented the worsening inequities resulting from governmental disregard of the evidence.[59]

Subsequently, in 2019, the Commission of the Pan American Health Organization (PAHO) on Equity and Health Inequalities in the Americas published its report *Sociedades Justas: Equidad en la Salud y Viva Digna/Just Societies: Health Equity and Dignified Lives.*[60] Akin to the 2008 WHO document, the PAHO report discussed socioeconomic health inequities.[61] Beyond this, it newly addressed the continued impacts of histories of settler-colonialism and enslavement on current Indigenous and Black health inequities, and likewise newly discussed the impacts of climate change and climate injustice.[62] Related radical civil society critiques bringing together these issues further paralleled and expanded on the official reports, as illustrated by the *Global Health Watch* series, whose reports have appeared in 2005, 2008, 2011, 2014, and 2017.[63]

As this brief history suggests, the "embodied truths" about the state of people's health thus comprise a unique currency for contesting harms and proposing salutary alternatives. Government and societal responses to these embodied truths—whether their dismissal and denial or embrace and use to inform action—speak deeply about the state of power relations, constituting one variant of "speaking truth to power." Stated another way, what I refer to as "the stories that bodies tell"[64] can reveal powerful truths about the connections between "the body natural" and "the body politic."

The catch is that these "embodied truths" are not self-evident and instead are always shaped and infused by theory. They are not simply "facts" that can be "read off" an individual's literal body or a body of statistical data. Partly tempering claims of truth is the recognition that scientific knowledge is dynamic and reflexive: inevitably, as scientists generate, test, refine, and reject hypotheses, using new and different technologies and analytic methods, both theoretical understanding of disease processes and systems of classifying disease can change, as can nosologies of death.[65] Beyond this, it deeply matters who is telling—and testing—the stories that bodies tell, and doing so from what vantage, about whose bodies, in what historical, societal, and ecological context. It equally matters whose bodies are invisible or ignored.

Calling attention to these assumptions and omissions is core to ensuring scientific rigor: it is about doing correct science, not politically correct science.[66] After all, a cardinal principle of scientific knowledge—and the basis for claims of valid and reliable evidence and explanations—is that it depends on the public testing of public ideas and data by independent scientists, whose methods, ideas, and data are open for public scrutiny.[67]

Here an aphorism from Donna Haraway (1944–), a renowned biologist and philosopher of science,[68] is useful: science and its evidence "are *made* but not *made up*."[69] Reflecting on contemporary controversies over scientific evidence, which span from outright rejection of science to critical questioning of bias in science, in 2019 she observed that for her and kindred critical scholars who have critiqued simplistic stances about science inherently being "objective" and bias-free:[70]

> Our view was *never* that truth is just a question of which perspective you see it from. . . . The idea that reality is a question of belief is a barely secularized legacy of the religious wars. In fact, reality is a matter of worlding and inhabiting. It is a matter of testing the holdingness of things. Do things *hold* or not?
>
> Take evolution. The notion that you would or would not "believe" in evolution already gives away the game. If you say, "Of course I believe in evolution," you have lost, because you have entered the semiotics of representationalism—and post-truth, frankly. You have entered an arena where these are all just matters of internal conviction and have nothing to do with the world. You have left the domain of worlding.

It is in this very material world, one existing long before humans evolved but now profoundly shaped by human actions informed by human ideas, for good and for bad, that the realities of embodied truths play out.

On this note, it is now time to review briefly the key features of the *ecosocial theory of disease distribution*. As will become evident, the interlinked ecosocial constructs of *embodiment*, *embodying (in)justice*, and *embodied truths* animate its core.

Ecosocial Theory of Disease Distribution: Situating Embodiment and Embodying (In)justice

In 1994, I introduced the *ecosocial theory of disease distribution*.[71] Its purpose is to explain societal distributions of disease and health and thereby generate knowledge that can inform action to improve both population health and health equity. From the start, and in its subsequent elaborations,[72] this theory has emphasized the importance of engaging with the *multilevel spatiotemporal processes of embodying (in)justice*, across the lifecourse and historical generations, as shaped by the political economy and political ecology of the societies in which people live.

The invitation is to start with our real and material world—and to ask how living beings and the populations of which they are a part reciprocally and dynamically engage with, incorporate, and shape this world, by virtue of the capacities that their biology affords.[73] In the case of people—who are simultaneously, not concurrently, quirky individuals, kin, and members of the population into which they were born and the societies in which they live and engage[74]—these lived realities of embodiment are shaped by interlinked societal and ecological systems, by individual and collective agency, and by structured chance.[75] Embodiment is thus simultaneously a lived reality and a tool for thinking—and one that can challenge dominant narratives of disembodied genes and decontextualized behaviors.[76] As I will argue here, no other public health construct does so much, so comprehensively, so concisely. It is why the ecosocial construct of *embodiment*—and the ideas of *embodying (in)justice* and *embodied truths*—together have the power they do, to inspire hypotheses, explanations, understandings, and action for the people's health.

Because I have written extensively about the ecosocial theory of disease distribution (or ecosocial theory, for short) in other publications,[77] here I offer a concise recap of the theory's key features, as shown in Figure 1.1—and then focus primarily on why grappling with embodiment in the ways the theory proposes is critical.

Ecosocial Theory of Disease Distribution: Why Every Word in the Name Matters

In brief, the *ecosocial theory of disease distribution* takes its name seriously.[78] Thus:

- *Ecosocial* (a term not used in public health until I introduced it in 1994, nor in any other literature with any elaboration) deliberately conveys, in one word (with no hyphen!), the fundamental interdependence of societal and ecological contexts, whereby societal systems and interactions necessarily depend on, shape, and are shaped by ecological systems—and vice versa. Because this theory is focused on population health and

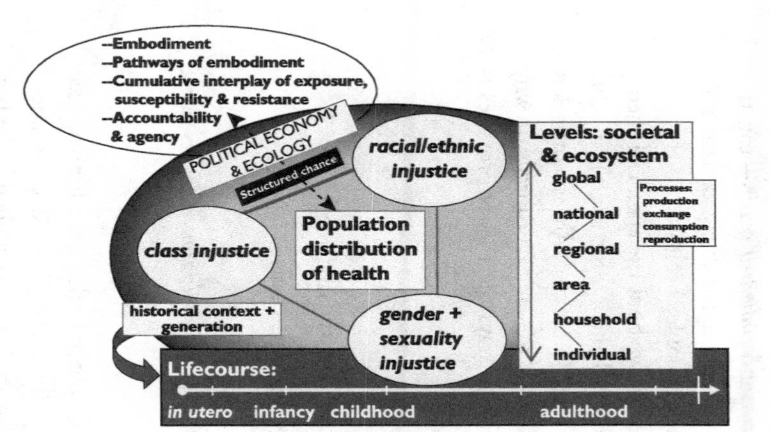

Figure 1.1 Ecosocial theory of disease distribution: levels, pathways, and power.

Sources: Krieger N. Epidemiology and the web of causation: has anyone seen the spider? *Soc Sci Med* 1994;39(7):887–903. doi: 10.1016/0277-9536(94)90202-x; Krieger N. Epidemiology and social sciences: towards a critical reengagement in the 21st century. *Epidemiol Rev* 2000; 22(1):155–163. doi: 10.1093/oxfordjournals.epirev.a018014; Krieger N. Theories for social epidemiology in the 21st century: an ecosocial perspective. *Int J Epidemiol* 2001; 30(4):668–677. doi: 10.1093/ije/30.4.668; Krieger N. A glossary for social epidemiology. *J Epidemiol Community Health* 2001; 55(10):693–700. doi: 10.1136/jech.55.10.693; Krieger N (ed). *Embodying Inequality: Epidemiologic Perspectives.* Amityville, NY: Baywood Publishing Co., 2004; Krieger N. Embodiment: a conceptual glossary for epidemiology. *J Epidemiol Community Health* 2005;59(5):350–355. doi: 10.1136/jech.2004.024562; Krieger N. Proximal, distal, and the politics of causation: what's level got to do with it? *Am J Public Health* 2008; 98(2):221–230. doi: 10.2105/AJPH.2007.111278. Epub 2008 Jan 2; Krieger N. *Epidemiology and The People's Health: Theory and Context.* New York: Oxford University Press, 2011; Krieger N. Measures of racism, sexism, heterosexism, and gender binarism for health equity research: from structural injustice to embodied harm-an ecosocial analysis. *Annu Rev Public Health* 2020;41:37–62. doi: 10.1146/annurev-publhealth-040119-094017. Epub 2019 Nov 25.

health inequities, and because societal systems and issues of injustice and equity are central to evaluating if differences in health status across social groups are unjust, avoidable, and in principle preventable, the "eco" modifies "social" (and not the other way around). Hence:

- ***Eco*** is meant literally, not metaphorically, and refers to the actual ecosystems (i.e., ecologies) that enable life to exist on our planet Earth (and, presumably, any other planet).[79] The construct of "eco" thus encompasses the lives of myriad species in their many evolved, evolving, and endlessly reproducing forms, past, present, and future. At issue are organisms and species, from one generation to the next, and one historical epoch to the next, jointly living and dying in real biophysical places that afford the possibilities for life, and which shape and are shaped by each organism's and each species' interactions with the living and abiotic world around them. The "eco" in ecosocial is not restricted to humans and instead encompasses complex cross-species and cross-level dynamic ecological systems that evolve. That said, ever since *Homo sapiens* evolved some 200,000+ years ago, as one new species joining 4.5 billion+ years of life on Earth, people have shaped and been shaped by their local ecosystems, with rising impacts on regional and increasingly planetary scales, especially since the 15th-century CE rise of global colonialism and commerce on a large scale.[80]
- ***Social*** in turn refers to the sociality of species life, involving the actions and interactions of living beings, within and across species, that affect the terms by which they and others live, reproduce, and die.[81] In the case of people, "social"

additionally encompasses "society."[82] It thus includes both forms of governance and the ideas people generate—and act on (with purpose in mind, and sometimes mindful of possible unintended consequences)—to explain and variously to structure, celebrate, honor, control, denigrate, or challenge their society's formal and informal rules plus impacts on the ecosystems of which they are a part.[83]

- ***Theory*** derives from the Greek word *theoria*, whose original meaning of "looking at," in relation to theater and spectacle, transmuted to mental schemes, and thus inner vision.[84] In the case of science, theory refers, as I have noted previously, to a "coherent and presumptively testable set of inter-related ideas that enable independent scientists to discover, describe, explain, and predict features of a commonly shared biophysical reality in which cause-and-effect exists."[85] Like all thinking, such theories inevitably rely on metaphor, to enable the "unknown" to be conceptualized in relation to the "known"—which is vital, given that our world, indeed universe, is more wonderful, unruly, and stranger than anything we can imagine.[86] Scientific theories are thus more than models; they encompass and spark hypotheses and frame the collection, analysis, and interpretation of data—and they are premised on science being a public way of knowing, involving the public testing of public knowledge.[87]
- ***Disease*** is a term I explicitly meant broadly, to refer not just to specific ailments, but as shorthand to encompass a diversity of somatic and mental phenomena that render an organism ill, disabled, or unhealthy; unable to partake in usual daily activities; and ultimately or immediately dead (whether via sickness, an injury, or an assault).[88] The contrast is thus to being alive, healthy, and in a state of well-being, which the theory also encompasses

(i.e., as health-relevant phenomena with a population distribution). While people have endlessly debated definitions of what constitutes "disease" and "health" (let alone "alive" and "dead"), in both ancient to current texts and oral traditions of diverse societies worldwide,[89] one striking commonality is that life, health, disease, injury, disability, and death are phenomena that manifest within individual organisms. The shorthand of "disease" thus refers to health-related phenomena occurring within individuals, which can then, whether literally or metaphorically, be aggregated to paint a picture of population health. Moreover, while the health status of individuals can depend on interactions between individuals, and also between individuals and their enmeshed societal and ecological systems, and thus be linked to and influenced by population rates of disease, nevertheless the health status experienced occurs within individual bodies.

- ***Distribution*** in turn links individual and population phenomena within a specified time and place. A noun that notably refers to both description and action, *distribution* can describe either (1) the static frequencies or probabilities with which specified characteristics of individual units occurs within a delimited population, and (2) the dynamic processes producing, that is, distributing, this allocation of characteristics.[90] In the case of population sciences, distributions are thus necessarily a multilevel phenomenon, involving individual units and the populations of which they are a part. It follows that description and analysis of distributions at the most abstract level require knowing the bounds of the values (minimum, maximum) and their clustering within these bounds (e.g., normal distribution, bimodal distributions, something else), within the population context in which the distribution occurs.[91]

- In the case of population health, measures of distribution typically are expressed as population rates (cases per total population in a specified place in a specified period of time) or population frequencies (proportions of a population, again in a specified place and time). They also, however, may pertain to frequencies or probabilities of events within an individual (e.g., respiratory rate, or pulse or heart rate, per unit of time).[92]
- No matter what the outcome, for any population distribution, it is critical to know the criteria used to delimit the population in which the distribution occurs, and in the case of population health, this requires characterization in relation to place or institution, social groups, and time.[93] There is, after all, no such thing as one adult human height distribution, even as human adults typically range between 4 feet and 7 feet tall (with, of course, outliers in both directions); the specific distribution depends on population, place, and time, whether 500 BCE or 2021 CE.[94]

Hence the name: the *ecosocial theory of disease distribution*—nothing more, and also, nothing less.

What Ecosocial Theory Is Not

The specifics of what is the *ecosocial theory disease distribution* can also be illuminated by being explicit about ***what it is not***. Thus, the ecosocial theory of disease distribution

- is *not* simply a theory of disease causation—since at issue is not only engaging with the social or biophysical mechanisms involved in the causation (or prevention) of individual cases of disease but also accounting for societal patterns of disease distribution, both present and past;
- is *not* simply a "model"—because it is a theory, and thus (1) presents a coherent and presumptively testable set of interrelated ideas that enable independent scientists to discover, describe, explain, and predict societal patterns of health and disease, and (2) specifies and structures the types of phenomena necessary for developing specific models to analyze particular population distributions and risk;
- is *not* simply a "framework"—because its purpose is not just to "frame" ideas but to generate testable hypotheses, explanations, and predictions about who and what drives population patterns of health and health inequities;
- is *not* simply a "social" theory of disease distribution because it engages with embodiment as a biological phenomenon and engages with the conjoint and temporally dynamic societal, biological, and ecological processes that shape population distributions of disease and health;
- is *not* simply a "social ecological" theory, as per the "social ecological theory" widely used in public health that was developed by Bronfenbrenner, since in this theory "ecology" refers solely to multilevel human phenomena, for example, children nested within families, households, neighborhoods, and schools, with no attention to actual ecosystems or other species;[95] and
- is *not* simply a "biosocial" theory because (1) such a construct ignores ecology, and (2) this term is haunted by its eugenic and

sociobiological past, whereby many scientists and journal titles still use it as shorthand for asserting biological determinism, even as others in diverse fields are trying to recast it to mean socioculturally shaped biological plasticity.[96]

Hence: the ecosocial theory of disease distribution is what its name says—literally—and its intent is to theorize the profound embodied connections that exist between people, politics, ecologies, and health, so as to understand and alter who and what drive population rates of disease and health inequities. It is an integrative theory, not a theory of "everything," that provides the principles and constructs for analyzing population distributions of health and promoting health equity in societal and ecological context—and stands in direct contrast to dominant theories that are individualistic and essentialist to their core.

Why Bother with Developing Ecosocial Theory?

The spark to my developing the ecosocial theory of disease distribution was my frustration, as I received my training in epidemiology in the mid- to late 1980s, with the narrow, individualistic, ahistorical, and highly biomedical bent of my field (I earned my master's degree in 1985 and my PhD in 1989, having previously obtained a BA in biochemistry in 1980). Noting that arguments can be an especially productive spur for critical thinking, I was at a journal club meeting soon after I received my PhD, and the proverbial "web of causation" was once again invoked to explain connections between so-called risk factors. I pointedly asked: who is the spider who has made such pernicious webs with such unjust distributions? This nascent crystallization of my argument with the ideas of my field led to my 1994 essay, "Epidemiology and the Web of Causation: Has Anyone Seen the Spider?,"[97] in which I formally introduced the ecosocial theory of disease distribution.

A key challenge, however, was that if the spiderless web of causation didn't cut it for me, what would? Criticism without suggestion of better alternatives is, after all, insufficient, if not irresponsible. What I came up with—which led me to rich theorizing about the idea of embodiment—was a fractal image that metaphorically and materially united the bush of evolution with the scaffolding of society that different social groups seek daily to reinforce or alter.[98] History, historical contingency, and chance are built into this image, as suggested by the branching shapes, as is deliberate social structure (and its intended and unintended consequences), as suggested by the scaffolding.

As I explained in the text, this image dynamically situates both (1) the bush of evolution within the changing ecologies in which evolution occurs, and (2) the scientists who research population health and the questions they do or do not ask. As I wrote then and would still argue now (albeit with greater recognition of the need to be clear that the social and biologic reflect and shape their ecological context):[99]

> This intertwining ensemble must be understood to exist at every level, sub-cellular to societal, repeating indefinitely, like a fractal object. Different epidemiologic profiles at the

> population-level would accordingly be seen as reflecting the interlinked and diverse patterns of exposure and susceptibility that are brought into play by the dynamic intertwining of these changing forms. It is an image that does ***not*** permit the cleavage of the social from the biologic, and the biologic from the social. It is an image that does not obscure agency. And it is an image that embraces history rather than hides it from view. . . . . this image makes clear that although the biologic may set the basis for the existence of humans and hence our social life, it is this social life that sets the path along which the biologic may flourish—or wilt.

I would further add now that another feature of this image is that it recognizes that causal arrows may fly in multiple directions, across and within levels, from macro to micro, but that does not mean they each do so with the same causal strength.[100] Recognition of the levels being considered—and ignored—is a first step to understanding who and what shape population patterns of health and health inequities.

KEY CONSTRUCTS OF THE ECOSOCIAL THEORY OF DISEASE DISTRIBUTION

As shown in Figure 1.1, the ecosocial theory of disease distribution theorizes population health—that is, *population health exposures, processes, and outcomes*—in relation to *levels, pathways, and power*, together situated in their *societal and ecological context,* driving the *processes and pathways of embodiment,* and especially *embodying (in)justice.* Connecting these levels, both societal and ecological, and also the phenomena occurring within levels, are the *processes of production, exchange, consumption, and reproduction*.

It is not accidental that these latter terms can refer both to biological and societal phenomena. Consider only: producing hormones within the body or producing them in commercial laboratories; exchanging oxygen and carbon dioxide in the lungs or exchanging US dollars for euros; consuming nutrients or consuming products; sexually reproducing or socially reproducing hierarchies of power or daily life in households via cooking, cleaning, and caring. The "eco" in both "ecology" and "economics" can be traced back to the ancient Greek word *oikos*, which refers to "household" generally, and "household management" more specifically.[101] When the biologist Ernst Haeckel (1834–1919) coined the term *ecology* in 1866, he stated his focus was on "the place each organism takes in the household of nature, in the economy of all nature," one in which organisms exist in "infinitely complicated relations" with inorganic and organic conditions of existence, including "among the other organisms its friends and enemies."[102] There is no way to theorize about population distributions of health without attending to *time*, *place*, and *historically structured* relationships, within and across species, and which, for people, includes structurally forged social groups.

From this standpoint, households, whether of humans or other species, are physical places and social spaces, they are units of biological and social reproduction, they require sustenance and produce waste, and in the case of people, their inhabitants and what they do depend on the political, social, economic, and ecologic context in which their households exist. It matters whether and which household inhabitants are—or are descended—from which permutations of people: enslaved versus free; undocumented versus documented; Indigenous versus immigrant versus "native" born; working class versus professional; impoverished versus wealthy. It matters if they are cis- or transwomen versus cis- or transmen; lesbian or gay or bisexual versus heterosexual; genderqueer or non-binary versus cis-gender. It matters if they are young versus old; disabled versus disability-free; alone or living with others. The various inhabitants of these households—and also homeless people, both sheltered and unsheltered—each and every day integrate, within their very bodies, their daily social and biological exposures and experiences, both in and outside their households. To ask theories of disease distribution to be equally integrative about how these realities are reflected in people's health status is thus to ask such theory to engage with the realities of life, health, disease, and death on Earth.

Engaging with the processes that produce population health requires one additional consideration: time. Processes, by definition, take place over time—and three aspects of time are critical. One is time in relation to an organism's *lifecourse*, including biological development and the constant interplay and modification of phenotype by lived experience, from birth to death. A second is the *historical generation* in which this life is lived (e.g., birth cohort). A third is the *etiologic period* (e.g., incubation period for infectious diseases, latency period for noninfectious chronic diseases), that is, the amount of time causally required from initiation of exposure for pathologic processes to result in change(s) in health status.[103] This etiologic period can range from practically instantaneous (e.g., being shot by a bullet to the head and dying) to a couple of weeks (e.g., the time from exposure to the measles virus to developing measles[104]) to several decades (e.g., from asbestos exposure to mesothelioma[105]). The chronicity of exposure also matters: both acute traumatic events and prolonged abuse can, over variable time periods, increase risk of subsequent chronic mental and physical health problems,[106] and acute and chronic high alcohol consumption can respectively lead to acute alcohol intoxication and, over decades, to cirrhosis.[107]

A Concrete Example: Who and What Drive Population Distributions of Lead Poisoning? When and Where?

As one example that underscores the centrality of societal and ecological context, including time, place, and social group, to population distributions of health and epidemiologic theorizing, consider the age-old case of lead poisoning, known as early as 4,000 years ago, as described in ancient Egyptian papyri.[108] The population distributions of the health impacts of acute and sustained lead exposure depend in part on the age at exposure (in utero, infancy, early childhood, adolescence, adulthood) as well as on the sources of exposure: leaded water pipes, leaded paint and paint chips, leaded gasoline, lead dust at work, or lead contamination of cosmetics, beverages, food, and medicine.[109] In the case of water and leaded pipes, they depend as well on what else is in the water (e.g., how "hard" or "soft" it is) and what other chemicals, including industrial pollutants, are present.[110] All of these exposures also depend on the technologies available: leaded gasoline for automobiles, driven by the invention of cars, introduced entirely new ways of widely dispersing lead unrelated to occupational exposures or water pipes.[111] The distributions of exposure and outcomes likewise depend on the state of knowledge about lead poisoning, the existence and enforcement of regulations to prevent lead exposure—and also who benefits from intentional use of lead or from seeking to undercut or ignore these regulations.[112]

Economic, racialized, and gender inequities in population distributions of lead exposure and attendant outcomes have thus varied by time and place, shaped by practice and policies affecting exposures at work, at home, and in the community—of not only people but also pets, other animals, and plants.[113] Within the United States, the distribution of lead poisoning among children has depended on whether they were born before versus after the 20th century CE introductions of leaded gasoline and leaded paint, the fights to ban them, and the passage and enforcement of versus disregard for lead exposure prevention and abatement policies for housing and drinking water.[114] Bringing this home, as it were, is the ongoing recent debacle of the explosive rise of lead contamination of water in Flint, Michigan, a Black-majority city (54%) in which 40% of the population in 2019 was below the poverty line.[115] Lead contamination of its water supply soared in 2014 after state and city officials, in a corrupt cost-cutting measure, switched the water source from Lake Huron to the Flint River without necessary corrosion control treatment, and then denied the severity of the ensuing crisis, leading to extensive litigation, criminal indictments, convictions, and community initiatives to redress the damage.[116]

The example of lead poisoning also illustrates why explanations of disease distribution cannot be reduced solely to biophysical explanations of disease mechanisms, since the latter do not account for why rates and population patterns of disease change, in complex ways, over time and place. As with any causal analysis, questions of "why" and questions of "how" both matter, so as to understand the causal processes that interventions must address.[117] The biophysical mechanisms of lead toxicity (i.e., the "how") are presumably the same as they were millennia ago, when described in medical texts of antiquity in Egypt, Greece, India, China, and Rome.[118] Enormous expansion of the knowledge of the biophysical mechanisms, facilitated by improved technologies, crucially has enabled detection of harms associated with increasingly detectable very low levels of exposure, and has also improved options for therapies such as chelation.[119] But knowledge about the "how" is not the same as knowledge about the "why." To understand and intervene on the distribution of lead poisoning and prevent its harms, it is also essential to ask and investigate: why is exposure occurring, who is it affecting, and at whose cost and whose benefit? And too: why have some efforts to prevent exposure succeeded and others failed, also at whose cost and whose benefit?[120]

As should be clear, societal and ecological context is key. There is not and can never be one answer to the question: what is, and who and what cause, the epidemiology—that is, the actual population health distribution—of the adverse impacts of lead exposure? But there can be systematic approaches to asking these questions, informed by theory, with due attention to who and what is shaping population health distribution and inequities in health—which are for this reason at the very center, conceptually, of the ecosocial theory of disease distribution—as literally depicted in Figure 1.1.

Delineating Ecosocial Theory's Core Constructs: From Embodiment to Agency and Accountability

Figure 1.1 also shows, in the upper left-hand corner, the four core and conjoined conceptual constructs of the ecosocial theory of disease distribution, whose real-world manifestations are always transmuted, via structured chance (as I explain later), through the extant political economy and political ecology of the population in which disease distribution occurs.[121] In Chapter 2, I use concrete examples to illustrate the utility of these constructs for public health research and practice; here, I briefly delineate the core concepts.[122] They are as follows:

1. *Embodiment*: referring to how we humans, and all other biological organisms, literally biologically incorporate, that is, embody, our societal and ecologic context. While this may seem to be an obvious truism, as noted earlier, this approach to conceptualizing embodiment stands in direct opposition to the dominant frameworks of "nature" versus "nurture" and "gene × environment interaction."
2. *Pathways of embodiment*: referring to the concrete and concurrent social and biophysical processes and mechanisms, from macro to micro, involved in organisms embodying their societal and ecological context, thereby producing population distributions of health, disease, and death. Identification of these pathways is guided by ecosocial theory's other core constructs and consideration of both political economy and

political ecology: both the "why" and "how" of causal processes matter.

3. *Cumulative interplay of exposure, susceptibility, and resistance*: in relation to the pathways of embodiment, at each and every level, and in relation to both lifecourse and historical generation. Social movements resisting injustice and organizing for equity, for example, are every much a part of resistance relevant to shaping population health profiles as are individual-level interactions or resilience.
4. *Agency and accountability*: referring both to who and what, at each and every level, are responsible for health inequities and also for the research to explain population health. Key questions are: who benefits from injustice and from research that ignores injustice, and, conversely, who is harmed by injustice and by the overt suppression, lack of funding, and self-censorship that can limit research on these issues?

As with any theory, none of ecosocial theory's core constructs are "stand alone": they are interdependent and should not be used in isolation.[123] The ecosocial constructs of "embodiment" and "embodying (in)justice" can thus not be invoked on their own, as solely phenomena to explain mechanisms in biophysical or psychosocial terms. Instead, it must engage with the theory's other core constructs—taking into account levels, pathways, and power—with the objective of explaining population distributions of disease, within and across societies and places, at a given point in time and over time, and including but not restricted to health inequities.[124]

Embodiment and the processes of embodying (in)justice, as conceptualized in the ecosocial theory of disease distribution, accordingly are multifaceted, multilevel, temporally dynamic constructs. Embodiment is a noun that conveys a process.[125] It interweaves the links between that which is being embodied and the body doing the embodying. It requires agency and action and engagement. It requires vitality. And it requires context—which is provided by its companion construct "embodying (in)justice."

From an ecosocial public health standpoint, embodiment and embodying (in)justice—in societal and ecological context—are what connect, causally and conceptually, exposures and outcomes, and also individual risk and population rates of health. At issue is who and what is responsible for who is embodying what exposures, with what health consequences, along with (but not solely) the biophysical mechanisms by which such embodiment occurs within individuals. Stated plainly: just as bodies daily integrate organisms' experiences and the individual and collective actions they take to engage with and shape their world, the constructs of embodiment and embodying (in)justice afford a robust integrative way of thinking that can bridge the conceptual divides of individual versus population and societal versus biophysical. In Chapter 3, implications for interventions are considered.

Using Ecosocial Theory to Reject Biological Essentialism and Embrace Embodied Integration

Two key corollaries to the ecosocial constructs of embodiment and embodying (in)justice are (1) *the need to reject biological essentialism* and (2) *the need to embrace, instead, embodied integration.*

A third is to recognize that all living beings on our planet are, in ecosocial terms, *emergent embodied phenotypes.*[126] In this increasingly –omic, nanoscale-centric, and hyper-biomedical research moment, it is critical to emphasize that the drivers of current and changing societal patterns of disease distribution, including health inequities, are exogenous to people's bodies and reside instead in the body politic.

Ecosocial Constructs: Emergent Embodied Phenotypes and Distinguishing Between Biological Expressions of Injustice Versus Unjust Interpretations of Biology

Consider, for example, how the ecosocial theory of disease distribution engages with embodiment and embodying (in)justice by distinguishing between what I newly term, building on earlier work,[127] *biological expressions of injustice* versus *unjust interpretations of biology.* The former refers to the biological consequences of embodying societal injustice and constitutes manifestations of emergent embodied phenotypes. The latter, by contrast, refers to how societal injustice distorts interpretations of biology—and especially how it can lead to "naturalizing" the social phenomenon of health inequities. These constructs build on concepts I introduced to the epidemiologic literature in the mid-1990s, referring then to *biological expressions of racism* and *racialized expressions of biology* and, related, *biological expressions of gender* and *gendered expressions of biology*.[128] The intent was, and remains, to counter how "race/ethnicity" and "sex/gender" routinely appear in epidemiologic analyses as if self-evident stand-alone ostensibly biological characteristics that were solely individual-level characteristics, and with scant to no attention to how racial/ethnic and gender inequality shape both these very categories and the health of people so categorized.[129]

With regard to "race," the epidemiological, public health, and even more voluminous biomedical literature was—and remains—steeped with the supposition that "race" is an innate biological category, whereby "races" are posited to be distinguished by systematic genetic differences, and these differences give rise to observed racial/ethnic differences in health.[130] Reams of scholarship exist regarding the origins of these ideas, both outside and within medicine and science, as tied to histories of the rise of the European transatlantic slave trade and worldwide colonialism.[131] The upshot has been to racialize biology, via a circular logic that presumes that observed racial/ethnic differences are due to alleged innate differences, and therefore any such observable differences constitute proof of said differences. Missing from the analysis is any consideration of how the very power relations that have constituted and enforced racial injustice and forged the very construct of "race" have durable biological consequences: both for those who have endured racial discrimination and attendant economic deprivation and, conversely, for those who have benefited from racialized economic privilege.

Distinguishing between *biological expressions of racism,* as reflected in racial/ethnic health inequities, versus *racialized expressions of biology,* referring to health status and other bodily characteristics interpreted as signifiers of alleged innate "racial" differences, has offered new options for addressing links between categories of racialized groups and biology in population health research. After all, if structural racism affects the material and social conditions in which people live, and people embody these conditions, the ensuing biological expressions of this embodiment will be present in individual bodies and manifest in on-average differences in health status across the specified racialized groups. How could it be otherwise?

The analytic implications are twofold.[132] One pertains to the need to develop empirical measures of racial injustice, at multiple levels, that are feasible to use in epidemiologic research, cognizant of cohort effects related to historical events affecting exposure (e.g., imposition of Jim Crow, passage of civil rights acts, etc.). The other concerns the need to become aware, for each outcome studied and its biological features, of the histories of scientific debates over its associations with the assigned race/ethnicity of the racialized groups—that is, as being due to racism or to "race."

Clarity regarding these constructs has critical implications for the rigor and accuracy of *all* population health research—since *no one* is exempt from membership in these types of socially assigned groups. It accordingly is important to underscore the ongoing challenges for improving the methods and intellectual rigor of research on these issues.[133] Current work, increasingly breaking through to being published in prominent medical and public health journals, explicitly challenges the long-dominant framing, where since at least the early 1800s it has been deemed "scientific" to posit "race"—understood as a stand-in for inborn traits—as explanatory, but "ideological" to raise issues of the health impact of racial injustice.[134] The deeper truth is that it is ideological and intellectually weak to ignore how racial injustice harms health and, in the absence of such considerations, unscientific to make biological claims about "race" being the cause of observed differences in on-average health between racialized groups.[135]

Similar issues in the mid-1990s swirled around the framing of "women's health" in epidemiology and public health (noting that "men's health," as such, was rarely discussed in the public health literature back then, and questioning of the gender binary was virtually nonexistent).[136] As with racism and health, these debates were informed by a long history, easily extending back millennia, regarding the extent to which differences in biology between women and men explained their individual and group differences in power, property, behaviors, and health—both for health outcomes potentially experienced by both sexes and for those uniquely related to reproduction (e.g., birth).[137] Here, however, the challenge was different. In contrast to "race/ethnicity" (and thus racialized groups) being solely a social construct with no legitimate biological basis, for sex/gender, different considerations existed regarding the social and biological issues involved. Specifically, in addition to the social construct of "gender," there were also the realities of humans being a sexually dimorphic species, meaning two different sexes are required for reproduction—which is separate from the question of where any particular individual lies in the distribution of diverse sex-linked biological traits.[138] That said, it is critical to underscore that people, regardless of their sex-linked biology, necessarily and overwhelmingly share a common biology by virtue of being members of the same species.[139]

Distinguishing between the constructs of *biological expressions of gender* and *gendered expressions of biology* likewise has been useful to sharpen analysis within population health science, with embodiment again analyzed in relation to levels, pathways, power, lifecourse, and historical generation.[140] The former construct has encouraged analysis of the impacts of social gender and structural gender inequality on the health of all genders, while the latter has allowed for critical questioning of how scientific sexism has shaped both scientific questions and affected interpretation of biology, including for comparative analysis of sex-linked biology across species.[141] Decentering human biology can be a way of looking again at human biology with fresh eyes—with new insights gained into what have been termed *flexible phenotypes*, as illustrated by how for some types of animals (e.g., teleost fish), individual organisms can change their reproductive sex repeatedly during their lifetime.[142] These biological realities bolster ecosocial theory's focus on population distributions when considering any particular exposure, trait, or health outcome.

As with racism and health, articulation of the ecosocial constructs involving embodying (in)justice, gender, and biology—that is, *biological expressions of gender* versus *gendered expressions of biology*—has likewise underscored how the concepts, methods, and measures for analysis of gender systems, gender inequality, and population distributions of health need to be strengthened.[143] They also led me to cease using the word *sex* by itself, and to refer instead to *sex-linked biology*, for two reasons. One was to avoid the common treatment of biological "sex" as one "thing," which ignores the complex layering of myriad biological characteristics and systems involved in the capacity to reproduce sexually (including chromosomes, genomes, epigenomes, hormones, and hormone receptors) and the distribution of these characteristics both within and across individuals and populations.[144] The second was because important aspects of sex-linked biology, such as estrogen receptors, can, among humans and other animals, occur in every organ system and can have biological roles not directly tied to sexual reproduction.[145] The larger empirical question could thus be rephrased—and remains—whether gender, sex-linked biology, neither, or both contribute to observed population patterns of specified health outcomes.[146]

More broadly, by hypothesizing that health inequities comprise the *emergent embodied phenotypes*[147] via which injustice is biologically expressed—whether the injustice be in relation to social class, racism, nativity, sexism, heterosexism, gender binarism, ableism, or other types, separately or combined[148]—ecosocial theory invites population health scientists to develop the concepts and methods to put these ideas to the test. It likewise invites critique of population health and biomedical research that neglects to consider how injustice may affect both the population health phenomena under study and the ideas and methods used to study it. And because it is a scientific theory, one that can be studied and applied like any other theory, ecosocial theory clarifies that the knowledge needed to formulate these kinds of hypotheses is possible for anyone who takes the time to learn and acquire the relevant expertise, above and beyond whatever their own lived experience might be.[149]

Ecosocial Theory, Embodying (In)justice, and Intersectionality: Disciplinary Connections and Differences and Contrasts to Metaphors of "Weathering" and "Under the Skin" and the Construct of the "Exposome"

The ecosocial theory of disease distribution and its constructs of embodiment and embodying (in)justice additionally invite integrative thinking about bodily and population health phenomena. By implication, embodiment means we are not a member of a particular racialized or ethnic group one day, have a particular gender identity on another, and on still another day have a particular sexual orientation: we are all of these at once—and the same holds for our social class, nationality, immigration status, and where we live and work.[150] With and within our bodies, people and other organisms daily integrate their experiences and exposures, social and biophysical, in each and every moment. Doing so is literally part and parcel of being alive. Consequently, causal theorizing about these integrative processes needs to be equally integrative, including for such integrative phenomena as embodiment, embodying (in)justice, and population distributions of health.

A recognition that people's experiences are simultaneously shaped by their multiple historically forged social positions and identities is also conveyed by *intersectionality*. This construct first surfaced in the academy in the legal and social sciences in the early 1990s,[151] notably as used by Kimberlé Crenshaw in 1991 in a legal essay on violence against women of color.[152] The construct itself built on a legacy of activist insights and organizing in the 1970s and 1980s, galvanized by Black feminists, around the interlinked issues and lived realities of sexism, racism, heterosexism, and class injustice.[153] This latter body of work has also aided my development of ecosocial theory.[154]

The construct of *intersectionality,* whose use has dramatically risen in numerous fields over the last 30 years, has given rise to productive discussion and debate as to what it means and what it entails for critical analysis and action, including within public health.[155] In 2020, Patricia Hill Collins, one of the first-wave Black feminist scholars explicitly engaged in developing this framework, writing with her coauthor, Sirma Bilge, noted that while there can be varied and sometimes contradictory answers as to what "intersectionality" means, nevertheless a common working definition would be:[156]

> Intersectionality investigates how intersecting power relations influence social relations across diverse societies as well as individual experiences in everyday life. As an analytic tool, intersectionality views categories of race, class, gender, sexuality, class (*sic*), nation, ability, ethnicity, and age—among others—as interrelated and mutually shaping one another. Intersectionality is a way of understanding and explaining complexity in the world, in people, and in human experiences.

The objective is to foster the ideas and practices that people can use to challenge interlocking and mutually reinforcing distinct systems of power and their impacts on both societies and individuals, with the intent of creating equitable societies.

Or, as Claudia Rankine, an African American poet, stated in her 2014 award-winning prose-poem *Citizen: An American Lyric*:[157]

> The world is wrong, You can't put the past behind you. It's buried in you, it's turned your flesh into its own cupboard. Not everything remembered is useful but it all comes from the world to be stored in you.

As this metaphor suggests, the totality of people's lived experiences, shaped by their societies, lodge and manifest in their bodies, such that to understand the state of the people's health, one must start by grappling with their histories, in societal context, within and across generations.

Thus, like the ecosocial theory of disease distribution, intersectionality is concerned with socially structured entanglements of multiple types of injustice. Where it differs is that intersectionality's intended use is for analysis and action regarding intersecting forms of structural injustice in human societies.[158] It was not and has never been intended to guide theorizing jointly concerned with social, biological, and ecological phenomena and the pathways and processes of literal biological embodiment and their translation, via emergent embodied phenotypes, to patterns of population health and health inequities.[159] Of note, while the ideas and methods of intersectionality have been and can be productively applied in public health research and practice,[160] "intersectionality" by itself is not—nor ever was intended to be—a population health theory. That said, deep resonance exists between the analytic approaches and objectives of "intersectionality" and ecosocial theory.

The same can be said for the sociocultural construct of "embodiment."[161] An old idea long the focus of profound philosophical and theological debates about ties between material existence (i.e., bodies) and consciousness (or spirit or soul),[162] since the 1980s interest in "embodiment" has surged in both the humanities and social sciences. Active discussion about and use of this idea occurs within and between anthropology, ethnography, sociology, history, critical feminist studies, critical race studies, queer studies, postcolonial studies, disability studies, literary and art criticism, and philosophy and theology, to name a few.[163] A primary focus has concerned understanding how, within specified societal contexts and constraints, people's ideas and their power to enact them shape what they and others do with and to their own bodies and the bodies of others—where, when, and with whom.[164] The primary interest is in the body as site of action and contestation in the world, not its biological being. That said, new scholarship in both sociology and anthropology (especially that of Margaret Lock[165]) is addressing biology as it analyzes embodiment in relation to societal systems, power, social and biophysical exposures, health practices, and health status.[166] However, as befitting for the social sciences, this sociocultural scholarship employs chiefly conceptual and qualitative analysis and is not intended to provide quantitative description or analysis of population distributions of health.[167]

So too with the construct of "habitus," a complex concept regarding what the influential French sociologist Pierre Bourdieu (1930–2002)[168] referred to as "the social made body."[169] Although Bourdieu never provided one single definition of "habitus,"[170] sociologically and philosophically he held that social processes of pedagogy and socialization mediate how the outside world transports itself into and transforms people's bodies, from infancy onward, leading to their internalizing dispositions that shape how they act, feel, think, and talk—which they then bodily manifest as their "habitus."[171] Examples include how people's societal context and their family's standing within it variously affects (consciously or unconsciously) their posture, gestures, facial expressions, and gait; how they dress and adorn their bodies; how and what they eat and drink; how they stimulate or dull their senses (including via recreation and cultural activities); and how they have sex and with whom.[172] Moreover, although Bourdieu recognized that "habitus" had health implications, his focus as a sociologist was on cultural and social phenomena, not disease distribution, epidemiology, or public health.

Precisely because of their different emphases, ecosocial and sociocultural understandings of "embodiment" can enrich each other in complementary ways.[173] The sociocultural analyses can provide rich or thick information regarding the ways in which people describe, interpret, and shape their lived experiences,[174] providing clues that ecosocial theory can use to delineate and test hypotheses about potential literal "pathways of embodiment" connecting this lived experience to population distributions of health outcomes. Conversely, ecosocial theory, attuned to the context-dependent population distributions of the stories that bodies tell—both separate from and in conjunction with what people are willing or able to self-report about their lives[175]—can reveal context-specific patterns of designated health outcomes and test hypotheses about the historically shaped experiences and exposures that produce them. Together, the distinct but related ecosocial and sociocultural approaches to analyzing "embodiment"—along with ecosocial theory's explicit focus on embodying (in)justice—can generate insights relevant to understanding the people's health and guiding action for health justice.

Three additional public health concepts concerned with exposures' impacts on bodies and health status are *weathering*,[176] *get under the skin*,[177] and the *exposome*.[178]

- The first, "weathering," is a metaphor introduced by Arline Geronimus in 1992 as part of a novel hypothesis about why "the black-white infant mortality differential is larger at older maternal ages than at younger ages."[179] It evokes how weather (i.e., wind, precipitation, temperature, etc.) can physically sculpt and also disintegrate substances, whether inorganic (e.g., rocks), dead (e.g., timber), or alive (e.g., skin);[180] by association, it conveys how people's social context affects their biological aging and risk of disease and death.[181]
- The second, "get under the skin," construed as a metaphor (and as opposed to literal use describing skin-penetrating parasites[182]), emerged in psychosocial epidemiology in the late 1990s;[183] rooted in health psychology and neurobiology, it serves as shorthand for biological mechanisms by which psychosocial stressors affect disease risk.[184]
- The third, "exposome," a term coined in 2005 by Christopher Wild to capture the "nongenetic" complement to "genome,"[185] refers to biological markers of environmental exposures (initially biophysical, then expanded to include psychosocial) whose "footprints" can be measured within individuals.[186]

While none of these three concepts by themselves constitute theories of disease distribution or explicitly focus attention on who (and not just what) drives health inequities, all attest to the growing interest in understanding how people's psychosocial and biophysical exposures affect their biology and disease risk.

Finally, starting in the early 2010s, several social worker scholars and practitioners have called for their field to adopt an "ecosocial" approach to their work, prompted by the growing crisis of climate change.[187] What this entails is for social workers to be more engaged with how people's ecological context affects their societies, their social welfare systems, and the hardships they face. Responding to kindred societal and ecological contexts that spurred my introducing the ecosocial theory of disease distribution back in 1994, the social work approach, however, is fundamentally more akin to a "social-ecological" approach (as described earlier), with the added component of a nonhuman ecological context. As befitting for its focus, the social work approach does not address analysis of disease distributions or biological embodiment of societal and ecological context; it should not be confused with the ecosocial theory of disease distribution.

Embodiment: Producing Population Distributions of Disease in Societal and Ecological Context, and a Causal Challenge to the Proximal/Distal and Upstream/Downstream Divides

The notion that biology is expressed in societal and ecological context, is sensuous, and is structured by history is not unique to the ecosocial theory of disease distribution. It is, rather, core to analysis of biology, development, and evolution for all species.[188] As emphasized in the collaborative work of the evolutionary biologist and paleontologist Niles Eldrege (b. 1943) and Marjorie Grene (1910–2009), a major 20th-century CE philosopher of biology, all living beings on Earth are always and inseparably social beings and biological organisms and members of species whose lives reflect and create entwined histories of ecologies, genealogies, and social interactions.[189] This is why the lives of all living beings, from start to end, necessarily are *emergent embodied phenotypes.*[190]

To this mix, humans have distinctively expanded modes of inheritance from being first and foremost biological (what is passed from parents to progeny via sexual or asexual reproduction, including but not limited to DNA and RNA[191]) to societal rules about passage of property from one generation to the next.[192] The relationship of people to place—understood socially and ecologically—and what this entails for the health of people and other living beings is likewise core to such disciplines as critical health geography[193] and political ecology.[194]

Ecosocial theory's focus on population distributions of disease and embodying (in)justice, however, entails specific theorizing about populations, distributions, exposures, and health outcomes, and the causal processes underlying their existence and relationships. To close this chapter, and as a segue to the concrete work of using ecosocial theory in Chapter 2, I accordingly address the interrelated concepts and realities of *populations* and *structured chance*, and why their embodied connections refute the misleading characterization of "distal" versus "proximal" and "upstream" versus "downstream" causes of health.[195] These are topics I have written about at length in other publications,[196] informed by the work of scholars in diverse disciplines, so my presentation here is telescoped to bring key points into clear view.

In brief, the ecosocial theory of disease distribution recognizes that the construct of "populations" is complex, with profound implications for causal analysis of "population distributions." At the crux are (1) who and what define the "population" and delimit who and what it includes *and* excludes, (2) whether these defining characteristics are viewed as fixed versus dynamic, and (3) whether populations comprise entities from which random samples of individuals can be taken to estimate population characteristics.[197] Deeply entangled in all of these considerations are issues of structure, agency, and chance—and, related, whether health inequities are analyzed as embodied biological expressions of injustice versus unjustly interpreted as resulting from innate biological differences or flawed versus virtuous individual or cultural choices.

Stated bluntly, the embodied stance of ecosocial theory rejects biological essentialism and strict determinism. It instead embraces historical contingency, structured chance, and possibilities of change—including for the very definitions of populations and diseases, as brought about by people debating ideas and seeking to alter how they live within their societies.[198] The contrast is to the dominant view, aptly captured by the famous apparatus of the Quincunx (Figure 1.2), designed in 1889 by Sir Francis Galton (1822–1911).[199] As noted earlier, Galton infamously coined the term *eugenics* in 1893 and he also invented the correlation coefficient in his quest to prove that population distributions, including of intelligence, were a product of "heredity," not "environment."[200]

Galton designed his Quincunx to be "an apparatus . . . that mimics in a very pretty way the conditions on which Deviation depends."[201] The device consisted of pegs placed on a vertical board and was designed so that identical pellets poured through a funnel that randomly bounced off the carefully placed pins would fall into different bins and produce a normal distribution, with the height of the column reflecting the probability of different pathways of descent (Figure 1.2). To Galton, the device beautifully demonstrated the properties of what he called the "Law of Frequency of Error," a "law" that he claimed "would have been personified by the Greeks and deified, if they had known of it. . . . each element, as it is sorted into place, finds, as it were, a pre-ordained niche, accurately adapted to fit it."[202] From this standpoint, the placement and shape of the pins and funnels

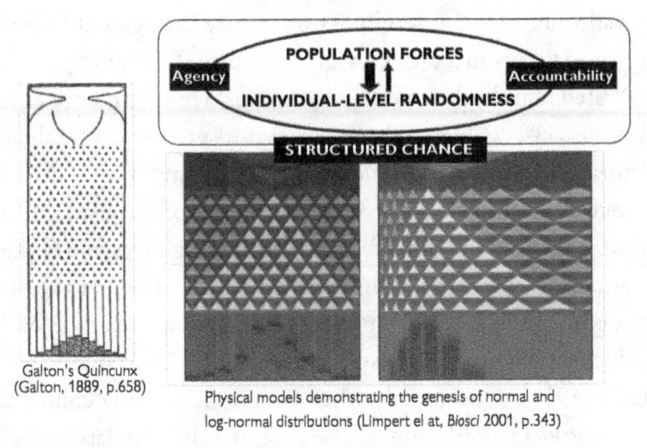

Figure 1.2 Producing population distributions: structured chances as represented by physical models.

Sources: Krieger N. "Who and what is a "population"? Historical debates, current controversies, and implications for understanding "population health" and rectifying health inequities". *Milbank Q* 2012; 90(4):634–681; figure: p. 658; Galton F. *Natural Inheritance*. London: Macmillan, 1889; p. 63, which is freely available at: https://galton.org/books/natural-inheritance/index.html; Limpert E, Stahel WA, Abbt M. Log-Normal distributions across the sciences: keys and clues. *BioSci* 2001; 51:341–352; figure: p. 343.

were a given in the background—and the resultant distributions were simply preordained.

Galton's fundamental assumption—common to the political elites of his time, and still influential to this day[203]—was that the global political, economic, and intellectual dominance of white Anglo-Saxon men of rank and means simply reflected their innate superiority, as a natural group.[204] This superiority was to women in their own "race" and class; men in their own "race" but who were working class; and men and women in all other "races," regardless of class. A second assumption, based on scant, biased, or otherwise dubious data,[205] was that variation in characteristics between these populations vastly exceeded within-group variation.[206]

Yes, outliers existed, but the average and distribution told the "real" story; in Galton's own words:[207]

> Let us then compare the negro race with the Anglo-Saxon, with respect to those qualities alone which are capable of producing judges, statesmen, commanders, men of literature and science, poets, artists, and divines. . . .
>
> First, the negro race has occasionally, but very rarely, produced such men as Toussaint L'Overture. . . .
>
> Second, the negro race is by no means wholly deficient in men capable of becoming good factors, thriving merchants, and otherwise considerably raised above the average of whites . . . a result which again points to the conclusion, that the average intellectual standard of the negro race is some two grades below our own.

Toussaint L'Overture notably commanded Galton's attention because of his leadership role in the first successful anti-colonial and anti-slavery revolution, which liberated Haiti from French rule (1791–1804), with enormous implications for not only France but also Great Britain and other European colonial powers and their reliance on slavery, as well as the US anti-slavery struggle.[208]

Innate difference, to Galton, is what also explained group differences in age at death, despite flux of individuals.[209] To convey the idea that populations' stability arose from the common essence of each transient member, he reached for a celestial metaphor: "The cloud and the population are composed of elements that resemble each other in the brevity of their existence, while the general features of the cloud and of the population are alike in that they abide."[210] This perspective clearly was indifferent to changes wrought by the wind (how long have you seen a cloud hold its shape?). Instead, to Galton, inner properties determined outer essence—and the external context was of secondary or no importance.

However, consider the alternative mechanical device also shown in Figure 1.2. It was built by several late 20th-century CE physicists, who altered the shape of the funnel and pins to generate a log normal curve (i.e., the logarithm of the values has a normal distribution).[211] Juxtaposing these two figures enabled me to introduce the ecosocial idea of *structured chance*,[212] via use of an illuminating mechanical metaphor for showing how altered structures can change population distributions, including of identical balls, by altering the probabilities or chances of what pins they will hit, what angle they will bounce off, and where they will end up. This concept recognizes that while randomness, or stochasticity, is integral to disease distributions and the outcomes for individuals, at the same time, the population distributions themselves are not simply random, but reflect the societal and ecological contexts that create the populations and structure exposure and risk.[213]

One implication is that observed embodied differences between populations are simply that: observed embodied differences in a given context, not clear-cut indicators of "innate difference," with structuring of chance playing out over time and different spatial scales as well. In the case of population health, what matters is not only time reckoned in terms of seconds, minutes, or people's lifecourse but also the historical time period in which they live.

Are comparisons of, say, Black versus white health status carried out when racialized slavery is legal, when formal legal equality exists but without reparations for enslavement, or (as is yet to be the case anywhere) when equity has existed for generations, in all domains of human rights (social, economic, civic, political, and cultural)?[214] Is Indigenous health analyzed and compared to that of other groups (e.g., settler-colonialists and their descendants, other immigrants) at a time of forcible expropriation of territory and forced settlement on reservations, when formal legal equality and formal recognition of sovereign treaties exist but without reparations for colonization, or, again, when equity and respect for human rights has existed for generations? Are comparisons of health status across categories of gender and sexual identities conducted when women as a class are denied suffrage and property rights, when homosexuality or same-sex marriage is illegal, when transgender as a category is denied legal recognition, or when equality in legal recognition and rights exist?

The juxtaposition of structures and distributions in Figure 1.2, when combined with the ecosocial approach to analyzing embodiment, additionally underscores problems with the conventional causal divides, in public health and other literatures, of "proximal" versus "distal," and "upstream" versus "downstream."[215] Levels coexist, interact, and exert their influence simultaneously: the so-called distal policies that affect, say, population distributions of wealth and impoverishment, or subsidies for fossil fuels or deregulation of environmental protections, are as directly embodied by individuals as are the more specific so-called proximal exposures these policies engender. Moreover, the embodied impacts can last long after the policies change, across the lifecourse and across generations. Reckoning with these cross-level temporal realities is what the daily embodied integration of societal and ecological context entails. Additionally, individuals (typically construed as "downstream") can engage in collective action to change so-called upstream systems and policies, a possibility obscured if causal power flows solely from "upstream" to "downstream." Indeed, only collective organizing by individuals has ever had the strength to change societal structures.[216] A frank reckoning with levels, pathways, and power, as articulated by the ecosocial theory of disease distribution, makes these connections and possibilities clear.

In sum: people and other living beings live in socially, spatially, and temporally structured multilevel realities—history and context always matter, as do the lived realities of embodying (in)justice. How could they not? The privileging of individual-level phenomena by dominant theories, at the heart of the "individualistic fallacy," flies in the face of actual lived experience, as has its denigration of use of contextual data as mere "ecological fallacy."[217] Rather, as I wrote when reflecting on being at the first People's Climate March in 2014: "Indeed, the real 'ecological fallacy' is to think epidemiologists or others could ever understand the people's health except in societal and ecological, and hence historical, context."[218] Putting these ideas to work is the task of Chapter 2.

2 EMBODYING (IN)JUSTICE AND EMBODIED TRUTHS

USING ECOSOCIAL THEORY TO ANALYZE POPULATION HEALTH DATA

As I started writing this chapter in late July 2020, a new study grabbed headlines globally on July 24,[1] offering yet more evidence as to the power of embodied truths. Its news coverage managed to break through the relentless and intensifying missives about COVID-19, including in the United States, whose daily cases per million residents in late July put it in the top 10 worst countries worldwide, with 10 of its states reporting higher rates of new cases per day than any other country in the world.[2] This high-profile study—titled "Genetic Consequences of the Transatlantic Slave Trade in the Americas"[3]—offers a good entryway into illustrating ecosocial theory at work, critically joining societal and biological analysis in a historical and ecological context.

After considering this example, I then explicate how I have used ecosocial theory and its core constructs, including embodiment and embodying (in)justice, as a guide for my thinking and research, including to generate and test novel hypotheses and develop new metrics to help answer the questions I have posed. I offer these examples as an invitation to strengthen the work for

Ecosocial Theory, Embodied Truths, and the People's Health. Nancy Krieger, Oxford University Press.
 DOI: 10.1093/oso/9780197510728.003.0002

health justice. The first concerns conceptual and empirical analyses of health inequities and the breast cancer estrogen receptor. The second pertains to the simultaneous realities and embodied health impacts of myriad types of discrimination and other physical and social hazards, present and past, across diverse domains and levels. The third focuses on expanding the array of structural measures of injustice for health research and action. The final section asks if the ecosocial theory of disease distribution and its focus on embodied truths and embodying (in)justice have been of use to other researchers, scholars, practitioners, and advocates—and a small hint: the answer is "yes."

Embodied Truths: From Enslavement to Jim Crow to Civil Rights and Ongoing Racialized Health Inequities in the United States

Enslavement, Relations of Power, and Embodied Ancestries

Cutting through the COVID-19 media coverage, the newsworthy investigation about genetic consequences of the transatlantic slave trade mentioned earlier[4] was published at a time when issues of structural racism, slavery, and state violence against people of color were at a high pitch in the United States.[5] It comprised, at the time of its release, the largest genetic study to date ($N = 50{,}281$ research participants) to investigate the geographic origins, forced relocations, and differential inferred ancestry—African, European, or Native American—of descendants of enslaved people throughout the Americas.

The participants were self-selected individuals who had submitted their DNA and genealogical history to a commercial firm (23andMe) to learn more about their ancestry, with the researchers supplementing these genetic and self-report data with data from slave ship manifests.[6] The net result was a study whose findings corroborated extant and extensive historical and anthropological research, as well as other genetic studies.[7] The new data unsurprisingly reaffirmed well-known regional differences across the Americas regarding the diverse African regions from which Europeans forcibly seized, enslaved, and transported 12.5+ million Africans, sending them to labor, live, and die in European rulers' forcibly expropriated so-called New World colonial territories ("new" only to the Europeans, not their Indigenous inhabitants).

The investigation also notably quantified the far greater genetic contribution of the enslaved African women compared to men, despite men comprising 60% of those enslaved, along with regional variation in contributions from European men. Putting new numbers on old facts, the study estimated that in Latin America, enslaved women contributed 17 times more to the genomes of the region's current persons of African descent than did the enslaved men; in the Latin Caribbean, they contributed 13 times more; and in the United States, they contributed 1.5 times more. In Central America, the Native American contribution of women was greatest and exceeded that of men by 28 times. Additionally, while the contribution of European men versus women was 3 times higher in the United States, it was 25 times higher in the British Caribbean. As the study authors discussed, these findings reflect continent-wide histories of rape combined with regional politics, whereby

South American colonial rulers sought to "dilute" African ancestry, whereas those in North America enforced segregation.[8]

As summarized by one expert commentator on the study, Dr. Alondra Nelson,[9] "The study illustrates how much physical and sexual violence were part of slavery—and how they are still built into our society. . . . It confirms the 'mistreatment, discrimination, sexual abuse, and violence that has persisted for generations' . . . that many people are protesting today."[10] As such, these compelling new genetic data substantiate the widespread role of reproductive violence in economies built on enslaved labor—long documented by the targets and their descendants, yet often denied by the perpetrators and their progeny.[11] By doing so, these genetic methods newly applied on a large scale are akin to how smartphones and social media have provided new tools to targets and witnesses of police violence to document violations long described by the afflicted but denied by the police and their protectors.[12] In both cases, though, use of these new technologies can be the proverbial two-edged sword, given the role of the new genetic methods in the rise of such new fields as sociogenomics, which seeks once again to focus on finding a genetic basis for social group differences.[13] So too can data from smartphones, social media, and other digital technologies be employed for state surveillance.[14]

However, turning the usual genetics-is-race determinism on its head, the new study's horrific embodied truths offer further material evidence exposing the fundamental falsehoods of "biological race."[15] As discussed in Chapter 1, this essentialist construct, whose origins lie in Europe with the Inquisition, the slave trade, and contingent claims of "blood" as the basis of white Christian European superiority and dominance versus the alleged inborn inferiority and subordination of others, has saturated mainstream European and colonial-based medicine and science since the 1500s CE.[16] Yet, as the new study's genetic data underscore, societal systems, not genetic inheritance, have been the key arbiters of whose progeny was free versus enslaved. Is it any wonder that this widespread racialized injustice, manifested in ways that are society specific, would produce the racialized health inequities long evident throughout the Americas,[17] *despite* the very different genealogies, geographies, and surrounding ecologies of the populations at issue?

The importance and power of using anti-essentialist theories of disease distribution, in conjunction with the latest methods and data, should be clear. It is for this reason that the ecosocial theory of disease distribution's approach to analyzing population health engages with and offers guidance for thinking about, and developing and testing hypotheses regarding, the societal, biological, and ecological processes at the heart of embodying (in)justice—with an eye toward producing knowledge that can aid efforts to end these inequities.

Embodying Jim Crow—and Its Abolition

Just one week before the publication of the genetic study, another newsworthy event transpired in the United States tied deeply to the histories invoked: the passing of the powerful US civil rights leader Congressman John Lewis (February 21, 1940–July 17, 2020).[18] In his final op-ed, written 2 days before he passed and published, as requested, on July 30, the day of his burial, Lewis recalled how he was 15, in 1955, when Emmet Till was lynched, and he realized this could have been him. Connecting across generations, he further said that who Emmet Till was to him and his peers, so too George Floyd is to the present generation, with Floyd's murder by police in Minneapolis on May 25, 2020, serving as a key spark to massive sustained protests in the United States and globally.[19]

Till's lynching was part and parcel of the vigilante and state-sponsored terror used to entrench the legally codified reign of Jim Crow, in which white supremacy was the law of the land in 21 US states and the District of Columbia.[20] It existed in parallel and in synergy with local legal and illegal racial discrimination in the technically non- Jim Crow North, whose political economy was entangled with that of the Jim Crow states, and whose legacy of state and local laws entrenching segregation and discrimination notably arose in conjunction with laws in the early 1800s that outlawed slavery.[21] Codified and enforced via passage of laws enacted in the 1880s and 1890s, Jim Crow in the US South constituted a backlash against the civil rights gains of the Reconstruction, following the defeat of the Confederacy in the US civil war in 1865.[22] These Southern Jim Crow laws were so numerous that they filled 524 pages in small print in the pathbreaking text *States' Laws on Race and Color*,[23] the first comprehensive compilation of state and local ordinances, published in 1950 by the pathbreaking civil rights activist and lawyer Pauli Murray.[24] These racially discriminatory laws—both in the US South and the rest of the country—remained in effect until the 1964 passage of the US Civil Rights Act, an act whose full realization and enforcement remains problematic to this day.[25]

Galvanized by Reverend Dr. Martin Luther King Jr.'s call for abolition of Jim Crow and for civil rights, along with his nonviolent philosophy, Lewis started his career of getting into what he called "good trouble." On March 7, 1965, at the age of 25, he led a critical march for voting rights in Selma, Alabama—on what came to be known as "Bloody Sunday"—and his vicious beating by the state troopers in riot gear that day on the Edmund Pettus Bridge was televised nationally.[26] Five months later, in response to the outpouring of support for civil rights, and building on generations of organizing for these rights, President Lyndon B. Johnson presented the US Voting Rights Act to Congress, which he signed into law on August 6, 1965.[27]

Also in 1965, under the auspices of the newly established Office of Economic Opportunity (OEO), the federal government authorized the first grant to set up the first federally funded community health centers in the United States—one in Mound Bayou, Mississippi, and one in Boston, Massachusetts.[28] Made possible by the surge for civil rights and desegregation of health services, plus the newly funded federal "war on poverty," these first community health centers had two intertwined goals. One was to treat patients, and the other was to work with community members and organizations to change conditions that harmed their health in the first place and prevented them from thriving—in relation to employment, housing, food insecurity, and more.[29] This is not the place to recount the history of these clinics (ably told elsewhere[30]) or their growth to a network of nearly 1,400 community health centers nationwide, currently "serving as the primary medical care home for over 29 million people in more than 12,000 rural and urban communities across America."[31] Rather, it is a jumping point to more deeply understand what it means to embody history and the harms caused by not recognizing this—as well as get a glimpse of what it might mean to embody justice.

Of note, an astoundingly still very small number of population health studies directly focus on Jim Crow and population distributions of health: as of the start of 2021, they total only 10.[32] Among these are four I have led,[33] one of which documented that fully two-thirds of African American children born between 1960 and 1964 were born in the Jim Crow states.[34] Together, these studies provide suggestive evidence of the immediate beneficial impacts of dismantling Jim Crow on infant mortality, premature mortality, and other outcomes related to hospital care.[35] For example, infant mortality rates prior to 1965 were notably higher among the US Black population in the Jim Crow compared to non–Jim Crow states. However, during the first several years after abolition of Jim Crow, the rates among the former swiftly converged to those of the latter—albeit still higher, for both groups, compared to the infant mortality rates among the US white populations in these two disparate regions,[36] reflecting the entrenched racial injustice throughout the United States, as discussed earlier.[37] These results should not be surprising—yet the point is not simply to analyze the immediate impact of Jim Crow's abolition, as I shall now explain.

In the winter of 1969 and the spring of 1970, the Mississippi community health center teamed up with a film crew to produce a documentary—called *Out in the Rural*—about its vision for and the realities of the community it served and who were part of its governance.[38] The film (22 minutes long and freely available online[39]) featured people across the lifecourse, including a newborn infant just a few days old, toddlers, young children, adolescents, young adults, all the way up to community elders. Capturing both extreme hardship and a strong spirit to thrive, the film starkly portrayed the extreme racial segregation, unemployment, and impoverishment tied to disinvestment by the former employers of sharecroppers, along with dilapidated worn housing barely a step up from shacks, a lack of sanitation, and inadequate transportation. Yet it did not focus solely on hardship. Instead, it simultaneously highlighted the community's determination—in partnership with the community health center—to rectify these problems. Steps taken included setting up an agricultural cooperative, fixing housing, building latrines, and building a community bookstore to feed the mind as well as spirit. Also depicted were the highly engaged staff of the health center, taking care of patients, meeting and organizing with and learning from the community members, while also teaching cooking classes and teaching science to young people from the community so that they could become health professionals.

I have shown this film to the students I teach for quite some time. Each time I show it, I ask them to think about what age the people shown in the film would be that year. Each year, they do the math—and each year they are surprised if not shocked. This is not ancient history. In 2020, when I started writing this chapter, and as per when I taught my course in fall 2020, the baby in the film would be turning 50 and the little ones would be in their 50s to 60s. Moreover, those who were born the same year as Representative John Lewis, 1940, who were then 30, would now be turning 80—unless, of course, their lives were cut short, given well-known inequities in premature mortality.[40]

If the notion of embodied history were commonplace, it would be no surprise and no stretch for students to grasp quickly the contemporary relevance of the film, gauged in relation to health status of people alive in the here and now. I find the same surprised reaction when I give lectures on my work on Jim Crow and contemporary health and present a slide (Table 2.1) that shows what age persons born in different birth cohorts, from 1920 through 2000,

Table 2.1 **Isn't Jim Crow Ancient History? Do the Math!**

Year of Birth	Age in 1965	Age in 2000	Age in 2020	Born to Parents Born Before 1965	May Have Become a Parent Before 1965
1920	45	80	100	yes	yes
1930	35	70	90	yes	yes
1940	25	60	80	yes	yes
1950	15	50	70	yes	yes
1960	5	40	60	yes	no
1970	–	30	50	yes	no
1980	–	20	40	yes	no
1990	–	10	30	maybe	no
2000	–	0	20	maybe	no

How old were you in 1965?

How old were your parents?

How old were your grandparents?

would have been in 1965, in 2000, and in the year I am giving the lecture. Two additional columns assess whether they were born to parents before 1965 or may have themselves become a parent before 1965. I ask the audience: How old were they in 1965 (including not yet born)? How old were their parents? How old were their grandparents? And I point out that, given my own birth in 1958, I was alive, as a child, when Jim Crow was legal and no constitutional amendment or federal laws nationally outlawed racial discrimination throughout the United States.[41]

This simple but not simplistic exercise brings home the significance of embodied history—and why it merits attention, given the massive numbers of people affected. It underscores why analyzing people and their health in the here and now (or in any historical generation) is always and necessarily an endeavor grounded in embodied history and requires considering the impacts of embodying (in)justice—and further requires considering who, in the same age range(s), is no longer alive, and why. Ignoring massive historically contingent sociopolitical exposures that can readily be presumed to have inequitable health impacts, as is the case with Jim Crow, is bound to lead to incomplete and biased knowledge inadequate to the task of improving population health and promoting health equity.[42]

A Thought Experiment: How Quickly Would Abolishing Racism End Racialized Health Inequities?

Consider the related thought experiment (or counterfactual, if you will): in an instant, imagine an "intervention" that eliminates structural racism, including all material racialized and economic inequities in living and working conditions, income and wealth, and access to health services, but leaves everyone's bodies and minds exactly as they are at the moment of this fantastic(al) change. Would health inequities suddenly vanish?

The short answer is: no. The more complex answer is: it depends. Thus, inequities in some temporally acute outcomes could potentially shrink quickly. Among the likely candidates would be childhood vaccination rates, food insecurity, psychological distress, disability or mortality due to treatable ailments or injuries, and heat-related deaths among the elderly.[43]

However, to the extent health insults have already been irreversibly embodied within a person, ecosocial theory and its approach to conceptualizing embodied truths and embodying (in)justice would predict that these inequities will continue to manifest as long as persons are alive who were born before this momentous change in circumstances. Such a prediction has huge implications for which sets of health metrics would or would not shift quickly, in which age groups, after such large changes in living and working conditions. Reckoning with embodying injustice means engaging with both historical cohorts and lifecourse and assessing the impacts of interventions, and the continued need for equitable redress regarding the embodied impacts, with this knowledge front and center.[44]

It also requires engaging with biology, including in relation to intergenerational transmission of privilege versus deprivation—which is not the same as biological inheritance.[45] A considerable body of research is investigating the injurious impacts of early-life economic and social adversity on subsequent health through the lifecourse, potentially involving period and cohort effects (e.g., differential impacts, by age, of, say, an economic depression, carried forward).[46] It is another matter entirely, however, to conjecture that these health inequities are biologically transmitted from parent to progeny—as is the case with a new round of hypotheses invoking epigenetic inheritance of methylation marks on gametes' DNA to explain persistent health inequities.[47]

Here it is crucial not to confuse two types of biological inheritance: (1) from one cell generation to another, within an organism, versus (2) inheritance of genetic material contained in gametes that join to produce the next generation.[48] Persistence of epigenetic modification from one cell generation to the next is well established and is core to biological development, and underlies how a zygote develops into an organism with different organ systems, despite all cells originating from the same fused gametes.[49] But generation of gametes is not the same as cell division within somatic cells, since meiosis largely strips gametes of their methylation marks—allowing for a fresh start, as it were.[50] As a result, direct biological inheritance of durable epigenetic marks, at least within mammals, has been found to be highly unlikely[51]—and the handful of examples (based on lab experiments with nonhuman species, e.g., rodents) that are repeatedly cited in favor of epigenetic imprinting and transgenerational inheritance[52] may well be the proverbial exceptions that prove the rule (or potentially errors of experimentation or interpretation).

Instead, transgenerational societal inheritance of adverse or beneficial conditions of life (i.e., the social reproduction of inequities, not genetic or epigenetic inheritance) is plausibly the most powerful driver of differential health profiles of social groups across generations. This is cause for not only despair but also optimism, since it implies that if conditions change for the better, so too will health inequities, within and across generations, on timescales that match the biological processes involved—but if social injustice continues, so will its contingent embodied truths.

Consequently, when I was asked by a colleague in political science if it would be acceptable to stop collecting data on race/ethnicity in health records, given growing recognition that "race" is a social, not biological, variable, and that racial discrimination has been formally outlawed, my answer was: no.[53] As long as people are alive whose embodied histories are shaped by racial injustice, data on race/ethnicity—to identify racialized groups—must be included, along with data on exposures and pathways through which racial injustice harms health.[54] When evidence shows that health inequities between racialized groups no longer exist, for each and every age group, it will be a landmark sign of embodying justice.

Of course, this is not a new observation. Consider the prediction made in 1858 by Dr. John S. Rock (1825–1866)—one of the first credentialed African American physicians, a powerful abolitionist, and in 1865 the first African American lawyer to be admitted to the Bar of the US Supreme Court.[55] At a speech to commemorate the death of Crispus Attucks—a Black man who escaped from enslavement, who was the first person to be killed for the American Revolution, dying at the hands of British troops at the 1770 Boston Massacre, and who was the son of Prince, an enslaved African and Nancy, an enslaved member of the Wampanoag nation —Rock declared once justice and true equality exist, "black will be a pretty color," and have no special salience for health inequities.[56] Until then, however, embodied history remains, and rigorous science requires reckoning with its implications for embodying (in)justice. As the Uruguayan writer Eduardo Galeano (1940–2015) sagely observed 2 years before he died: "History never really says goodbye. History says, see you later."[57]

Embodying (In)justice and Putting Ecosocial Theory to Work: Breast Cancer as an Illustrative Example

What, then, does it mean for public health research and practice to take seriously embodying (in)justice and embodied truths as framed by ecosocial theory? These are questions I ask myself for each and every project I undertake as a social epidemiologist, however small or big. The theory's tenets, core constructs, and attention to levels, pathways, and power, in historical, societal, and ecological context, serve as an ever-present mental checklist, to evaluate every type of idea and data that feed into the work. This systematic critical questioning takes place at every stage of the process:

- critically engaging with new or old evidence and ideas from myriad disciplines and sources;
- predicting new as-of-yet unobserved findings;
- generating specific hypotheses and devising the data and methods to test them;
- evaluating and interpreting the results; and
- reflexively working with others to guide action based on this knowledge.

The intent is to generate new knowledge and new possibilities for improving population health and advancing health equity.

Curiosity is key, as is a comparative and integrative approach—in relation to not only people in societal and historical context but also the broader world in which people live. Ecosocial theory mandates paying attention to the basic fact that we humans live in a dynamic world (and universe) that has long preceded us (and undoubtedly will outlast us), one that is literally more wonderful, contingent, quirky, contradictory, and expansive than human beings can ever imagine.

But as ecosocial theory also acknowledges, people's creative collective capacity to develop and test theories to aid understanding of this world is concomitantly matched by our collective abilities to forge and act on political credos—which can variously gut or expand options for equitable and sustainable ways of living.[58] Stated another way, and as reflected in scientific theory, the human capacity for imagination is both wonderful and terrible, a gift and a curse: it can lead to expansive inclusive innovative thinking that upholds the dignity of all persons and values our connections to people and other life on this planet—or to restrictive exclusionary reactionary thinking that denigrates and dehumanizes other people and treats everything on this planet as entities for economic gain.

One key test of a scientific theory is whether it can generate testable and verifiable predictions—that is, lead to new evidence and knowledge about the world, above and beyond the challenge of offering testable explanations for already existing observations.[59] I accordingly discuss how the ecosocial theory of disease distribution and its construct of embodying (in)justice has enabled me to make and test novel predictions leading to new actionable knowledge, as well as informing the work of others. The intent is to concretize what this way of thinking means for research and action, for any outcome, not just the ones that are the focus of my own work.

The Case of the Breast Cancer Estrogen Receptor: Using the Ecosocial Theory of Disease Distribution to Conceptualize Emergent Embodied Phenotypes and Predict Historically Contingent Health Inequities

One longstanding interest of mine concerns the epidemiology of breast cancer—and especially racialized and socioeconomic inequities in incidence, survival, and death.[60] Breast cancer is a disease that invites historical thinking attuned to context, for four reasons:[61]

- One is that population rates of incidence, survival, and mortality have exhibited different and complex trends over time, which have varied by place and social group, thereby necessitating attention to historical generation, societal conditions, and ecological environs.
- Another, bringing into play individual lifecourse, is that the litany of identified etiological exposures span from in utero to menarche to menopause and thereafter, both exogenous and also shaped by the timing and occurrence (or not) of pregnancy, and the impact of these exposures on potentially tumor initiation and tumor growth.
- Third, the breast itself is a uniquely dynamic organ and is the sole organ that continues to develop biologically and change, in ways that can affect breast cancer risk, from conception

to puberty to pregnancy (if one occurs) to menopause and thereafter.

- Fourth, breasts are typically a bilateral organ, and primary breast cancers can occur (once or repeatedly) in either one or both breasts, thereby indicating that "static" individual person-level "risk factors" (such as age at menarche) only partially contribute to tumor risk, and raising questions about potential within-breast localized factors and also structured chance.

Such multilevel dynamic context-dependent phenomena invite multilevel dynamic context-dependent theorizing, as provided by the ecosocial theory of disease distribution.

Background to developing these ideas was my work in the mid-1980s on social class and Black/white differences in breast cancer survival—at a time when cancer registry data included only data on race/ethnicity and none on socioeconomic position.[62] This omission, combined with an absence of socioeconomic data in breast cancer research using other sources of data, contributed to the pronounced racialization of US data and analyses regarding cancer outcomes.[63] In the case of breast cancer, this led to researchers either implicitly or explicitly assuming that some unmeasured genetic "racial" difference accounted for the shorter survival and higher mortality among the Black women.[64]

The solution was to geocode (back then, by hand!) the residential address of where the cases lived at diagnosis or death and to link the records to their census tract and block group's corresponding socioeconomic and racial/ethnic composition.[65] Although this strategy for using census tract data in conjunction with health was first developed in the 1920s, after census tracts were first introduced by the US Census Bureau in 1910 (and referred to as "sanitary areas" due to their public health usage),[66] this methodology still remained underutilized in public health well over a half century later.[67]

Characterizing both the cases and the population from which they arose in relation to their area-based socioeconomic indicators enabled analysis of breast cancer survival jointly in relation to race/ethnicity and socioeconomic position.[68] Relevant comparisons included not only Black versus white women, controlling for socioeconomic position, but also Black versus white women within socioeconomic strata (both impoverished and affluent), and socioeconomic gradients among Black women and among white women. As I have subsequently demonstrated through my *Public Health Disparities Geocoding Project*,[69] the methodology of geocoding and linkage to area-based social metrics can be used for myriad outcomes, from birth to death, in relation to incidence, prevalence, survival, and mortality rates, and also comparison of case characteristics. In such analyses, these area-based measures have *never* been simply a "proxy" for "individual" characteristics. Instead, as conceptualized using ecosocial theory, with its emphasis on myriad levels, these area-based measures have always constituted contextual measures, and the importance of their contribution to shaping population health has only become more manifest via use of multilevel models that employ both individual and area-based social metrics.[70]

As new technologies made it increasingly possible, logistically and economically, to conduct population-based large-scale epidemiologic studies using molecular biomarkers, epidemiological research on breast cancer began to focus on whether tumors at diagnosis were estrogen receptor (ER) positive (ER+) versus ER negative (ER–), or borderline.[71] The ER status of the tumor matters, since only ER+ tumors can be treated by medications that interfere with the tumors' responsiveness to estrogen-mediated growth signals.[72] Once these treatments became the standard for medical care, survival for women with ER+ tumors became notably greater compared to women with ER– tumors.[73]

Subsequently, assays also became widely available for the progesterone receptor (PR) and, later, the protein HER2, a growth-promoting hormone on the surface of all breast cells.[74] An important discovery was that breast tumors negative for ER, PR, and HER2—termed "triple negative" tumors—had the worst survival, since they were least responsive to available medications.[75] Starting in 2000 CE, molecular phenotyping of breast cancers further refined classification of breast cancer subtypes;[76] however, among these subtypes, ER status has remained a key axis for classification and treatment.[77]

The initial research on ER status and breast cancer outcomes followed the predictable racialization,[78] compounded by the lack of socioeconomic data in cancer registries and medical records (the source of cancer registry data).[79] In the late 1990s, when US studies (including some of my own) started to investigate explicitly the socioeconomic patterning of the breast cancer ER, overall and in relation to race/ethnicity, two findings stood out.[80] First, risk of having an ER– tumor was greatest among those worst off, by whatever simple metrics were used—education, income, occupation.[81] Second, "controlling" for these socioeconomic metrics typically contributed to but did not fully explain US Black versus white differences in ER status or breast cancer survival (the main foci of research at that time).[82] Yet, rather than prompt concerns about inadequate measures of socioeconomic position, let alone no data on other aspects of structural racism (e.g., residential segregation[83]), the dominant approach emphasized genetic hypotheses.[84] To this day, major breast cancer review articles continue to discuss the genetics of women whom they term to be "of African descent" or "African ancestry" lumped into one category[85]—as if "Africa" was one discrete place with a homogenous population, and not the largest continent in the world with enormous heterogeneity, including in breast cancer ER prevalence.[86]

Confronted by the state of this literature, I first used the construct of embodied histories to publish, in 2011, the study to test formally for—and find evidence of—temporal variation in the Black/white odds ratio for being diagnosed with an ER+ tumor.[87] From what I could glean, this anti-essentialist study, providing evidence of historical contingency of inequities, appeared also to be the first investigation to test formally for temporal trends in racialized inequities in biomarker expression of any kind. Building on this work, in 2013 I published an article titled: "History, Biology, and Health Inequities: Emergent Embodied Phenotypes and the Illustrative Case of the Breast Cancer Estrogen Receptor."[88]

In this article I explicitly used the ecosocial theory of disease distribution to generate (1) a systematic approach to thinking about history in relation to the breast cancer ER, (2) a new construct—that of the *emergent embodied phenotype*, and (3) new predictions, which I tested in subsequent research.[89] Drawing on new insights from ecological evolutionary developmental biology (eco-evo-devo) and ecosocial theory to question dominant gene-centric and ultimately static approaches to conceptualizing biology, I evaluated the history of the breast cancer ER, conceptually and empirically, in relation to four types of history: (1) societal, (2) individual (lifecourse), (3) tumor (cellular pathology), and (4) evolutionary.

From this case example, I developed four sets of questions population health scientists could ask about any biological trait considered in their research:[90]

- **Question 1: Societal history.** What data exist on historical trends in the average population rates of—and health inequities in—the embodied biomarker or outcome (e.g., between and within countries and regions, defined geopolitically and in relation to societal divisions involving property, power, resources, and discrimination, including socioeconomic position, race/ethnicity, Indigenous status, gender, sexuality, disability, nativity, and immigrant status)?
- **Question 2: Individual (life course) history.** What is the "natural"—and "unnatural"—history of the embodied biomarker or outcome across a person's life course? Does its expression change over time for a given course of illness, or across repeat bouts of an illness? Does its expression vary by the societal groups considered in Question 1 (i.e., display health inequities)?
- **Question 3: Pathological/cellular history.** What is the "natural"—and "unnatural"—history of the embodied biomarker analyzed at the level of the tissue(s) involved? Does its expression change over the course of the disease? Or vary by the societal groups considered in Question 1 (i.e., display health inequities)?

- **Question 4: Evolutionary history.** What is known—and debated—about the evolutionary history of the embodied biomarker or outcome under analysis? What insight does this history provide regarding the likely dynamics of expression, within and across individuals, historical generations, and societal groups?

As with the four core ecosocial constructs, the point is to consider all four questions, not just ask them selectively, one at a time.

These questions guided me in a rigorous (and fascinating) review of the literature, one that led me to posit that the ER is "a flexible characteristic of cells, tumors, individuals, and populations, with magnitudes of health inequities tellingly changing over time."[91] I found that the evidence available at the time supported my argument against essentialist interpretations of ER status.[92]

- First, regarding societal history, mortality rates for premenopausal breast cancer (for which ER– tumors predominate) did not, contrary to now, show strong class gradients in the United Kingdom or Black/white inequities in the United States prior to the 1960s.
- Second, regarding individual or lifecourse history, numerous studies reported on high levels of discordance for ER status for women diagnosed with a first and second primary breast tumor, as high as 70% for women whose first primary tumor was ER+ and as high as 48% for those for whom it was ER–, even among BRCA1 and BRCA2 carriers.
- Third, regarding cellular/pathological history, evidence indicated that the ER status of a given breast tumor can change

over time (in part reflecting tumor clonal heterogeneity), and also that ER– tumors could experimentally be induced to re-express the ER.

- Lastly, regarding evolutionary history, the literature held that the ancestral estrogen-related receptor likely evolved to have high sensitivity to extracellular signals, since it apparently first functioned as a xenobiotic sensor, able to detect if substances exogenous to the organism could be eaten or might require detoxification.

Together, these lines of evidence supported analyzing the epidemiology of breast cancer ER status as a dynamic and historically contingent phenomenon, rather than a fixed inborn characteristic. My larger inference was that "our science will likely be better served by conceptualizing disease and its biomarkers, along with changing magnitudes of health inequities, as embodied history—that is, emergent embodied phenotype, not innate biology."[93] Such thinking allows for changing what is wrongly construed as a "constant" into a dynamic contextualized variable.

Breast Cancer ER Meets Jim Crow: New Hypotheses and New Clues

Building on these ideas, I was able to generate a new hypothesis, not simply try to explain the extant evidence. Drawing on my other work regarding Jim Crow, the 1960s war on poverty, and population health,[94] and at a time when no studies had investigated Jim Crow and cancer, my first prediction was that Jim Crow birthplace would be associated with risk of breast cancer ER status among US Black women, but not US white women. My second prediction was that in both groups, trends in breast cancer ER expression would vary by birth cohort, albeit for the Black women in ways that varied by Jim Crow birthplace. In both cases, my a priori hypotheses were that these embodied histories would reflect the biological impact of severe early-life economic deprivation and racial discrimination bound up with Jim Crow, in contrast to the then-predominant hypotheses about the genetic contributions to Black versus white risk of ER– breast tumors, especially among younger women.[95]

Notably, my predictions were upheld, not refuted, by the two studies I published, one in 2017,[96] the other in 2018.[97] As of early 2021, they still stand as the only studies, to my knowledge, on cancer and Jim Crow, despite its likely large impacts on current population health, as discussed earlier, and only two (including one that I led) have explored the impact of residential segregation and ethnic enclaves on breast cancer ER status.[98]

The first of these Jim Crow studies used US national cancer registry data we obtained from the Surveillance, Epidemiology, and End Results (SEER) 13 registry group (excluding Alaska) for 47,157 US-born Black non-Hispanic and 348,514 US-born white non-Hispanic women, aged 25 to 84 inclusive, diagnosed with primary invasive breast cancer between January 1, 1992, and December 31, 2012.[99] Among these Black and white women, 92% and 96% were born when Jim Crow was in effect (i.e., before 1965), and 46% and 59% were age 20 or older in 1965, respectively. The key finding was that Jim Crow birthplace was associated with increased odds of ER– breast cancer only among the Black, not white, women, with the effect strongest for women born before 1965. Specifically, among Black women, the odds ratio (OR) for an ER− tumor, comparing women born in a Jim Crow versus not Jim Crow state, equaled 1.09 (95% confidence interval [CI] 1.06, 1.13), on par with the OR comparing women in the worst versus best census tract socioeconomic quintiles (1.15; 95% CI 1.07, 1.23). The Black versus white OR for ER− was higher among women born in Jim Crow versus non–Jim Crow states (1.41 [95% CI 1.13, 1.46] vs. 1.27 [95% CI 1.24, 1.31]). The existence of a unique Jim Crow effect for US Black women for breast cancer ER status brings home the reality of breast cancer ER− as an emergent embodied phenotype and underscores why analysis of racialized inequities must be historically contextualized.

The second study drew from this same database to ask a different question, informed by the evolutionary measure known as the "haldane," which measures the pace of change in traits in relation to biological generation, and used principally for analysis of ancient as well as contemporary nonhuman species.[100] In 2013, I had introduced the haldane into the epidemiologic literature in studies that looked at trends, by race/ethnicity and socioeconomic position, in body size[101] and age at menarche,[102] both relevant to risk of breast cancer and ER status at diagnosis. In this follow-up study, we grouped the cases according to birth cohort and used the haldane to quantify the rate of change. Key findings were that the percentage of ER+ cases rose, according to birth cohort (1915–1919 to 1975–1979), only among the women diagnosed before age 55 years; that the changes according to biological generation were greater for the Black women than for white women; and that among the Black women, they were greatest for those born in the states with Jim Crow laws (versus non–Jim Crow states), with this group the only one to exhibit high haldane values (>|0.3|). The net impact, among women under age 55 years, was to close the ER+ prevalence gap between Black and white women *by half.* Such rapid phenotypic change can plausibly be driven only by factors exogenous to populations' genomes.[103]

Supporting an interpretation regarding extreme childhood material deprivation are the two extant studies that have examined breast cancer ER status in relation to women's childhood experiences with food rationing and famine. Both found strong associations with increased risk of ER– tumors: one focused on UK women who were children in the 1940s, during World War II,[104] and the other on Chinese women who were children during the Great Famine of 1959–1961.[105] Since then, newer research has investigated the impact of exposure to extreme poverty and famine, both in utero and in early childhood, on breast cancer ER status in relation to both body size and metabolic syndrome, both hypothesized to affect expression of receptors for growth hormones.[106]

Troublingly, US public health research, including on breast cancer, nevertheless has paid scant regard to the likely contemporary impacts of the extraordinarily high rates of impoverishment and food insecurity among Black Americans in the US South in the decades prior to the abolition of Jim Crow.[107] As stated in the 1968 presidential commission report on *Rural Poverty in the United States,* regarding the available 1960 data:[108]

> Thus, while the incidence of poverty among all residences of the rural South is very high, the incidence among rural Negroes is shockingly so: 78 percent of all southern rural Negro families have incomes below $3,000 per year as

> compared to 39 percent of southern whites. The differential between urban white and Negro families (i.e., those living in communities of more than 2,500 population) in the South is somewhat smaller, 18 versus 53 percent.

Again, children born in 1960 would be only 60 now, in 2020. To understand their health, as well as who and what drive health inequities in breast cancer ER status—not to mention other outcomes!—ecosocial theorizing about embodying (in)justice in historical context is needed, and can lead to new hypotheses relevant to revealing the structural basis of current health inequities.

More Examples of Thinking with Ecosocial Theory to Integrate and Analyze Embodied Health Impacts of Myriad Social and Biophysical Exposures, Present and Past, Across Diverse Domains and Levels

Of course, *embodiment* is but one of the four core constructs of the ecosocial theory of disease distribution involved in the processes of *embodying (in)justice*. It features prominently because, as discussed in Chapter 1, embodiment is at the nexus of the concurrent and integrated lived realities of health, disease, and well-being, in societal and ecological context, at both the individual and population levels. In brief, no bodies, no embodiment, no life—or health, disease, or death.

Consider, then, two more sets of examples of how I think with ecosocial theory to do my analytical work, whether conceptual, methodological, or empirical substantive research. One concerns the embodied health impacts of embodying (in)justice in relation to the multiple types of socially structured exposures and experiences people variously are subjected to, generate, shape, and respond to at work and home, which they daily integrate in ways that affect their health and well-being. The other concerns developing metrics that reveal, rather than conceal, who and what structure inequitable neighborhood conditions bound up in producing health inequities.

United for Health and the "Inverse Hazard Law"

In the early 2000s, I was part of a team, led by my colleague Elizabeth Barbeau, to conduct a study—which we called "United for Health"—that we expressly devised to look at the health impacts of social and physical hazards at work. We designed the study to be able to analyze these impacts both overall and in relation to the workers' race/ethnicity, gender, sexuality, nativity, and socioeconomic position, along with their exposure to social hazards outside of work, including unsafe sex and intimate partner violence (IPV). At the time we did this study, scant research simultaneously investigated these different yet entangled social and physical aspects of work, let alone took into account workers' larger societal context, including as manifested in their sexual and relationship conditions.[109] Yet workers, as people, live these experiences and contexts concurrently, not sequentially or compartmentalized.

We conducted our study in partnership with two amalgamated unions in the Greater Boston area, whose members were primarily low-wage workers.[110] In Box 2.1, I include the introduction from our paper that explained how the study's approach was informed by the ecosocial theory of disease distribution and its conceptualization of integrated embodiment,[111] and also include the conceptual model we developed to operationalize our approach.[112] As will be evident, use of this theory was crucial for focusing the investigation on the selected hazards and relevant covariates.

For this study, we recruited 1,202 union members from 14 worksites between March 2003 and August 2004 who were between the ages of 25 and 64, employed for at least 2 months at the worksite, and could complete the survey in either English or Spanish. Participants worked in a range of industries, including meat processing, electrical lighting manufacturing, retail grocery, and school bus driving, and we categorized the participants (and that of any other heads of household they identified) in relation to their occupational class (defined in relational terms, i.e., supervisory or nonsupervisory employee, self-employed, or employing others) and also educational level.[113] We obtained self-report data on hourly wage, which we assessed in relation to that year's living wage estimate for the Greater Boston area, along with their household income and number and age of persons in the household, to calculate their poverty level in relation to the US definition of poverty. Other self-report data pertained to race/ethnicity, nativity, gender, sexual orientation, exposure to unsafe sex and IPV, and psychological distress.[114] We additionally directly measured the participants' height, weight, and blood pressure, using validated protocols. Questions on social hazards at work (using multi-item

Box 2.1

United for Health Study

This box provides explanatory text and the conceptual model illustrating how we used the ecosocial theory of disease distribution to study the integrated embodied health impacts of social and physical hazards at work, overall and in relation to the workers' race/ethnicity, gender, nativity, and socioeconomic position, along with their exposure to social hazards outside of work, as shaped by the sociopolitical realities of their societal context (Greater Boston, Massachusetts; recruitment: 2003–2004).

1. Introduction to Krieger N, Waterman PD, Hartman C, Bates LM, Stoddard AM, Quinn MM, Sorensen G, Barbeau EM. Social hazards on the job: workplace abuse, sexual harassment, and racial discrimination—a study of black, Latino, and white low-income women and men workers (US). *Int J Health Services* 2006; 36:51–85 (quote: pp. 51–53)

Workplace hazards. Typically, these are defined—and their health impacts analyzed—in relation to important physical exposures at work (e.g., chemicals, dust, fumes, noise, and ergonomic strain) and also unsafe conditions (e.g., inadequate guards on machines, slippery surfaces, and poor lighting) (1). Also warranting concern are psychosocial job characteristics that can potentially harm health, including job strain (2) (whether defined in terms of job control-demand-support (3) or effort-reward imbalance (4)); a related strand of psychosocial research focuses on links between disease and social status defined in relation to rank in an occupational hierarchy (5).

Yet while work is fundamentally a locus of production of commodities and value, via contractually recognized chains of command that involve the purchase and sale of people's physical and mental labor (as well as the extraction of surplus value) (6–8), it is also more than this. Work necessarily is also a site of social engagement, within and across job categories (6–10). For any given person in any given job at any given worksite, working entails interactions with permutations of peers, subordinates, and supervisors or superiors—and, in some jobs, interacting also with external clients or customers, including the public at large. Moreover, the "any given person" in any particular job is not simply a lump sum of "human capital" with a specified amount of experience, skill, and education. Instead, each worker is necessarily embedded in her or his societal context, and thus simultaneously embodies and brings to the work her or his social position in relation to key societal divisions involving property and power, including class, gender, sexuality, race/ethnicity, nationality, and citizen status, to name a few (11,12). From this embodied perspective, as elaborated by ecosocial theory (11–13), work is a locus not only of economic production but also of social reproduction of social relationships of the society at large.

In other words, "the worker" is not simply a "worker." Within a US context, this person is or self-identifies as or is labeled as a woman or a man (or perhaps transsexual or transgendered); heterosexual or lesbian, gay, or bisexual; white, Black, Latino/a, Asian, Pacific Islander, American Indian, or some other race/ethnicity not included in the major categories used by the US Census;

US-born or foreign-born; and legal citizen or resident, or illegal and undocumented resident. It manifestly follows that worksites, like any other social domain, will be arenas in which these social relations are expressed and contested. The net implication is that, in a context of societal inequality, additional workplace hazards can plausibly include racial discrimination, sexual harassment, and workplace abuse—with the first two also encompassing experiences that occur both in and outside of work (7,9–11,14–20). A fuller analysis of workers' health and workplace hazards thus translates to a concern with not only job-specific hazards but also the broader societal context in which workers live their lives and do their work (21).

To date, however, relatively little public health research has empirically assessed the prevalence and health impact of social hazards at work, singly and combined, especially simultaneously in relation to class, race/ethnicity, gender, and sexuality (14–20). Although there is a growing body of work on sexual harassment, workplace abuse, and health (19,20,22–25), acknowledged shortcomings include a lack of research on these hazards among multi-racial/ethnic low-wage working-class women and men, whether singly, together, or also in relation to racial discrimination (20). Conversely, among new work investigating links between racial discrimination and health (14,26), little has characterized the types and extent of self-reported experiences of racial discrimination among specifically working-class populations, and few studies provide data on racial/ethnic groups other than the African American and white population (14,26,27). Extant evidence, however, suggests that workplace abuse is most likely to be experienced by subordinate workers, both men and women (19). Sexual harassment in turn has been documented to affect chiefly women workers, especially women of color, with the caveat that among the few studies conducted among predominantly working-class populations, most found women and men reported similar levels (19,20). Research has also shown that self-reported experiences of racial discrimination are most likely to be reported by African Americans and least likely to be reported by white Americans (14,26). The net result is a marked dearth of data on the social patterning of social hazards at work, singly, combined, or in conjunction with physical hazards.

To address these knowledge gaps, we established the United for Health study to assess the prevalence of social and physical hazards at work and ascertain their combined health impact on workers' health among a multi-racial/ethnic working-class population of women and men (27,28). This article reports, as a first step, the distribution of three key social hazards—workplace abuse, sexual harassment, and racial discrimination—by the workers' race/ethnicity, gender, and wage level, defined in relation to a living wage. Based on the limited extant evidence, we hypothesized that low-wage workers of color would be most likely to experience the greatest combined burden of these three social hazards, with the women especially at risk for exposure to sexual harassment.

2. Conceptual model, from Barbeau EM, Hartman C, Quinn MM, Stoddard AM, Krieger N. Methods for recruiting white, black, and Hispanic working class women and men to a study of physical and social hazards at work: the United for Health Study. *Int J Health Services* 2007; 37:127–144 (model: p. 130)

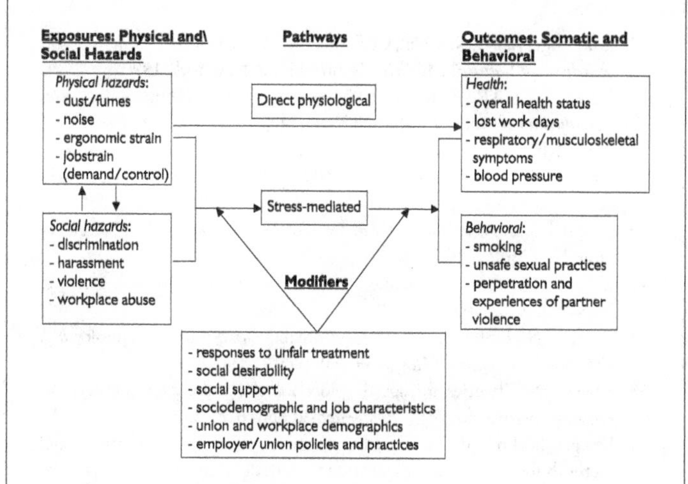

References

1. Levy, B. S., and Wegman, D. H. (eds.). *Occupational Health: Recognizing and Preventing Work-Related Disease and Injury,* Ed. 4. Lippincott Williams and Wilkins, Philadelphia, 2000.
2. Kasl, S. V. The influence of the work environment on cardiovascular health: A historical, conceptual, and methodological perspective. *J. Occup. Health Psychol.* 1:42–56, 1996.
3. Theorell, T., and Karasek, R. A. Current issues relating to psychosocial job strain and cardiovascular disease research. *J. Occup. Health Psychol.* 1:9–26, 1996.
4. Siegrist, J. Adverse health effects of high-effort/low-reward conditions. *J. Occup. Health Psychol.* 1:27–41, 1996.
5. Singh-Manoux, A., Adler, N. E., and Marmot, M. G. Subjective social status: Its determinants and its association with measures of ill-health in the Whitehall II study. *Soc. Sci. Med.* 56:1321–1333, 2003.
6. Wright, E. O. *Classes Count: Comparative Studies in Class Analysis.* Cambridge University Press, Cambridge, 1997.

7. Wooding, J., and Levenstein, C. *The Point of Production: Work Environment in Advanced Industrial Societies.* Guilford Press, New York, 1999.
8. Rose, D., and O'Reilly, K. (eds.). *Constructing Classes: Towards a New Social Classification for the UK.* Office of National Statistics, London, 1997.
9. Baxandall, R., and Gordon, L. (eds.). *America's Working Women: A Documentary History, 1600 to the Present,* rev. and updated. Norton, New York, 1995.
10. Ehrenreich, B. *Nickel and Dimed: On (Not) Getting by in America.* Henry Holt, New York, 2002.
11. Krieger, N. (ed.). *Embodying Inequality: Epidemiologic Perspectives.* Baywood, Amityville, NY, 2004.
12. Krieger, N. Embodiment: A conceptual glossary for epidemiology. *J. Epidemiol. Community Health* 59:350–355, 2005.
13. Krieger, N. Theories for social epidemiology in the 21st century: An ecosocial perspective. *Int. J. Epidemiol.* 30:668–677, 2001.
14. Krieger, N. Embodying inequality: A review of concepts, measures, and methods for studying health consequences of discrimination. *Int. J. Health Serv.* 29:295–352, 1999; updated as Krieger, N. Discrimination and health. In *Social Epidemiology,* ed. L. Berman and I. Kawachi, pp. 36–75. Oxford University Press, Oxford, 2000.
15. Yen, I. H., et al. Racial discrimination and alcohol-related behavior in urban transit operators: Findings from the San Francisco Muni Health and Safety Study. *Public Health Rep.* 114:448–458, 1999.
16. Baker, C. N. Blue-collar feminism: The link between male domination and sexual harassment. In *The Company of Men: Male Dominance and Sexual Harassment,* ed. J. E. Gruber and P. Morgan, pp. 242–270. Northeastern University Press, Boston, 2005.
17. Wyatt, G. E., and Riederele, M. The prevalence and context of sexual harassment among African American and white American women. *J. Interpers. Violence* 10:309–321, 1995.
18. Nelson, N. L., and Probst, T. M. Multiple minority individuals: Multiplying the risk of workplace harassment and discrimination. In *The Psychology of Prejudice and Discrimination: Volume 2, Ethnicity and Multiracial Identity,* ed. J. L. Chin, pp. 193–217. Praeger, Westport, CT, 2004.
19. Richman, J. A., et al. Sexual harassment and generalized workplace abuse among university employees: Prevalence and mental health correlates. *Am. J. Public Health* 89:358–363, 1999.

20. DeFour, D. C., et al. The interface of race, sex, sexual orientation, and ethnicity in understanding sexual harassment. In *Academic and Workplace Sexual Harassment: A Handbook of Social Science, Legal, Cultural, and Management Perspectives,* ed. M. Paludi and C. Paludi, pp. 31–45. Praeger, Westport, CT, 2003.
21. Quinn, M. M. Occupational health, public health, worker health. *Am. J. Public Health* 93:526, 2003.
22. Paludi, M., and Paludi, C. (eds.). *Academic and Workplace Sexual Harassment: A Handbook of Social Science, Legal, Cultural, and Management Perspectives.* Praeger, Westport, CT, 2003.
23. Gruber, J. E., and Morgan, P. *In the Company of Men: Male Dominance and Sexual Harassment.* Northeastern University Press, Boston, 2005.
24. Fendrich, M., Woodword, P., and Richman, J. A. The structure of harassment and abuse in the workplace: A factorial comparison of two measures. *Violence Vict.* 17:491–505, 2002.
25. Lim, S., and Cortina, L. M. Interpersonal mistreatment in the workplace: The interface and impact of general incivility and sexual harassment. *J. Appl. Psychol.* 90:483–496, 2005.
26. Williams, D. R., Neighbors, H. W., and Jackson, J. S. Racial/ethnic discrimination and health: Findings from community studies. *Am. J. Public Health* 93:200–208, 2003.
27. Krieger, N., et al. Experiences of discrimination: Validity and reliability of a self-report measure for population health research on racism and health. *Soc. Sci. Med.* 61:1576–1596, 2005.
28. Barbeau, E. M., et al. Methods for recruiting white, black, and Hispanic working class women and men to a study of physical and social hazards at work: The United for Health study. *Int. J. Health Serv.* 37(1):127–144, 2007.

validated instruments) pertained to racial discrimination, sexual harassment, and workplace abuse;[115] a single-item question about exposure to discrimination based on sexual orientation or identity was also included.[116] Questions on physical hazards (using multi-item validated instruments) concerned exposure to dust, chemicals, noise, work-related musculoskeletal hazards, and job strain (demand/control), supplemented by additional exposure data obtained during a walk-through conducted by the study team's industrial hygienists.[117]

My purpose here is not to review in depth the findings from the roster of publications stemming from this project.[118] Nor is it to address the well-known challenges of using self-report measures that reflect what people are willing or able to self-report, as I and others have discussed elsewhere at length.[119] It is instead to underscore the centrality of integrated embodiment of exposures, predating the current focus of contemporary quantitative public health research on "intersectionality."[120]

Briefly, then, 39% of the study population self-identified as being Black, 23% as Latinx, 25% as white non-Hispanic, and the remaining 13% including Asian Americans, American Indians, and additional racial/ethnic groups. A similar proportion across these first three racial/ethnic groups identified as being women (32%, 41%, and 36%, respectively); overall, 14% identified as being either lesbian, gay, bisexual, or transgender (LGBT), and within the total study population, 13% reported having same-sex partners but did not identify as being LGBT. Nearly a third of the workers earned less than a living wage, with wages, income, and education level lower among the Black and Latino as compared to the white non-Hispanic participants, who were by contrast the most likely to be in supervisory positions. Additionally, immigrants comprised 48% of the participants, including 58% of the Black and 66% of the Latino workers, albeit in ways that varied by race/ethnicity and gender. About 90% of the white non-Hispanic women and men and 70% of the Black women were born in the United States, whereas 60% to 70% of the Latinx women and men and also the Black men were foreign born. Notably, compared to the total population of the Greater Boston area, the participants were respectively 8.4 and 3.8 times more likely to be Black and Latino, 4.5 times more likely to be living in poverty, 1.6 times more likely to have less than a high school education, and 4 times more likely to be foreign born. These contrasts alone suggest that associations observed within our study population would yield conservative estimates compared to what might have been observed if we had included people in more privileged occupations and households.

Within the United for Health study population, fully 85% of the participants reported at least one high exposure to an occupational hazard (dust, chemical, noise, ergonomic strain) in the past year, 46% reported three or more high exposures, and 17% reported five or more high exposures. Workers of color were at higher risk of high exposures, with little variation by gender. Additionally, 85% of the workers reported exposure to at least one social hazard (racial discrimination, sexual harassment, workplace abuse). Moreover, exposure to all three reached 20% to 30% among the Black workers, among whom the two most common combinations, together reported by over half the Black women and men, were, first, racial discrimination combined with workplace abuse, followed by all three types combined, with the latter most common among Black women. Among the Latinx workers, the two most common combinations were racial discrimination and workplace abuse, both reported by over half the women and men, followed by workplace abuse alone. By contrast, among the white non-Hispanic workers, the most common category was workplace abuse alone, reported by slightly over 40% of the women and men. Additionally, LGBT workers reported twice as much sexual harassment compared to their cisgender heterosexual counterparts.

With regard to relationship hazards, two-thirds of participants reported having had a sexual partner in the last 6 months, among whom the proportion reporting safe sex ranged from a low of 5% and 9% among white men and women to a high of 28% and 24% among the Black men and women. The majority of persons classified as having unsafe sex were persons in married heterosexual relationships; also included was a smaller group with multiple and often risky partners. In each racial/ethnic group, about one-third of the men reported having ever been a perpetrator and about one-third of the women reported having been a target of IPV, with 20% stating this had occurred in the past year. Moreover, bringing work back into the picture, about 11% of the women reported that, during the past year, they had experienced employment-related IPV (e.g., a partner coming to work to threaten or harass them), and this proportion did vary by race/ethnicity, ranging from 7% among the Black women to 25% among the white women.

What were the implications of these exposures and experiences for the workers' health? Focusing on the results for psychological distress,[121] in models containing all three types of hazards (occupational, social, and relationship), along with the array of covariates, several hazards from each of the three hazard domains, analyzed separately, continued to be independently associated with psychological distress, albeit in ways that varied by gender. Thus, among both women and men, these independent associations occurred for both workplace abuse and racial discrimination; among men, they persisted for being a perpetrator of IPV; and among women, they remained evident for poverty, high exposure to occupational hazards, and smoking. Second, in the model containing all three types of hazards, some exposures that had mattered in models containing only one type of hazard no longer were statistically significant, including high exposure to occupational hazards among men and being a target of IPV among women. In a substudy, we also found that self-reported exposure due to discrimination based on sexual orientation was associated with increased psychological distress among the LGBT workers.[122] We also found, in a second substudy, that while US-born Black Americans self-reported more experiences of racial discrimination than their foreign-born counterparts (with the latter's exposure increasing with years of residence in the United States), in both groups this exposure was strongly and similarly associated with psychological distress.[123]

Thus, supporting our a priori hypothesis, these findings suggest that important confounding due to omitted variables could bias exposure-outcome associations in analyses examining only singly the occupational, relationship, or social hazards. The larger implication is that it is critical to reckon with the joint and embodied reality of diverse types of hazards involving how people live *and* work. In other words, had we considered only relationship hazards, or only occupational hazards, or only social hazards, we would have generated an incomplete and even distorted picture of which exposures were associated with increased risk of psychological distress. Thinking through embodiment and what it means to embody (in)justice is what enabled us to avoid these problems, including omitted variable bias, and the erroneous causal inferences to which they give rise.

The results of our project led us to formulate an "inverse hazard law," which we defined as "the accumulation of health hazards tends to vary inversely with the power and resources of the populations affected."[124] We modeled this law after the influential and highly cited "inverse care law," famously penned by Julian Tudor Hart in 1971: "The availability of good medical care tends to vary inversely with the need for the population served"[125]—and which, though seemingly a self-evident statement, had not been previously concisely articulated in this way. Our formulation enabled us to capture how, clustered together and embodied conjointly, these health hazards can include economic deprivation, discrimination, and hazardous living and working conditions, together harming the health of societal groups exposed to—as compared to those who are buffered from, and often benefit from—these inequities.

One major implication of this "inverse hazard law"—and the realities of embodying (in)justice—is that to understand health inequities, research is needed that contrasts exposures and health status population-wide, not just among those most inequitably exposed.[126] As ecosocial theory and its distributional approach emphasize, research on those termed the most "marginalized" in intersectional analyses is of course critical. But a focus solely on these groups is insufficient. This is because the extent to which these groups are burdened by health inequities can be rendered visible only by situating their exposures and health profiles in their societal context and comparing them to those groups in society with the most power and most resources—and rendering the latter visible is crucial for accountability.[127]

Structural Exposures, Agency, and Accountability

As ecosocial theory further clarifies, however, measuring exposures at only one level can go only so far—whether at the individual level (e.g., self-reports), within individuals (e.g., biomarkers), or at the population level. In 1999, I published the first epidemiologic review article on discrimination and health, titled "Embodying Inequality: A Review of Concepts, Measures, and Methods for Studying Health Consequences of Discrimination."[128] At that time, within the tiny public health literature focused explicitly on discrimination and health, the focus of the extant 20 studies with individual-level data was on self-reported measures, mainly in relation to race/ethnicity, along with 7 that included data on residential racial segregation, almost always measured at the city level or higher.[129] In 2014, I published an update to this review,[130] and while the number of studies focused on individual-level measures of discrimination had soared (to the point where I was reviewing review articles!), the number using structural measures remained tiny in comparison. Other recent reviews have likewise noted that individual-level measures of exposure predominate, which cannot capture the structural drivers of social injustice.[131]

To help take this work to another level, as it were, in 2020 I prepared an invited review article for *Annual Review of Public Health* on "Measures of Racism, Sexism, Heterosexism, and Gender Binarism for Health Equity Research: From Structural Injustice to Embodied Harm—An Ecosocial Analysis."[132] To orient this review, I used ecosocial theory and its focus on embodying (in)justice to develop a novel schema (Table 2.2) that went beyond the now-conventional differentiation between structural, individual-level, and internalized measures,[133] in two ways.

First, informed by rich theorizing in political sociology,[134] I distinguished between three types of structural measures:[135]

- The first pertains to explicit "rules of the game," that is, unjust laws and policies that explicitly name the targeted group (e.g., Jim Crow).
- The second concerns nonexplicit "rules of the game," which are intended to affect but do not name the targeted group because it is no longer legal to do so (e.g., voter suppression achieved by reducing polling sites in counties with high proportions of populations of color albeit by using nonracial criteria).[136]
- The third, the most commonly used in public health research, comprises area-based or institutional indicators of injustice, which do not directly measure either explicit or nonexplicit "rules of the game," but which presumably reflect their operation (e.g., racial composition of neighborhoods or employees).

Table 2.2 **Analyzing Health Inequities: Levels, Timeframes, and Comparison Groups**

Level	Type of Measure	Example of Metric
Structural	Explicit "rules of the game": unjust laws, policies, and rules	• Laws, policies, and rules that explicitly discriminate adversely against the targeted group and privilege the dominant group
	Nonexplicit "rules of the game": unjust laws, policies, and rules	• Laws, policies, and rules that are intentionally designed to evade anti-discrimination laws in order to effectively, but not explicitly, discriminate against the targeted group and to privilege the dominant group
	Area-based or institutional indicators of injustice (legacies & current)	• Indicators: • Differences in social outcomes across targeted vs. privileged group • Population-based data on attitudes and belief in support of injustice • Audit studies of institutional or public unjust practices

(*continued*)

Table 2.2 *Continued*

Level	Type of Measure	Example of Metric
Individual (exogenous)	Explicit self-report Implicit measurement Experimental	• Explicit survey questions, or Implicit association Tests (IATs), or experimental scenarios involving exposure to unjust conditions
Internalized	Explicit self-report Implicit measurement	• Explicit survey questions, or Implicit Association Tests (IATs), or experimental scenarios involving internalization of unjust conditions
• **Comparison groups: within targeted group; between targeted and privileged group** • **Temporal issues: before vs. after change in rule; cross-sectional across places with different rules; etiologic period**		

Source: Krieger N. Measures of Racism, Sexism, Heterosexism, and Gender Binarism for Health Equity Research: From Structural Injustice to Embodied Harm-An Ecosocial Analysis. Annu Rev Public Health. 2020 Apr 2;41:37-62. doi: 10.1146/annurev-publhealth-040119-094017. Epub 2019 Nov 25

I also clarified the relevance of historical generation and lifecourse for identifying relevant pathways of embodiment. This approach newly illuminated why contemporary research on structural racism can rarely use contemporary explicit "rules of the game" measures, since such discrimination is formally illegal, but it can use older such measures (e.g., Jim Crow laws). By contrast, contemporary research focused on anti-LGBTQ discrimination can employ explicit "rules of the game" measures precisely because this type of explicit, legally sanctioned discrimination still exists. By organizing the review of extant literature in relation to this schema, one grounded in the ecosocial theory of disease distribution and its constructs involving embodying (in)justice, I could demonstrate why each type of measure requires different methodological approaches. In the review I also found that, once again, most research tended to focus on individual-level measures, leading me to propose new ideas for future research foci: conceptual, methodological, and substantive, especially in relation to structural measures.[137]

Making good on my recommendations, my current work has been investigating both new area-based structural measures of spatial social polarization and the literal "rules of the game" origins of contemporary residential economic and racial segregation. My intent has been to bring into view those who currently benefit from a status quo in which working families are below the poverty line, while the proverbial middle is hollowed out.[138] One manifestation of this problem is residential economic segregation, which in the United States is heavily racialized. The value of using measures of segregation is because they capture and portray inequity as a social relationship—because segregation by definition involves two or more social groups. One cannot focus on, say, solely impoverished persons—instead, in any such measure, affluent persons, as holders of economic power, are part of the picture as well.

However, upon delving deeper into the public health and social science literature, I confronted two problems. First, work seemed to focus primarily on racial segregation, often controlling for socioeconomic position, or, less commonly, looked at economic segregation, often not taking into account race/ethnicity.[139] Moreover, the two phenomena were not studied together, in part because technically measures of these two types of segregation typically cannot be put into the same model, given their high—albeit not total—correlation. Second, I found that most studies measured segregation at the city level or above,[140] precisely because spatial social polarization within cities means that most conventional measures cannot be used for smaller areas. For example, the Gini index of income inequality is the same for a neighborhood where everyone is impoverished and where everyone is affluent, and the same problem affects the main measure used for racial segregation, the Index of Dissimilarity.

My searching for better measures, largely absent in public health research, led me to the social science literature, where I came across an intriguing measure, developed in 2001 by Douglas Massey, one of the leading US scholars on racial segregation.[141] He termed his metric the "Index of Concentration at the Extremes,"[142] or "ICE" (which, I grant, in the United States is now an unfortunate acronym in this day and age of politically stoked anti-immigrant vitriol, since it is also the acronym for US Immigration and Customs

Enforcement[143]). Massey designed the ICE to be a measure of spatial social polarization, one that could be computed at multiple scales and levels of geography.[144]

Ranging from −1 to 1, the ICE is calculated by a provocative and deceptively simple formula:

$$ICE_i = (A_i - P_i) / T_i$$

where A_i, P_i, and T_i correspond, respectively, to the number of persons in the *i*th geographic area who are categorized as belonging to the most privileged extreme or the most deprived extreme and the total population whose privilege level was measured. The ICE thus ranges from −1 to 1, delineating areas in which 100% of the population is in the most extreme group for deprivation to 100% in the most extreme group for privilege.

The ICE thus uniquely captures, in a single metric, the extent to which an area's population is—or is not—concentrated at one or the other end of a spectrum of privilege and deprivation. At the time I started exploring using the ICE, in 2014, it had been used mainly in social science studies, plus a handful of health studies—and computed solely in relation to socioeconomic measures.[145] My novel contribution has been to extend the use of the ICE to measure both (1) racial segregation and (2) racialized economic segregation, and to do so in a multilevel framework.[146] No such joint measure of racialized economic segregation had previously been used in the social science or population health literature.

Of note, the ICE can be computed in relation to any two groups conceptualized as being at the extremes of a distribution, whether in relation to groups categorized in relation to measures of, say, socioeconomic position, race/ethnicity, nativity, or other sociodemographic measures, or even political affiliation.[147] Additionally, the ICE for racialized economic segregation can be computed in relation to diverse racial/ethnic groups, whether contrasting specific groups (e.g., Latinx vs. white non-Hispanic) or more generally contrasting people of color to white non-Hispanics. However, mindful that the most extreme residential racial segregation in the United States has been between the US non-Hispanic Black and non-Hispanic white populations,[148] my research has primarily employed an ICE for racialized economic segregation that uses this contrast. This ICE sets A_i = number of non-Hispanic white persons in the top income households (80th percentile) in neighborhood *i;* P_i = number of non-Hispanic Black persons in the bottom income households (20th percentile) in neighborhood *i*; and T_i = total population across all income percentiles in neighborhood *I*.[149]

I first introduced the ICE measures for racialized economic segregation in a study published in 2015, conducted in the Greater Boston area, which focused on an area-based exposure: air pollution.[150] My teams and I have subsequently used different ICE measures to study a wide range of outcomes, including infant mortality and premature mortality, cancer incidence and stage at diagnosis, weapons-related fatal and nonfatal assaults, and COVID-19 cases, deaths, and excess mortality.[151] In virtually all cases, the ICE for racialized economic segregation has detected the strongest gradients, compared to ICE measures that used only race/ethnicity or only income, and has also has detected stronger gradients than the poverty level. We have also shown, in studies using ICE metrics at both the census tract and city/town levels, that steeper gradients occurred, in single-level models, for the census tract compared to city/town measures, and, in multilevel models, the effects remained strongest for the census tract level and were attenuated for the city/level.[152] The importance of this finding is that it underscores that the conventional approach of using only city-level measures of residential segregation likely underestimates the impact of segregation on health.[153] Attesting to growing interest in this approach, the ICE measures we have developed are increasingly being used by other researchers to study the impact of structural racism and spatial social polarization on diverse health outcomes.[154]

Who and what, however, have driven these patterns of intense and intensifying spatial social polarization and their attendant health inequities? Here again the ecosocial theory of disease distribution has underscored the need to identify structural measures for accountability. My most recent empirical work along these lines has been to start conducting research on the impact of historical redlining on health inequities, whereby "redlining" refers to US federal policies, commencing in the 1930s and inscribed on maps, which imposed—and expanded nationwide—residential racial segregation and racialized patterns of community investment and divestment.[155] The role of government, not just private individuals, as the causal agent responsible for producing and reinforcing this residential segregation sharply reveals the structures animating structural injustice—and the basis for claims for government to repair the harms it has caused.[156]

Of note, government-instigated redlining has long been a topic of research and advocacy (e.g., for reparations) in urban planning, education (given the impact of property value on available funds for US public schools), and other fields outside of public health, especially once the original 1930s redlining maps were discovered in the US government archives in the early 1980s.[157] However, attention to the contemporary health impacts of historical redlining remains nascent, in part because these maps were digitized and made publicly available only in the mid-2010s.[158] The two studies I have led thus far in this area, on risk of preterm birth[159] and risk of late stage at cancer diagnosis,[160] were both published in 2020. Together with the nine other extant published empirical investigations (as of the start of 2021) on associations between historical redlining and current health injustice (plus one preprint), they add to the tiny literature in this area that has only just begun to emerge, with the first study published in 2017.[161] This work is likely to be further boosted by new digital resources, just released in late 2020, which show the associations, for 200 US cities, between area's redlining score, their current Centers for Disease Control and Prevention (CDC) Social Vulnerability Index, and their sociodemographic and health characteristics.[162]

This research is not simply "academic." One of the groups involved with producing these new digital resources on historical redlining and health inequities is the National Community Reinvestment Coalition.[163] Its mission, with its grassroots partners, is to "work with community leaders, policymakers and financial institutions to champion fairness and end discrimination in lending, housing and business" in order to "create opportunities for people to build wealth."[164] As I have learned from my urban planning colleagues, research linking historical redlining to contemporary adverse outcomes, including health inequities, has important implications for numerous policies addressing affordable housing and racial segregation. Examples include allocation of Low-Income Housing Tax Credits,[165] Community Reinvestment Act regulations,[166] Housing Choice Vouchers,[167] remedies under the Fair Housing Act,[168] and local policies regarding community preferences or displacement preferences for subsidized housing.[169] In particular, the 2015 federal Affirmatively Furthering Fair Housing Rule (AFFHR)[170]—passed during the Obama administration but suspended and then rescinded by the Trump administration[171]—included an assessment tool that explicitly took into account health. Among the relevant factors it said could be considered were "the type and number of hazards, the degree of concentration or dispersion (including in older housing stock), and *health effects such as asthma, cancer clusters, obesity*" (italics added for emphasis).[172]

The power of this rule, and the reason it engendered opposition from the Trump administration, is that it required US Department of Housing and Urban Development (HUD) grantees to conduct an Assessment of Fair Housing and to carry out robust public engagement in order to create specific measurable goals and actions to address obstacles to fair housing.[173] These goals were tied to future planning and assessments necessary to receive future HUD funding. For example, under the AFFHR rule, a municipality without an accepted Assessment of Fair Housing would not have been eligible to receive any of the roughly $4.6 billion in block grant funding that HUD disburses to more than 1,200 state and local governments annually.[174] Before the rule's suspension in 2018,[175] HUD had refused to accept roughly one-third of the Assessments of Fair Housing submitted since the rule went into effect in 2015.[176] This high rejection rate had put pressure on cities to create truly rigorous and innovative analyses of obstacles to achieving and goals for realizing fair housing, including data on such embodied facts as afforded by causally linking current health inequities to past government policies that enforced and expanded residential racial segregation.

Underscoring the likely relevance of the historical redlining and health equity analyses going forward, the incoming (as I write this) Biden-Harris administration has pledged to reinstate the Obama administration's AFFHR rule.[177] The US National Fair Housing Alliance's policy roadmap for the new administration has called for reinstating the prior AFFHR rule,[178] and, not surprisingly, the now-outgoing president, a real estate developer, and his conservative allies are vehemently opposed.[179] Rendering the toll of this form of structural racism visible through the embodied truths of health injustice can potentially contribute new evidence that can help with shifting the causal narrative toward health justice.[180] In other words, stay tuned!

To Be of Use: Ecosocial Analysis of Embodiment Applied to Other Exposures and Outcomes—Examples

Of course, the examples in this chapter are simply that: selected illustrations of how I have used and refined the ecosocial theory of disease distribution and its approach to embodiment. I could just as easily have employed other examples, whether primarily conceptual or involving my empirical epidemiological research on the range of other exposures and outcomes, using diverse analytic methods.[181]

Instead, to close this chapter, consider Table 2.3, which lists selected examples from the past decade of how others, in a range of academic disciplines and agencies, worldwide, have employed the ecosocial theory of disease distribution to conceptualize and analyze embodiment of (in)justice across a wide range of exposures and outcomes. This table also includes selected quotes from several articles explicitly addressing how ecosocial theory and its constructs of embodiment and embodying (in)justice have guided their work.

To be of use, after all, is the point. This phrase deliberately echoes that of the title of the first epidemiologic textbook in the English language, written by Jeremy Morris (1910–2009), a leading figure in the post–World War II rise of chronic disease epidemiology who also kept concerns about social inequalities in health center stage.[182]

Table 2.3 **Selected Recent Examples of the Explicit Use of the Ecosocial Theory of Disease Distribution and Its Approach to Conceptualizing Embodiment and Embodying (In)justice by Public Health Researchers, Practitioners, Policymakers, and Advocates as a Tool for Thinking and as a Guide to Conduct of Empirical Analyses, 2009–2020 (not including Krieger an author)**

Focus	**Example**
Theorizing about determinants of population health and health inequities	• In relation to conceptual frameworks for public health and health equity: Solar O, Irwin A. *A Conceptual Framework for Action on the Social Determinants of Health*. Social Determinants of Health Discussion Paper 2 (Policy and Practice). Geneva: World Health Organization, 2010. https://www.who.int/social_determinants/publications/9789241500852/en/ Wemrell M, Merlo J, Mulinari S, Homborg A-C. Contemporary epidemiology: a review of critical discussions within the discipline and a call for further dialogue with social theory. *Sociol Compass* 2016;10:153–171. Jayasinghe S. Conceptualising population health: from mechanistic thinking to complexity science. *Emerg Themes Epidemiol* 2011;8(1):2. doi:10.1186/1742-7622-8-2 Almeida-Filho N. Towards a unified theory of health-disease: II. Holopathogenesis. *Rev Saude Publica* 2014;48(2):192–205. Robinson WR, Bailey ZD. Invited commentary: what social epidemiology brings to the table—reconciling social epidemiology and causal inference. *Am J Epidemiol* 2020;189(3):171–174.

	Ozer EJ, Abraczinskas M, Duarte C, Mathur R, Ballard PJ, Gibbs L, Olivas ET, Bewa MJ, Afifi R. Youth participatory approaches and health equity: conceptualization and integrative review. *Am J Community Psychol* 2020;66:267–268. Harvey M. How do we explain the social, political, and economic determinants of health? A call for the inclusion of social theories of health inequality within US-based public health pedagogy. *Pedagogy Health Promotion* 2020;6(4):246–252. Breilh J. *Critical Epidemiology and the People's Health*. New York: Oxford University Press (2021).
	• In relation to history: Kramer MR. Why history? Explanation and accountability. *Am J Public Health* 2020;110(7):933–934.
	• In relation to place and health, including urban health and environmental justice: Petteway R, Mujahid M, Allen A. Understanding embodiment in place-based research: approaches, limitations, opportunities. *J Urban Health* 2019;96:289–299. Elliott SJ. 50 years of medical health geography(ies) of health and wellbeing. *Soc Sci Med* 2018;196:206–208. Corburn J. Urban place and health equity: critical issues and practices. *Int J Environ Res Public Health* 2017;14(2):117. doi:10.3390/ijerph14020117 Corburn J. Concepts for studying urban environmental justice. *Curr Environ Health Rep* 2017;4(1):61–67.

(*continued*)

Table 2.3 *Continued*

Focus	Example
	Dzudzek I, Strüver A. Urban health justice: what critical urban geography can learn from ecosocial epidemiology for researching embodied inequalities. *Geographische Zeitschrift* 2020;108(4):249–271. Pineo H, Audia C, Black D, French M, Gemmell E, Lovasi GS, Milner J, Montes F, Niu Y, Pérez-Ferrer C, Siri J. Building a methodological foundation for impactful urban planetary health science. *J Urban Health* 2020. https://doi.org/10.1007/s11524-020-00463-5
	• In relation to biological pathways of embodiment and health: Vineas P, Delpierre C, Castagné R, Fiorito G, McCrory C, Kivimaki M, Stringhini S, Carmeli C, Kelly-Irving M. Health inequalities: embodied evidence across biological layers. *Soc Sci Med* 2020;246:112781. Meloni M, Cromby J, Fitzgerald D, Lloyd S. Introducing the new biosocial landscape. In: Meloni M, Cromby J, Fitzgerald D, Lloyd S (eds). *The Palgrave Handbook of Biology and Society*. London: Palgrave Macmillan, 2017; 1–22. Kelly-Irving M, Delpierre C. The embodiment dynamic over the life course: a case for examining cancer aetiology. In: Meloni M, Cromby J, Fitzgerald D, Lloyd S (eds). *The Palgrave Handbook of Biology and Society*. London: Palgrave Macmillan, 2017; 519–540. Stringhini S, Vineas P. Epigenetic signature of socioeconomic status across the lifecourse. In: Meloni M, Cromby J, Fitzgerald D, Lloyd S (eds). *The Palgrave Handbook of Biology and Society*. London: Palgrave Macmillan, 2017; 541–560.

	• In relation to anthropology, embodiment, and health: Lock M. Toxic environments and the embedded psyche. *Med Anthro Q* 2019;34(1):21–40. Niewöhner J, Lock M. Situating local biologies: anthropological perspectives on environment/human entanglements. *Biosocieties* 2018;13:681–697. Lock M. Recovering the body. *Annu Rev Anthropol* 2017;46:1–14. Lock M. Comprehending the body in the era of the epigenome. *Current Anthropol* 2015;56(2):151–177.
	• In relation to policy: Vitrai J. How should we change the culture of health? A note on the margin of an outstanding debate. *J Public Health-Heidelberg* 2018;26(4):385–389.
	• In relation to health impacts of embodying racism: Gravlee CC. How race becomes biology: embodiment of social inequality. *Am J Phys Anthropol* 2009;139:47–57. Kuzawa CW, Sweet E. Epigenetics and the embodiment of race: developmental origins of US racial disparities in cardiovascular health. *Am J Hum Biol* 2009;21(1):2–15. Meloni M. Race in an epigenetic time: thinking biology in the plural. *Br J Sociol* 2017;68(3):389–409. Gravlee CC. Systemic racism, chronic health inequities, and COVID-19: a syndemic in the making? [published online ahead of print August 4, 2020]. *Am J Hum Biol* 2020;32(5):e23482. doi:10.1002/ajhb.23482.

(*continued*)

Table 2.3 *Continued*

Focus	Example
	• In relation to Indigenous health: Walters KL, Mohammed SA, Evans-Campbell T, Beltrán RE, Chae DH, Duran B. Bodies don't just tell stories, they tell histories. *Du Bois Rev* 2011;8:179–189. Conching AKS, Thayer Z. Biological pathways for historical trauma to affect health: a conceptual model focusing on epigenetic modifications. *Soc Sci Med* 2019;230:74–82.
	• In relation to communicable diseases and ecosystem health: Kenyon C. Emergence of zoonoses such as COVID-19 reveals the need for health sciences to embrace an explicit eco-social conceptual framework of health and disease. *Epidemics* 2020;33:100410.
	• In relation to noncommunicable diseases (NCDs), including relevant health behaviors: Adjaye-Gbewonyo K, Vaughan M. Reframing NCDs? An analysis of current debates. *Global Health Action* 2019;12(1):1611043. Budreviciute A, Damiati S, Sabir DK, Onder K, Schuller-Goetzburg P, Plakys G, Katileviciute A, Khoja S, Kodzius R. Management and prevention strategies for non-communicable diseases (NCDs) and their risk factors. *Frontiers Public Health*. 2020;8:788. Bloomfield K. Understanding the alcohol-harm paradox: what next? *Lancet Public Health* 2020;5(6):e300–e301. Moutran-Barroso HG. Epistemología, salud y género: diálogos entre la cardiología y las ciencias sociales. *Rev Colombiana Bioética* 2020;15(1).e3091.

	• In relation to reproductive health: Doll KM. Investigating Black-white disparities in gynecologic oncology: theories, conceptual models, and applications. *Gyn Oncol* 2018;149(1):78–83. Eichelberger KY, Alson JG, Doll KM. Should race be used as a variable in research on preterm birth? *AMA J Ethics* 2018;20(1):296–302. Smith MV, Lincoln AK. Integrating social epidemiology into public health research and practice for maternal depression. *Am J Public Health* 2011;101(6):990–994. Kramer M. Race, place, and space: ecosocial theory and spatiotemporal patterns of pregnancy outcomes. In: Howell FM, Porter JR, Matthews SA (eds). *Recapturing Space: New Middle-Range Theory in Spatial Demography*. Spatial Demography Book Series 1. Switzerland: Springer International Publishing, 2016; 275–299. Litt JS, Fraiman YS, Pursley DM. Health equity and the social determinants: putting newborn health in context. *Pediatrics* 2020;145(6): e20200817. https://doi.org/10.1542/peds.2020-0817
	• In relation to policing, incarceration, and restorative justice: Purtle J. Felon disenfranchisement in the United States: a health equity perspective. *Am J Public Health* 2013;103(4):632–637. Gaber N, Wright A. Protecting urban health and safety: balancing care and harm in the era of mass incarceration. *J Urban Health* 2016;93(Suppl 1):68–77. Duarte CDP, Salas-Hernández L, Griffin JS. Policy determinants of inequitable exposure to the criminal legal system and their health consequences among young people. *Am J Public Health* 2020;110(S1):S43–S49.

(*continued*)

Table 2.3 *Continued*

Focus	Example
	• In relation to lesbian, gay, bisexual, transgender, queer health: Marchia J. Lesbian, gay, and bisexual human rights as a global health issue. *Sociol Compass* 2018;12(12):UNSP e12642. Page KR, Martinez O, Nieves-Lugo K, Zea MC, Grieb SD, Yamanis TJ, Spear K, Davis WW. Promoting pre-exposure prophylaxis to prevent HIV infections among sexual and gender minority Hispanics/Latinxs. *AIDS Educ Prev* 2017;29(5):389–400.
	• In relation to disabilities and health: Filipe AM, Bogossian A, Zulla R, Nicholas D, Lach LM. Developing a Canadian framework for social determinants of health and well-being among children with neurodisabilities and their families: an ecosocial perspective. *Disabil Rehabil* 2020: 1–12. doi:10.1080/09638288.2020.1754926
	• In relation to occupational and environmental health: Kalweit A, Herrick RF, Flynn MA, et al. Eliminating take-home exposures: recognizing the role of occupational health and safety in broader community health. *Ann Work Expo Health* 2020;64(3):236–249. Okechukwu CA, Souza K, Davis KD, de Castro AB. Discrimination, harassment, abuse, and bullying in the workplace: contribution of workplace injustice to occupational health disparities. *Am J Ind Med* 2014;57(5):573–586. Hsieh YC, Apostolopoulos Y, Hatzudis K, Sönmez S. Social, occupational, and spatial exposures and mental health disparities of working-class Latinas in the US. *J Immigr Minor Health* 2016;18(3):589–599.

	• In relation to food insecurity: Canuto R, Fanton M, Lira PIC. Iniquidades sociais no consumo alimentar no Brasil: uma revisão crítica dos inquéritos nacionais [Social inequities in food consumption in Brazil: a critical review of the national surveys]. *Cien Saude Colet* 2019;24(9):3193–3212. Smith-Nonini S. The illegal and the dead: are Mexicans renewable energy? *Med Anthropol* 2011;30(5):454–474.
Methods for intersectional research and participatory research	Bauer GR. Incorporating intersectionality theory into population health research methodology: challenges and the potential to advance health equity. *Soc Sci Med* 2014;110:10–17. Evans CR. Modeling the intersectionality of processes in the social production of health inequalities. *Soc Sci Med* 2019;226:249–253. Petteway R, Mujahid M, Allen A, Morello-Frosch R. Towards a people's social epidemiology: envisioning a more inclusive and equitable future for social epi research and practice in the 21st century. *Int J Environ Res Public Health* 2019;16(20):3983. doi:10.3390/ijerph16203983 Wang X, Whittaker J, Kellom K, Garcia S, Marshall D, Dechert T, Matone M. Integrating the built and social environment into health assessments for maternal and child health: creating a planning-friendly index. *Int J Environ Res Public Health* 2020;17(24):9224.

(*continued*)

Table 2.3 *Continued*

Focus	Example
Application: conduct of empirical research on population health and embodying (in)justice	• Biological pathways of embodiment and social inequalities in health: Vineas P, Delpierre C, Castagné R, Fiorito G, McCrory C, Kivimaki M, Stringhini S, Carmeli C, Kelly-Irving M. Health inequalities: embodied evidence across biological layers. *Soc Sci Med* 2020;246:112781. • Health and place Petteway RJ, Mujahid M, Allen A Morello-Frosch R. The body language of place: a new method for mapping intergenerational "geographies of embodiment" in place-health research. *Soc Sci Med* 2019;223:51–63. Iyer HS, Valeri L, James P, Chen JT, Hart JE, Laden F, Holmes MD, Rebbeck TR. The contribution of residential greenness to mortality among men with prostate cancer: a registry-based cohort study of Black and white men. *Environ Epidemiol* 2020;4(2):e087. doi:10.1097/EE9.0000000000000087 • Population health surveillance Meagher-Stewart D, Edwards N, Aston M, Young L. Population health surveillance practice of public health nurses. *Public Health Nurs* 2009;26(6):553–560. • Discrimination and health Cormack D, Stanley J, Harris R. Multiple forms of discrimination and relationships with health and wellbeing: findings from national cross-sectional surveys in Aotearoa/New Zealand. *Int J Equity Health* 2018 Feb 17;17(1):26. doi:10.1186/s12939-018-0735-y

	Baldwin AM, Dodge B, Schick V, Sanders SA, Fortenberry JD. Sexual minority women's satisfaction with health care providers and state-level structural support: investigating the impact of lesbian, gay, bisexual, and transgender nondiscrimination legislation. *Womens Health Issues* 2017;27(3):271–278.
	Daoud N, Soskolne V, Mindell JS, Roth MA, Manor O. Ethnic inequalities in health between Arabs and Jews in Israel: the relative contribution of individual-level factors and the living environment. *Int J Public Health* 2018;63(3):313–323.
	Bell AN, Juvonen J. Gender discrimination, perceived school unfairness, depressive symptoms, and sleep duration among middle school girls. *Child Dev* 2020;91(6):1865–1876.
	• Economic trends and trends in population health
	Hammarström A, Virtanen P. The importance of financial recession for mental health among students: short- and long-term analyses from an ecosocial perspective. *J Public Health Res* 2019;8(2):1504. doi:10.4081/jphr.2019.1504
	• Well-being
	Kangmennaang J, Elliott SJ. "Wellbeing is shown in our appearance, the food we eat, what we wear, and what we buy": embodying wellbeing in Ghana. *Health Place* 2019;55:177–187.
	Kangmennaang J, Smale B, Elliott SJ. "When you think your neighbour's cooking pot is better than yours": a mixed-methods exploration of inequality and wellbeing in Ghana. *Soc Sci Med* 2019;242:112577. doi:10.1016/j.socscimed.2019.112577
	Onyango EO, Elliott SJ. Bleeding bodies, untrustworthy bodies: a social constructionist approach to health and wellbeing of young people in Kenya. *Int J Env Res Public Health* 2020;17(20):7555.

(*continued*)

Table 2.3 *Continued*

Focus	Example
	• Reproductive health Stanhope KK, Hogue CR, Suglia SF, Leon JS, Kramer MR. Restrictive sub-federal immigration policy climates and very preterm birth risk among US-born and foreign-born Hispanic mothers in the United States, 2005–2016. *Health Place* 2019;60:102209. Schwarz J, Dumbaugh M, Bapolisi W, Ndorere MS, Mwamini MC, Bisimwa G, Merten S. "So that's why I'm scared of these methods": Locating contraceptive side effects in embodied life circumstances in Burundi and eastern Democratic Republic of the Congo. *Soc Sci Med* 2019;220:264–272. Eni R, Phillips-Beck W, Mehta P. At the edges of embodiment: determinants of breastfeeding for first nations women. *Breastfeed Med.* 2014;9(4):203–214. Kola'A O, Ishola G, Bankole A. Relationship between religion and unintended childbearing in Nigeria: a cross-regional perspective. *Genus* 2020;76(1):1–20. Nardone AL, Casey JA, Rudolph KE, Karasek D, Mujahid M, Morello-Frosch R. Associations between historical redlining and birth outcomes from 2006 through 2015 in California. *PLoS One* 2020;15(8):e0237241. Agénor M, Pérez AE, Peitzmeier SM, Borrero S. Racial/ethnic disparities in human papillomavirus vaccination initiation and completion among US women in the post-Affordable Care Act era. *Ethn Health* 2020;25(3):393–407. Cubbin C, Kim Y, Vohra-Gupta S, Margerison C. Longitudinal measures of neighborhood poverty and income inequality are associated with adverse birth outcomes in Texas. *Soc Sci Med* 2020;245:112665. https://doi.org/10.1016/j.socscimed.2019.112665

	• Policing, incarceration, and restorative justice, including in relation to illicit injection drug use and HIV/AIDS Todić J, Cubbin C, Armour M, Rountree M, González T. Reframing school-based restorative justice as a structural population health intervention. *Health Place* 2020;62:102289. doi:10.1016/j.healthplace.2020.102289 Henry BF. Adverse experiences, mental health, and substance use disorders as social determinants of incarceration. *J Community Psychol* 2020;48(3):744–762. doi:10.1002/jcop.22289 Phillips JC, Domingue JL, Petty M, Coker MA, Howard T, Margolese S. HIV care nurses' knowledge of HIV criminalization: a feasibility study. *J Assoc Nurses AIDS Care* 2016;27(6):755–767. Harawa NT, Amani B, Rohde Bowers J, Sayles JN, Cunningham W. Understanding interactions of formerly incarcerated HIV-positive men and transgender women with substance use treatment, medical, and criminal justice systems. *Int J Drug Policy* 2017;48:63–71. Phillips JC. Antiretroviral therapy adherence: testing a social context model among Black men who use illicit drugs. *J Assoc Nurses AIDS Care* 2011;22(2):100–127. Phillips JC, Webel A, Rose CD, Corless IB, Sullivan KM, Voss J, Wantland D, Nokes K, Brion J, Chen WT, Iipinge S, Eller LS, Tyer-Viola L, Rivero-Méndez M, Nicholas PK, Johnson MO, Maryland M, Kemppainen J, Portillo CJ, Chaiphibalsarisdi P, Kirksey KM, Sefcik E, Reid P, Cuca Y, Huang E, Holzemer WL. Associations between the legal context of HIV, perceived social capital, and HIV antiretroviral adherence in North America. *BMC Public Health* 2013;13:736. doi:10.1186/1471-2458-13-736

(*continued*)

Table 2.3 *Continued*

Focus	Example
	Friedman SR, West BS, Pouget ER, Hall HI, Cantrell J, Tempalski B, Chatterjee S, Hu X, Cooper HL, Galea S, Des Jarlais DC. Metropolitan social environments and pre-HAART/HAART era changes in mortality rates (per 10,000 adult residents) among injection drug users living with AIDS. *PLoS One* 2013;8(2):e57201. doi:10.1371/journal.pone.0057201 Henry BF. Typologies of adversity in childhood and adulthood as determinants of mental health and substance use disorders of adults incarcerated in US prisons. *Child Abuse Neglect* 2020;99:104251. https://doi.org/10.1016/j.chiabu.2019.104251Get • Opioid use Persmark A, Wemrell M, Evans CR, Subramanian SV, Leckie G, Merlo J. Intersectional inequalities and the U.S. opioid crisis: challenging dominant narratives and revealing heterogeneities. *Critical Public Health* 2020;30(4):398–414. • Interpersonal violence Dias NG, Fraga S, Soares J, Hatzidimitriadou E, Ioannidi-Kapolou E, Lindert J, Sundin Ö, Toth O, Barros H, Ribeiro AI. Contextual determinants of intimate partner violence: a multi-level analysis in six European cities. *Int J Public Health* 2020;65(9):1669–1679. • Political violence, occupation, and health Sousa C, Marshall DJ. Political violence and mental health: effects of neoliberalism and the role of international social work practice. *Int Social Work* 2017;60(4):787–799.

	• Infectious disease Dasgupta R. Mapping cholera vulnerability in Delhi: an ecosocial perspective. *Asian J Water Env Pollution* 2010;7(1):19–26. Mulligan K, Elliott SJ, Schuster-Wallace C. The place of health and the health of place: dengue fever and urban governance in Putrajaya, Malaysia. *Health Place* 2012;18(3):613–620. Cordes J, Castro MC. Spatial analysis of COVID-19 clusters and contextual factors in New York City [published online ahead of print June 21, 2020]. *Spat Spatiotemporal Epidemiol* 2020;34:100355. doi:10.1016/j.sste.2020.100355. • Environmental health Gilbertson M. Index of congenital Minamata disease in Canadian areas of concern in the Great Lakes: an eco-social epidemiological approach. *J Environ Sci Health C Environ Carcinog Ecotoxicol Rev* 2009;27(4):246–275. doi:10.1080/10590500903310120. PMID: 19953398 Bisung E, Elliott SJ. "Everyone is exhausted and frustrated": exploring psychosocial impacts of the lack of access to safe water and adequate sanitation in Usoma, Kenya. *J Water Sanit Hyg Devel* 2016;6:205–214. Alvarez CH, Evans CR. Intersectional environmental justice and population health inequalities: a novel approach [published online ahead of print December 2, 2020]. *Soc Sci Med* 2020;113559. doi:10.1016/j.socscimed.2020.113559. • Occupational health Stuesse A. When they're done with you: legal violence and structural vulnerability among injured immigrant poultry workers. *Anthro Work Review* 2018;39(2):79–93.

(*continued*)

Table 2.3 *Continued*

Focus	Example
	• Food insecurity Hurtado-Bermúdez LJ, Vélez-Torres I, Méndez F. No land for food: prevalence of food insecurity in ethnic communities enclosed by sugarcane monocrop in Colombia. *Int J Public Health* 2020;65(7):1087–1096. • Health services van der Zande MM, Exley C, Wilson SA, Harris RV. Disentangling a web of causation: an ethnographic study of interlinked patient barriers to planned dental visiting, and strategies to overcome them. *Community Dentistry Oral Epidemiol* 2020. doi:10.1111/cdoe.12586
Quotes from selected articles regarding the utility of ecosocial theory and its approach to analyzing embodying (in)justice	a. Niewöhner J, Lock M. Situating local biologies: anthropological perspectives on environment/ human entanglements. *Biosocieties* 2018;13:681–697. p. 693: "Here epidemiologist Nancy Krieger's concept of the emergent embodied phenotype is helpful. Krieger contributes to an ecosocial theory in which bodies express ecology over time, based on the assumption that single genotypes can result in multiple phenotypes. Specific phenotypes emerge in reaction to ecological conditions, a complex process of embodiment itself situated within evolutionary, social, biographical, and cellular time or history, sometimes referred to as eco-evo-devo (Gilbert et al. 2015). Krieger uses this approach to study the development of health inequities in populations over time and how these become embodied. Importantly, her approach has a reflexive component, asking about the societal construction of disease categories and the related availability of data for any given period of time (see also Aronowitz, 2008).

	While not based on a process ontology as such, ecosocial theory offers an approach to study the distribution of disease within a given population that takes into account social and historical dynamics . . . which enables her to understand medical and biological categories as the result of epochal shifts in power/knowledge, not least in public health itself." b. Vineas P, Delpierre C, Castagné R, Fiorito G, McCrory C, Kivimaki M, Stringhini S, Carmeli C, Kelly-Irving M. Health inequalities: embodied evidence across biological layers. *Soc Sci Med* 2020;246:112781. pp. 1–2: "The multi-layered social environment within which humans exist and live ultimately affects cells, organs, and biological systems (Blane et al., 2013). This concept, known as *embodiment,* was initially developed by Nancy Krieger at the beginning of the century (Krieger, 2005). . . . In this paper we apply a biologically multi-layered approach to address the consistency of findings across measures of socio-economic conditions and health from populations to molecules, under the assumption that each layer (death, functional outcomes, DNA, RNA, proteins, infections) is characterized by different types of bias and confounding. The paper is based on a large consortium of cohorts, Lifepath, that has brought together data on socio-economic position and health from up to 1.7 million individuals, including extensive data on biomarkers." . . . "The findings show consistent associations of social disparities with unfavourable health outcomes spanning inflammatory biomarkers, DNA or RNA-based markers, infection, indicators of physical functioning and mortality. Although each of these associations has a different set of confounders, a dose-response relationship is neverthele ss consistently observed, thus showing the power of our multi-layered approach." . . . "This new evidence supports biological embodiment of social disadvantage."

(*continued*)

Table 2.3 *Continued*

Focus	Example
	c. Petteway R, Mujahid M, Allen A. Understanding embodiment in place-based research: approaches, limitations, opportunities. *J Urban Health* 2019;96:289–299. p. 290: "In public health, perhaps the most developed and useful conception of embodiment is that articulated by Krieger [22, 26]. As a foundational construct of ecosocial theory [27], embodiment is understood as the process through which the outside physical and social world becomes embedded into our biology. . . . Embodiment, then, is both continuous and dynamic, as well as both objective and subjective. Our bodies keep tally of our lived experiences—our physical and social encounters—and the health and well-being of our bodies can accordingly bear witness to the contexts and conditions of such experiences and encounters. Moreover, these contexts and conditions of embodiment are shaped and organized by societal arrangements of power, privilege, and opportunity—both current and historic. The processes and mechanisms of embodiment—the so called pathways of embodiment [22], forged through an interplay of our inner biology and the outer social world—are beholden to and an expression of such social, economic, and political arrangements. Inequalities in health across populations, then, present as "embodied expressions of social inequality [28]." p. 295: "As articulated by Krieger [22], embodiment and *pathways of embodiment* should be understood in light of and cannot be divorced from notions of *agency and accountability*. We should accordingly ask ourselves if the methodological and procedural choices we make enhance and facilitate community agency, or preclude and mask it. . . . For place-embodiment research this means evolving to include mixed-method, participatory, spatially and temporally dynamic, and intergenerational approaches that listen to what people say, in addition to what their bodies say."

d. Kramer MR. Why history? Explanation and accountability. *Am J Public Health* 2020;110(7):933–934.

p. 933: "In this issue of AJPH, Krieger et al. (p. 1046) make an important contribution . . . by asking whether the historical process of mortgage redlining in specific New York City neighborhoods predicts the risk for preterm birth among women residing in those neighborhoods and delivering liveborn, singleton infants in 2013 through 2017. . . . By contrast to much of the neighborhood effects research focusing on temporally proximate or contemporary exposures and outcomes, Krieger et al. use the 1938 maps created by the federally sponsored Home Owners Loan Corporation (HOLC) of investment 'risk' guiding mortgage lenders as predictive exposures. . . . The authors find modest empirical evidence that the 1938 HOLC categories predicted contemporary risk for preterm birth, independent of important individual demographic, socioeconomic, and health risk factors."

p. 934: "Perhaps for some, the identification of a historical neighborhood indicator as a predictor for a contemporary health outcome is an academic curiosity: interesting but not particularly actionable given that racialized lending practices have theoretically been outlawed by the Fair Housing Act of 1968, and it is presumably the contemporary neighborhood conditions that most directly affect population health outcomes, including preterm birth. However, this perspective that history is only in the past dangerously risks missing the value added from seeing contemporary health as part of a process rather than a current, static, and isolated event or state of being. . . . Rather than lacking in actionable insight, historical processes are useful because of their potential to inform public health action. . . . [H]istorically naïve public health disease-prevention efforts that do not account for the spatial and historical contingencies patterning health may fail to address root causes of health and health inequity. History provides critical explanation but also helps hold accountable actors, decisions, and processes continually shaping health and life chances."

Titled *Uses of Epidemiology*[183] and published in 1957, the book's aim was to present "epidemiology as a way of learning, of asking questions, and getting answers that raise further questions: that is, as a *method*" (italics in the original).[184]. What Morris wrote about epidemiology holds for its theories as well (as broadened to encompass the full range of population health sciences): "Epidemiology is only one way of asking some questions in medicine, one way of asking others (and no way at all to ask many)."[185]

The questions I leave you with are: How can the ecosocial theory of disease distribution and its constructs be of use to you? And how might these ideas and their application be extended? It is to latter questions that I turn in Chapter 3.

3 | CHALLENGES

EMBODIED TRUTHS, VISION, AND ADVANCING HEALTH JUSTICE

I began composing this last chapter in the waning days of August 2020 and continued writing as time permitted in the fall of 2020, acutely aware of my context (as should be expected, in accord with the ecosocial theory of disease distribution!)—with pauses due both to the intensifying COVID-19 pandemic in the United States and to mobilizing around the US elections. Especially relevant to issues of embodiment and health justice, my writing has also coincided with a heightened public awareness about—and new calls for action to address—the horrendous toll exacted by structural racism on physical and mental health, economic well-being, and environmental degradation, in the United States and worldwide, as connected also to issues of climate change.[1] With this awareness has come a regrettably predictable backlash, extending from the highest offices in US government on down to lethal violence instigated by white supremacists at protests against police violence.[2]

Heightening the visibility of these issues, on an almost daily basis since spring 2020, major newspaper and social media have been featuring stories about health inequities and tying them to unjust policies and practices, past and present, that affect the

Ecosocial Theory, Embodied Truths, and the People's Health. Nancy Krieger, Oxford University Press.
 DOI: 10.1093/oso/9780197510728.003.0003

conditions in which people live and work and the resources they and their communities have—or lack—to thrive.[3] Major US institutions—across government, business, academia, sports, news media, religion, the arts, and more—have been issuing first-ever pronouncements and pledges to end discrimination, with several also offering first-ever apologies for their roles in both ignoring and reinforcing these injustices and some even taking initial steps towards meaningful reparations.[4] By the end of August 2020, over 70 local jurisdictions in the United States newly declared racism to be a public health crisis—up from zero in 2018—and several US states were considering doing so.[5] As of late December, the numbers more than doubled: attesting to high levels of local activism, 89 US cities and towns, 67 US counties, and 3 US states (plus a fourth with legislation passed in the House and waiting for the Senate and governor to act) have declared racism to be a public health crisis.[6] Major professional organizations have also issued—many but not all for the first time ever—analogous statements declaring that racism is a public health crisis, including the American Public Health Association, the American Medical Association, the American College of Emergency Physicians, the American Psychological Association, the Society for Epidemiologic Research, and the American Heart Association, among others.[7] Whether and how these declarations will translate to concrete and meaningful action remains to be seen,[8] but they do mark a newfound public acknowledgment by government and professional organizations alike of their role in and responsibilities for rectifying health inequities resulting from structural racism.

That said, the profound connections between injustice and injuries to health have long been known—for eons—by those affected, including the myriad groups and individuals who have been organizing for justice, including health justice, for generation after generation.[9] Nor are these connections news to those in public health, health care, and other professional and academic fields who have made issues of health inequities—and the necessity of rectifying them—central to their work, whether as researchers, practitioners, teachers, or advocates, drawing on their expertise and also potentially their own experiences as members of inequitably burdened groups.[10]

When what ought to be commonplace knowledge leaps from being (1) publicly "invisible" to (2) palpably "obvious" and then (3) targeted for push-back and suppression, it should serve as an important reminder that privilege in part has always been revealed by what one can afford to ignore—and has the power to repress.[11]

Sparking this current outburst of awareness is a confluence of new and old crises, at once separate yet connected. In the United States, where I live, an abbreviated summary includes[12]

- the devastating health and economic impacts of the COVID-19 pandemic, which are compounding already skyrocketing private accumulations of wealth and massive wealth and income inequities, in the United States and globally;[13]
- heightened visibility of police brutality, police budgets, and mass incarceration, nationwide;[14]
- growing attacks on the norms and institutions necessary for democratic governance, whether by voter suppression, casting doubt on the integrity of the voting process, or distorting governance by gerrymandering (whereby incumbents divisively

draw district boundaries to pick their voters, rather than let voters pick their representatives);[15]

- ongoing attempts to manipulate the US decennial census (whose counts determine, by law, the allocation of political representation and distribution of government funding);[16]
- intensifying social polarization and amplification of misinformation, stoked and monetized by social media, pitting authoritarian, white supremacist, alt-right, nativist, racist, misogynist, anti-LGBT, anti-Semitic, fundamentalist, anti-regulatory, and anti-welfare state ideologies against democratic and human rights values and practices;[17] and
- rapidly accelerating climate change and gutting of regulation of environmental pollution, both fueled by insatiable economies reliant on fossil fuel, extractive industries, and mass consumption to increase corporate profits and private wealth.[18]

Of course, none of these crises are unique to the United States, but US issues continue to reverberate globally, reflecting its continued (even if declining) power as a major military, political, economic, and cultural force.

Notably, what I call *embodied truths* (see Chapter 1)—that is, embodied harms to population rates of health, disease, and death—feature in, cross-cut, and integrate concerns and analyses across every single set of these issues. They have become a common currency by which diverse "single issue" or "single focus" disciplines, professions, researchers, campaigns, and groups are newly understanding and articulating interlinked harms, fostering new collective strategizing for preventive action and restorative repair. When it can be shown that preventable harm to people's health benefits some at the expense of those harmed, and that these health inequities are repeated across multiple exposures and health outcomes, it potentially opens to the door to not only specific legal actions but also broader coalitions focused on primary prevention and transformative change.[19]

Illustrating this point, in a December 2020 landmark ruling, a UK death certificate for the first time ever listed air pollution as the primary cause of death for a young 9-year-old Black child, Ella Adoo-Kissi-Debrah, who died of a fatal attack of asthma, having lived near a highly congested traffic circle in southeast London.[20] Tying together issues of environmental racism and legal culpability, lawyers in and outside the United Kingdom posited this precedent could "open a new door to lawsuits by pollution victims and their families."[21] This new approach to listing cause of death has already caught the attention of a new network of pediatricians in the United States focused on raising awareness about the effects of climate change on children's health and health inequities among parents, the broader public, and politicians.[22] The issues at stake reverberate across centuries. Back in 1839, a famous (but now mainly forgotten) dispute over whether "starvation" could be listed as a cause of death embroiled William Farr, an eminent epidemiologist and statistician in the recently created office of the Registrar General of Births, Deaths, and Marriages, who argued "yes," and Edwin Chadwick, chief administrator of the Poor Law Commission, who argued "no."[23] To Chadwick, who won the debate and later led the seminal 1843 public health *Report on the Sanitary Condition of the Labouring Population*,[24] people died of specific diseases, not social conditions; individual immorality was the root cause of both poverty and sickness; and proper sanitation, as opposed to changes in wages and poor relief, could fix both problems.[25] Imagine if Farr had prevailed.

In this final chapter I will thus contend that the embodied truths of inequitable harm—coupled with recognition that emergent embodied phenotypes can express not only health inequities but also health equity—provide an integrative basis for multisectoral action for health justice, using the levers of research, organizing, litigation, policy change, social change, and political transformation. Embodiment is no metaphor. It is a living—and actionable—fact.

To make my case, I first briefly discuss the causal narratives of two current case examples that together highlight the salience of embodied truths for health justice: one concerns *police violence*, the other *global warming and climate change*. I then ask if the ecosocial theory of disease distribution and its approach to conceptualizing embodiment could spark or strengthen new possibilities for research, practice, knowledge, and action on presently poorly understood problems. Casting the net wide, I consider four very different examples, pertaining to

(1) *fossil fuel extraction and sexually transmitted infectious disease*;
(2) *health benefits of organic food—for whom?*;
(3) *public monuments, symbols, and the people's health*; and
(4) *light, vision, and the health of people and other species.*

Looping back on the deep connections between vision and theory,[26] I conclude by reflecting on why rigorous engagement with embodiment and its implications for embodying (in)justice, as conceptualized by ecosocial theory, is crucial for public health, population health sciences, and the work of health justice.

Embodied Truths and Causal Narratives in the Service of Health Justice: Police Violence and Climate Change

Concerns about harms—and specifically inequitable harms—due to police violence and to climate change are nothing new.[27] Long before the establishment of municipal police forces in the United States in the mid-19th century CE (Boston being the first city to establish such a force, in 1837), use of violence was a matter of business for privately employed security forces and for slave catcher patrols.[28] With the advent of municipal uniformed police, use of excessive violence has been repeatedly documented, variously by journalistic accounts, legal claims, and government reports, with such force targeted, in overlapping permutations, against Black Americans, striking workers, and Latinx, American Indian, LGBT, and mentally impaired populations.[29] In the case of climate change, concerns about its adverse impacts on people's health, above and beyond harms to other species, date back to the 1970s.[30]

However, as the case examples I consider show, it was only in the 2010s that both issues gained framing and traction as public health problems, featuring prominently the embodied harms affecting people's health, individually and at the population level. In this reframing, issues of causal accountability, costs, and remedies began to be analyzed as a matter of public health and health equity, no longer solely the concern, in the case of police violence, of criminologists, sociologists, and lawyers, or, in the case of climate change, of physical, natural, or environmental scientists.

Causal Stories and Causal Narratives

Before considering the specifics for each of these examples, it is worth pausing to reflect on an influential, highly cited article published in 1989 by Deborah Stone, "Causal Stories and the Formation of Policy Agendas."[31] Stone's noteworthy contribution was her analysis of the central role of causal ideas for transforming vexing and complex "issues" into tractable "problems" amenable to human intervention and action, whether involving policies and their formulation, enactment, and enforcement or other political changes. Bringing causal reasoning to the fore, it served as an important complement to another influential approach, articulated by John W. Kingdon in 1984, which analyzed policy change in relation to the opening and shutting of "policy windows" as tied to the confluence of three policy "streams" (i.e., problem streams, policy streams, and politics streams) plus chance.[32] By contrast, Stone emphasized the importance of what she termed "causal stories," which necessarily "have both an empirical and a moral dimension" and "move situations from the realm of fate to the realm of human agency."[33]

Setting up a 2×2 table as a heuristic, Stone classified people's understanding or portrayal of "causes" as (1) being either (a) devoid of intended purpose versus (b) purposeful, and (2) with either (a) intended versus (b) unintended consequences. From this standpoint, a tree branch that unexpectedly breaks with a sudden gust of wind and falls on someone who happens to be passing by is a story of unintended purpose with unintended consequences; by contrast, passage of discriminatory laws with discriminatory intent (e.g., Jim Crow laws) is a causal story of intended purpose and intended consequences.

Apart from a few examples concerning fiscal policy, the bulk of the cases considered by Stone notably pertained to human health. Examples included: injuries due to dangerous products or dangerous workplaces; illnesses due to adverse occupational exposures, environmental pollution, or chemical weapons; the prevalence of hookworm disease in the early 20th century CE; and deaths due to gun violence, drunk driving, and smoking. Observing that both law (especially tort law) and science come into play with adjudicating causal responsibilities and remedies for harms people experience, Stone delineated four key features of what causal stories accomplish. Specifically, they can[34]

1. "either challenge or protect an existing social order";
2. "assign responsibility to particular political actors so that someone will have to stop an activity, compensate its victims, or possibly face punishment";
3. "legitimate and empower particular actors as 'fixers' of the problem"; and
4. "create new political alliances among people who are shown to stand in the same victim relationship to the causal agent."

Crucially, the causal connections between these four aspects are the lived realities of embodied harm—the embodied truths simultaneously manifested in individuals' lives and in population rates and risks. Both matter: the singular instance of a case of harm and the larger pattern of which it is a part.

A corollary is that the existence of harm and population data documenting this harm are by themselves insufficient to convert bad conditions into a "problem" that can galvanize action. Instead, "strategic portrayal of causal stories"[35] is essential. Also important is understanding how strategic causal stories "implicitly call for a redistribution of power by demanding that causal agents cease producing harm and by suggesting the types of people who should be entrusted with reform" and, beyond this, "can restructure political alliances by creating common categories of victims."[36]

Only, such "victims" can become protagonists, precisely by virtue of the new political alliances forged. Illustrating this point are the new alliances currently arising in relation to both police violence and climate change by virtue of turning the focus to embodied harms (i.e., the lived realities of embodiment and embodied truths) at the population, and not solely individual, level.

Police Violence: New Options for Prevention Opened by Expanding from Framing as Chiefly a Legal Issue to Also a Public Health Issue

Search PubMed for articles on "police AND brutality" and, as of September 26, 2020, among the 68 articles identified, the first (not directly relevant) was published in 1985.[37] Thereafter, 1 or 2 articles per year appeared sporadically through 2015, when 4 articles were published, and fully 80% (i.e., 55) were published during or after 2016, mostly focused on the United States.[38] Search on "police AND racism" and, among the 153 articles identified, also mainly concerned with the United States, the first (also not directly relevant) was published in 1991.[39] One or 2 articles per year then appeared through 2012, after which 6 were published in 2013, 9 each in 2014 and 2015, and thereafter an average of 25 articles appeared per year, or 84% of the total.[40] By contrast, a Web of Science search [on topic search (TS) = (police AND (brutality OR racism)] across all databases, including social sciences, criminology, and other fields outside of public health or medicine, yielded 1,211 studies, dating back to 1945, with 113 articles published in 2016 and rising, in 2020, to 169, only two-thirds of the way through the year.[41]

The "waking up" of US public health literature in the mid-2010s to the longstanding problem of US police violence can be traced to two interlinked phenomena. First and foremost was the emergence of a powerful social movement, Black Lives Matter, which was founded in 2013 by three radical Black women—Alicia Garza, Patrisse Cullors, and Opal Tometi.[42] The spark was the acquittal of George Zimmerman, who, in an act of vigilante murder, had shot and killed Trayvon Martin, a 17-year-old African American, in Sanford, Florida in 2012.[43] Further galvanizing its growth were two high-profile police killings in 2014: Michael Brown, an 18-year-old African American, in Ferguson, Missouri, killed by shooting, and Eric Garner, a 44-year-old African American, in New York City, killed by a chokehold.[44] Also contributing was the fast-rising ability for people to document and share these killings on social media, in real time, along with expanded capacity of activists and journalists to compile online databases that newly brought into public focus, in real time, the frequency, location, and circumstances of these killings. Examples include Fatal

Encounters, a freely accessible website set up in 2012 by an independent journalist and whose records extend back to 2000;[45] Fatal Force, available to subscribers of the *Washington Post*,[46] with data back to 2015; and, with data for 2015–2016 only, *The Guardian's* freely available website The Counted.[47] The contrast to the incomplete and often unavailable or delayed data at the national, state, and local level was stark.[48] As stated in 2015 by the director of the FBI, James Comey: "It is unacceptable that the Washington Post and the Guardian newspaper from the UK are becoming the lead source of information about violent encounters between [US] police and civilians. . . . It's ridiculous—embarrassing and ridiculous."[49]

The new population data on the extent of US police killings—documenting upward of 1,100 such deaths per year, averaging ~3 per day[50]—have served as an important complement to the evidence brought forth for each specific case. Premised on the view that these data are public health data—and not solely criminal justice data[51]—these population health data have spurred work both to validate the open-source counting of police-involved deaths[52] and to improve public health monitoring of not only these deaths[53] but also police-involved injuries.[54] One contribution of this work has been to quantify the extent to which US vital statistics, US Department of Justice reports, and police department reports have vastly underreported the extent of these deaths, with estimates on the order of 50% or more.[55] Also gaining new traction is attention to potential "spillover" effects, affecting the health and well-being of entire communities, beyond the death of those killed and its impact on their immediate family members.[56] Taking into account relevant contextual and individual-level covariates (including crime rates), preliminary suggestive associations documented at the community level with police violence include increases in risk of high blood pressure,[57] adverse mental health,[58] and sexually transmitted infectious disease.[59] New research using regression discontinuity models has also pointed to harmful delays in Black persons going to emergency rooms for needed care in the wake of media reports of police killing an unarmed Black person.[60]

Additionally, these population health data have vividly exposed the high prevalence of cases in which police have used excessive and at times deadly force against persons who were mentally disabled at the time of the encounter.[61] This evidence has buttressed calls by public health professionals, advocates, and activists alike to redirect funds to enable staffing and deployment of response teams with mental health expertise to defuse situations and provide care.[62] These new coalitions of groups concerned with criminal justice and with public health have also heightened attention to and sparked debate over the decades-long disproportionate and escalating share of funding for police relative to social services and public health in local and state budgets.[63] They have also pointed to how public health departments can address police violence as a public health issue.[64] Framing structural police violence (as opposed to an individualistic blaming of so-called "bad apples") as a public health issue,[65] one with profound implications for health equity, is opening new possibilities for action. While thus far studied most extensively in the United States, investigations carried out in other continents and geopolitical contexts, spanning from Africa to East Asia to Europe, point to the global salience of these issues.[66]

In other words, the embodied integrated truths of harm caused by excessive use of policing and police force, translated into stories that bodies tell through population health data, alongside individual cases, is setting the basis for an integrative causal narrative that is creating new alliances. These new coalitions are, in Stone's prescient words, calling "for a redistribution of power by demanding that causal agents cease producing harm" and "suggesting the types of people who should be entrusted with reform."[67] Specifically, these new alliances have brought into the mix people concerned with health equity and reparative justice, offering new options for redress and repair, in ways that build on and move beyond a focus solely on criminal justice and carceral systems, informed by a broader vision of health justice.

Climate Justice: New Options for Prevention and Repair Opened by Expanding from a Framing as Chiefly an Environmental Issue to Also a Public Health Issue

Climate change, climate justice, and health offers a parallel case. Search on the Web of Science, using all databases across the social and natural sciences, for the topic TS = (global AND warming) or TS = (climate AND change), and, as of November 27, 2020, the number of records retrieved respectively equaled 104,621[68] and 589,249.[69] Of these, however, respectively only 8% (8,356)[70] and 7.3% (43,423)[71] belonged to the category of "public environmental occupational health"[72]—and only starting in the 2010s did a handful of articles began jointly addressing issues of environmental justice, climate justice, and public health.[73] PubMed searches using these same search terms yielded similar patterns: one or two articles per year on global warming or climate change began appearing in the mid-1970s, with the pace picking up considerably after 2000.[74] Use instead the phrase "(climate justice) and (public health)," and relevant articles began appearing only in the 2010s, just a few per year, but began to reach up to and beyond 40 per year starting in 2018.[75] The implication is that the literature jointly focused on climate change, climate justice, and public health comprises only a tiny, albeit fast-increasing, fraction of the scientific literature on climate change.[76]

Confirming the trends observed in these types of literature searches, the first assessment report by the Intergovernmental Panel on Climate Change (IPCC, based within the United Nations and founded in 1988[77]), published in 1990, included no discussion of potential impacts on human health.[78] The main foci instead were on physical aspects of climate change (temperature, ocean level), impacts on agriculture and forestry, and the physics and economics of energy sources.[79] The IPCC's second assessment report, issued in 1995,[80] continued with these foci, It also, however, touched briefly (in two pages) on potential impacts on human health, but noted that efforts to quantify these impacts were constrained by both the complexity of the task and the meager literature on the topic.[81]

Only in the third assessment report, published in 2001, did the IPCC first include a full chapter focused on "Human Health."[82] This chapter, 32 pages long, included 7.5 pages of references in small font in dual columns. Noting the "caveats and challenges" to "research on the relationship between climate change and health" (p. 456), it then systematically worked its way through myriad aspects warranting attention. Topics addressed were: "sensitivity, vulnerability, and adaptation"; "thermal stress (heat waves, cold spells); "extreme events and weather disasters"; "air pollution"; "infectious diseases"; "coastal water issues"; "food yields and nutrition"; "demographic and economic disruption"; "adaptation options"; "secondary health benefits of mitigation policies"; "research and information needs, including monitoring"; and "cross-cutting issues," in relation to "costing the health impacts of climate change" and "development, sustainability, and equity."

Since then, subsequent IPCC assessment reports, issued in 2007[83] and 2014,[84] have included expanded chapters on climate change and human health, each longer than the other—and each increasingly more attentive to how social inequities affect vulnerability to and may exacerbate impacts of climate change on health. The theme of potentially improving health and health equity, as a co-benefit of shifting from fossil fuels to renewable energy, gained prominence in the 2014 chapter (pp. 737–742). The outline for the upcoming sixth assessment (to be finalized by 2022) also includes a health chapter, one further expanded to consider the "psychological, social, and cultural dimensions" of how climate change can affect human health.[85] Hans-Otto Pörtner, co-chair of the Working Group II of the IPPC, notably stated, in November 2020: "Given current events and the increasing urgency to build back better after COVID-19, we anticipate that there will be great interest in several areas of our report such as the health chapter and the chapters on ecosystems, food, water, cities and climate-resilient development."[86]

The 2020 report of the *Lancet Countdown* on health and climate change likewise underscored the centrality of equity issues linking health and climate change, observing that "climate change interacts with existing social and economic inequalities and exacerbates longstanding trends within and between countries."[87] The report further stated: "An examination of the causes of climate change revealed similar issues, and many carbon-intensive practices and policies lead to poor air quality, poor food quality, and poor housing quality, which disproportionately harm the health of disadvantaged populations."[88] It also connected the rising scientific publications linking climate change and health (increasing by "a factor of eight from 2007 to 2019"[89]) to mounting public discourse on these issues, stating:[90]

> From 2018 to 2019, the coverage of health and climate change in the media increased by 96% worldwide, outpacing the increased coverage of climate change overall, and reaching the highest observed point to date (indicator 5.1). Just as it did with advancements in sanitation and hygiene and with tobacco control, growing and sustained engagement from the health profession during the past 5 years is now beginning to fill a crucial gap in the global response to climate change.

Further arguing that the societal and economic responses to COVID-19 and climate change should be aligned with each other, the *Lancet Countdown* report concluded that this joint work "must take advantage of the moment to combine public health and climate change policies in a way that addresses inequality directly."[91]

The rising interest in emphasizing the human health and equity dimensions of climate change is likewise evident in the 2020 websites of major US groups tackling climate change.[92] One of the oldest groups, the National Wildlife Federation, in 2020 released its first ever "Equity & Justice 2020 Strategic Plan," which is as much about their plans to focus on climate change, environmental justice, and health as it is about how they have to reorient who comprises their staff and the training they receive. They accordingly state:[93]

> Climate change is causing chaos in ecosystems, with temperatures and rainfall less and less predictable and more extreme. Many communities have experienced disinvestment, are faced with unsafe drinking water, and unclean air. We know that healthy communities mean healthy habitats for fish and wildlife. But our work is harder because our country is growing farther apart at the very moment we must work closely together. The environmental movement as a whole has not yet done its part to bring people together, especially across race. Conservation organizations like the National Wildlife Federation are over-represented by white staff and leadership when compared to our cities, towns, and our country as a whole.

The Sierra Club, long known for its emphasis on wildlife and environmental issues, but not human health, now features as one of its three primary "issues" the topic of "people and justice," with a focus on creating a "healthy world for all."[94] Similarly, the National Resource Defense Council, founded in 1970 with a focus on environmental issues only, now includes among its six primary programs "Healthy People & Thriving Communities."[95] It states that it is "working to ensure that everyone—not just some—benefits from a clean and healthy environment" and highlights the need to "foster the development of vibrant, healthy, and sustainable communities—especially working with those who have suffered from disinvestment and inequitable distribution of resources."[96] Pulling these different themes together, the newest organization—the Sunrise Movement, founded in 2017—describes itself as "building an army of young people to make climate change an urgent priority across America, end the corrupting influence of fossil fuel executives on our politics, and elect leaders who stand up for the health and wellbeing of all people."[97]

The turn to human health and health equity in research and advocacy regarding climate change is no accident. Partly, it reflects the greater engagement of public health agencies and organizations and population health scientists with this topic, as likely aided, in the US context, by the willingness of the Obama administration (2008–2016) to fund work in this area. For example, in 2009, the Institute of Medicine Roundtable on Environmental Health Sciences, Research, and Medicine organized a "Workshop on a Research Agenda for Managing the Health Risks of Climate Change," from which participants established an ad hoc federal Interagency Working Group on Climate Change and Health. This group involved the National Institute of Environmental Health Sciences (NIEHS), the National Oceanic and Atmospheric Administration (NOAA), the Centers for Disease Control and Prevention (CDC), and the Environmental Protection Agency (EPA). Its first report, issued in 2011, was a call for research, and was titled *A Human Health Perspective on Climate Change: A Report Outlining the Research Needs on the Human Health Effects of Climate Change.*[98]

A follow-up assessment, published in 2016, responded to "the rising demand for data that can be used to characterize how climate change affects health."[99] That same year, the American Public Health Association (APHA) articulated its first-ever "Climate Change and Health Strategic Plan,"[100] which undergirded its 2017 annual conference theme of "Year of Climate Change and Health"[101] and led to its establishment of the APHA Center for Climate, Health and Equity,[102] whose inaugural advisory board was appointed in June 2020.[103] Moreover, since 2019, the US National Resource Defense Council has been releasing a series of reports on the public health impacts related to climate change in different states, including how they exacerbate health inequities and health costs,[104] and journalists are increasingly reporting on how health departments lack the resources and power to curb these harms.[105]

But scientific insights and advances were not the only reason for the new emphasis on human health outcomes. Also salient was growing recognition, among both scientists and advocates, that to get political traction and increase support of the larger public for measures to mitigate climate change, it was essential to recast its impacts in human terms. One study in the United Kingdom, for example, found that participants were most mobilized by images linking climate change to air quality and the impacts of air pollution on people's health, as compared to images involving floods, heat, or infectious diseases.[106]

Starting in the early 2010s, this shift to engaging with human impacts and human health became popularized by the phrase "People not polar bears"[107]—not because there was no longer concern about polar bears, but because it was not enough to be concerned solely about them. Polar bears had, after all, become iconic symbols of the threat of global warming for destruction of habitats. However, for the majority of the world's people—other than Indigenous peoples living in the Artic, who were clear that disrupted environments portend terrible omens for not just their own but all life on this planet[108]—polar bears and melting ice flows seemed, however inaccurately, remote from daily life. They could readily be viewed as a luxury to worry about, a stance that likely suited those in the fossil fuel industries and their allies wielding power in the industrialized countries most responsible for carbon dioxide emissions just fine—with the added benefit, to industry, that polar bears, as such, could not sue for damages. Indeed, in 2013, the US and Russian governments, seeking to block curtailment of arctic drilling and mining, attempted to reframe the polar bear problem as one chiefly tied to Indigenous hunting practices, not climate change, thus changing the focus of legal liability.[109]

Documentation of the petrochemical industry's opposition to curbing use of fossil fuels is extensive, as is its adoption of many of the same obfuscating tactics employed by Big Tobacco since the 1960s to avoid regulation or banning of its product, as well as by other industries responsible for dangerous products or pollutants.[110] In all of these cases, once legal liability looms for damage to people's health—via a damage that cannot be excused away with either victim-blaming narratives or narratives of "we didn't know" (when the evidence shows they did)—the ability of people to sue can contribute to reducing harmful exposures and getting legislators to legislate to prevent further harm. Not that any such victories are inevitable: litigation is expensive and can drag on for years, and it can take multiple cases before the cost of damages and changes to regulations begin to have an impact. Consider only the decades of litigation, still ongoing, still unresolved, regarding the generations of health harms linked to the concentration of oil refineries and petrochemical plants in Louisiana's "Cancer Alley"—yet some curbs, however inadequate, on emissions have been established, and the fight continues.[111]

To date, the new arena of climate litigation has focused chiefly on deception and harm to properties, and not the health harms experienced by the plaintiffs.[112] However, the latest appeals, filed in March 2020, in support of *Juliana v. United States*, which is youth-led climate litigation primarily centered around their claim of a constitutional right to a stable environment (aka a right to a future),[113] now also emphasize threats to their health and safety, supported by public health "friend of the court" filings.[114] Similar lawsuits emphasizing the public health threat of climate change are underway in several US states, with support from public health and medical organizations and professionals.[115] Moreover, one state, Colorado, is seeking to include health harms from climate change under the rubric of current legislation that "protects public health and the environment from dangerous gas and oil operations."[116] In August 2020, the Union of Concerned Scientists (UCS) notably set up a UCS Science Hub for Climate Litigation, stating that "Litigation by Black, Indigenous, Latinx, and other marginalized communities to prevent human rights violations and hold corporations and governments accountable for toxic pollution and other abuses could yield important lessons for climate litigation."[117]

Whether this litigation will make a difference remains to be seen. Nevertheless, the reframing of climate change as a threat to human health and health equity is demonstrably helping to create new coalitions and initiatives. These new ensembles have the potential to expand greatly the support of the public, politicians, and ultimately the body politic, nationally and globally, for the changes needed not only to avert climate crisis but also to embark on the path of building sustainable and equitable societies. Once again, embodied truths—and evidence of bodily harms—matter, jointly at *both* the individual and population levels.

An Ecosocial Envisioning of Embodiment and New Paths for Research and Action for a Thriving World

The point of scientific theories is not simply to explain the occurrence of phenomena in the present (or past)—it is also to help identify data gaps and predict what no one has yet observed, whether about phenomena in the present or past, reflecting questions still unasked, or about phenomena that could transpire in the future. Here I briefly discuss four examples to illustrate what an ecosocial focus on integrated embodied truths might offer, vis-à-vis evidence relevant to envisioning new paths for research and action for a more equitable and sustainable, aka thriving, world (Figure 3.1).

Example 1: Fossil Fuel Extraction and Sexually Transmitted Infectious Disease

The health impacts of fossil fuel extraction are chiefly analyzed, understandably, in relation to the biophysical harms affecting people and other organisms that are caused by exposure to toxic

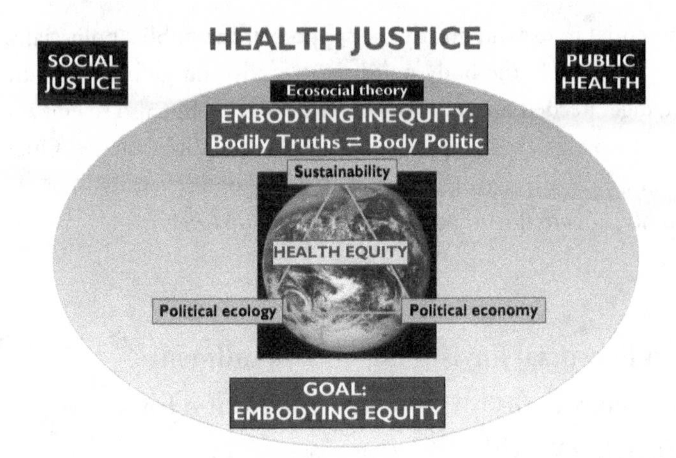

Figure 3.1 Health justice: ecosocial theory.

Image Credit: NASA.

substances in fossil fuels or the materials employed to extract them, as well as by climate change due to the burning of fossil fuels.[118] However, ecosocial theory would prompt the additional question: do the embodied impacts of fossil fuel extraction extend beyond these biophysical hazards and include the societal hazards of these extractive processes? These latter processes are, after all, driven by people—first and foremost the owners of and investors in fossil fuels, and, secondarily, the people employed to enable and carry out the extractive work.[119]

To date, scant literature has focused on the inequitable human toll, reckoned in diseases *not* directly linked to environmental hazards, of the typically boom-and-bust cycles of extractive industries, including psychosocial harms linked to community and landscape disruption.[120] Recently, however, a tiny but growing number of quantitative public health studies have focused on links between fracking and increased risk of sexually transmitted infectious diseases (STIs), especially syphilis, gonorrhea, and chlamydia.[121] The causal processes at issue pertain to how boom-and-bust economies often jointly lead to increases in transient worker populations and predominantly heterosexual sex work, alongside limited public health or health care infrastructure for prevention, treatment, and contact tracing.[122] The rise of fracking-associated US "man camps" has also been linked to increases in violence against women, especially Indigenous women, given the common siting of these locations on reservation lands, which are hampered by legal constraints on tribal prosecution of non-Indigenous assailants.[123]

One US study, published in 2017, used the methodology of difference-in-difference to examine trends in the gonorrhea incidence among counties that did versus did not experience an increase in fracking in the US Marcellus Shale region, which extends from West Virginia to Ohio, Pennsylvania, and New York.[124] The principal finding was that "fracking activity is associated with a 20 per cent increase in gonorrhea" (p. 477), which the investigators attributed to the combination of an influx of transient workers and the inability of overwhelmed rural counties to carry out public health screening and prevention campaigns.[125] A 2020 national study, using this same methodology, likewise found a 24% increase in gonorrhea risk.[126] In 2018, a study in Pennsylvania observed that fracking was associated with higher rates of both gonorrhea and chlamydia,[127] while a study in Ohio reported increases in gonorrhea and chlamydia rates, comparing counties with high activity versus no activity for shale gas extraction.[128] A 2020 study empirically documented higher gonorrhea rates in North Dakota counties in the oil-producing compared to non-oil-producing regions in the boom years (2006–2014), with contributing factors identified by the study's qualitative component including scant public health infrastructure, lack of available housing, and a "socially conservative environment"—which inhibited use of services for STI testing and reproductive health.[129] Studies linking oil booms and extractive mining industries to increased risk of STIs in other countries, ranging from Nigeria to Canada, reveal similar devastating impacts.[130]

An ecosocial analysis would not only foster such analyses but also further flag that the "outcome" of an STI can simultaneously be an embodied determinant of future birth outcomes, whereby infection with both gonorrhea and chlamydia during pregnancy is associated with increased risk of preterm birth and still birth.[131] To date, I have not identified any studies that have explored this likely link—and I predict this will be observed, a consequence of causal pathways of embodiment.

The larger implication is that tallying the toll of fossil fuel extraction requires reckoning with the actual ways in which fossil fuel extraction takes place, carried out by real people in real places, with impacts on community as well as environmental health, affecting both the present and future generations. It also requires development of interventions, at multiple levels, ranging from policies to support renewable energy and curtail harmful extractive industries to repairing the ruptured social and personal relations arising from these harms.[132] Once again, an integrated approach to embodiment and health justice matters.

Example 2: Health Benefits of Organic Food—for Whom?

Consider as well the case of the health benefits of organic food. Currently, the overwhelming focus of research, as documented by Cochrane reviews, concerns the adverse impacts of (1) pesticide residues on human consumers' nonorganic foods (with weak or inconsistent evidence of harm) and (2) pesticide pollution on nonhuman species (with robust evidence of harm to myriad mammals, birds, fish, reptiles, and insects, above and beyond those targeted as pests).[133] A search of Web of Science in late December 2020 across all databases likewise located 104,029 articles on the harmful ecosystem impacts of pesticides, 5,288 articles on organic food and consumer health, but only 1,536 articles focused on organic food and occupational health.[134]

As even this simple literature search reveals, who and what are missing from the dominant picture? Agricultural workers and their families.[135] Illustrating the problem, a 2017 article in the high-impact *Annual Review of Public Health,* focused on "Organic Food in the Diet: Exposure and Health Implications," considered consumer health at length (a major focus of marketing research[136]), briefly mentioned environmental health, and was silent regarding occupational health.[137] Another 2017 review article, expansively titled "Human Health Implications of Organic Food and Organic Agriculture: A Comprehensive Review," simply noted that "the focus of this article is on public health, not on occupational health of agricultural workers or local residents, although these issues are considered as part of the epidemiological evidence on pesticide effects."[138] Yet, as documented by a 2018 review article on plant-based diets and pesticides, "occupational exposure and pesticide self-poisoning directly imperil the health of 28 million people per year" who are landless agriculture workers, added to which are myriad "resource-poor smallholders in the tropics, where protective gear is rarely used."[139]

The net implication is that more comprehensive and actionable answers to the question "who benefits from organic food?" would be provided by jointly considering the interwoven pesticide-linked threats to the health of both people—as workers and/or consumers (since workers are consumers and consumers can be workers)—and other living beings in the affected ecosystems. In the United States, examples of emerging models of government agencies newly linking evidence and action across these domains include the Consumer, Environmental & Worker Protection Division for Alameda County, California,[140] and the 2016 expansion of New York City's Department of Consumer Protection to become the Department of Consumer and Worker Protection.[141] Further suggesting a more comprehensive causal narrative could bolster coalitions among advocates for worker, consumer, and environmental health, evidence indicates consumers can be—and have been—motivated to alter purchasing choices and participate in effective consumer boycotts and buycotts based on concerns about harms to others (both workers and the environment), not just harms to themselves.[142] The scarcity of scientific research, let alone health and consumer education materials, that jointly addresses and documents the linked embodied truths for these three sectors—workers, consumers, and the ecosystem at large—speaks to the urgency of the need for apt integrative frameworks, such as that afforded by the ecosocial theory of disease distribution.

Example 3: Public Monuments, Symbols, and the People's Health: Reckoning with Embodied Reverberations of Oppressive Histories and Present Realities

Hiding in plain sight is yet another type of exposure likely relevant to population health and health inequities, on which scant population health research exists: the embodied impacts of the presence and removal of oppressive public monuments and symbols and their replacement by more inclusive public art and government symbols.[143] This is distinct from, yet related to, the considerable literature on the violence of colonizer and state actions, in pursuit of power, territory, and wealth, to destroy a subjugated population's culture, language, religion, rituals, historical knowledge, kinship ties, and modes of dress, hairstyles, and public behavior afforded its individual members, conjoined with enforced political disenfranchisement and economic exploitation.[144]

Bringing this question of monuments, symbols, and the people's health to the fore was the surge of massive anti-racist protests, in the United States and globally, that erupted in the wake of the brutal US police murder on May 25, 2020, of George Floyd, a 46-year-old African American man, horrifically killed by a white police officer.[145] Floyd died, as I have written previously, "because he could not breathe, because police officer Derek Chauvin knelt on his neck for an agonizing 8 minutes and 46 seconds—in open view, as videoed for all to see, while three other police standing nearby failed to intervene."[146] During the myriad ensuing demonstrations and mobilizations—which built on generations of anger, as discussed earlier—US protestors toppled public Confederate monuments (mainly erected in the late 19th and early

20th century CE[147]) and also statues of Christopher Columbus and other Europeans who led the colonization of what became the United States.[148] In other countries, protestors tore down public statues of people who reaped fortunes from the slave trade or from colonizing the demonstrators' own or other countries.[149] Even the US military finally banned display of Confederate flags on its bases.[150]

Not only toppling but also replacement came to the fore. For example, on July 22, 2020, the US House of Representatives voted to remove from the Capitol "all statues of individuals who voluntarily served" the Confederacy[151]—to be replaced with telling alternatives. In addition to Confederate generals, also served notice was Supreme Court Chief Justice Roger B. Taney (1777–1864), an enslaver who entrenched white supremacy via the 1857 *Dredd Scott v. Sanford* decision.[152] This case ruled that no Black person in the United States, whether enslaved or free, could then or ever have the rights of US citizens—and asserted they were "so far inferior, that they had no rights which the white man was bound to respect."[153] After Taney's death, the abolitionist Senator Charles Sumner tried to block appropriations for the customary bust accorded deceased Chief Justices.[154] The now-chosen replacement? A newly commissioned bust of Supreme Court Justice Thurgood Marshall (1908–1993), its first Black member and a staunch supporter of civil rights.[155]

Another replacement took place on November 3, 2020, when the voters of Mississippi approved a new state flag, which back in 1894 had been redesigned to include the Confederate emblem, four years after the revised 1890 state constitution disenfranchised Black residents and imposed Jim Crow; a prior referendum to change the flag, in 2001, lost 2:1.[156] The new flag, to fly starting in January 2021, features a magnolia flower, the state's symbol, surrounded by a circle of 20 stars, symbolizing its status as the 20th state to join the union, and at the top and center of the circle is an additional golden star created by five diamond shapes, to represent Mississippi's Indigenous Native American tribes.[157]

A related vote reversal took place on June 27, 2020, when Princeton University's Board of Trustees voted, after years of protest, to remove Woodrow Wilson's name from its public policy school and a residential college, having previously voted against such a change in 2006.[158] Wilson had served as president of Princeton from 1902 to 1910, after which he became US president in 1913 and used his power to racially segregate the nation's civil service, reversing decades of their racial integration.[159] Following suit, the Board of Trustees of Stanford University voted on October 7, 2020, to remove from all buildings the name of its first president, David Starr Jordan (who served from 1891 to 1913), on account of his highly influential support for eugenics.[160] Continuing suit, on January 15, 2021, Caltech (the California Institute of Technology) removed the names of Robert A. Millikan (founding president and first Nobel laureate) and five other eugenics proponents from "buildings, honors, and assets."[161]

Similarly, in the United Kingdom, on June 19, 2020, the University College London (UCL) announced it was denaming spaces and buildings named after two prominent eugenicists: Francis Galton (who, as noted earlier, coined the term *eugenics*) and Karl Pearson, one of the influential founders of biostatistics (whose name still features prominently in the "Pearson correlation coefficient").[162] It followed this decision with its first ever public apology, issued on January 7, 2021, for the institution's role in the "development, propagation, and legitimization of eugenics."[163] Its apology (see Box 3.1) condemns eugenics and both its historical and ongoing impact, starkly stating, "This dangerous ideology . . . provided justification for some of the most appalling crimes in human history: genocide, forced euthanasia, colonialism and other forms of mass murder and oppression based on racial and ableist hierarchy."[164] Moreover, recognizing that "the right to freedom of expression is not unfettered," the UCL further affirmed that, while operating in "a climate of academic freedom," it would "commit to closing down any opportunity for this legacy to continue unacknowledged and unchallenged."[165]

In the midst of this swirl of protest, in summer 2020 a legal team contacted me to ask if there was any empirical public health evidence (as opposed to opinions) about the harms of being exposed to oppressive public symbols, since the existence of such empirical studies could help bolster their case.[166] I agreed to look—and could find none. That said, news coverage about the demonstrations and votes included numerous interviews with people expressing strong feelings about the removal and replacement of these symbols of white supremacy.[167] Rich literature from the humanities, arts, and social sciences has likewise explored public views on—but not the health impacts of—removing oppressive public symbols and replacing them with more inclusive public displays that reveal entwined histories of domination and resistance.[168] A newly launched website, Invisible Hate, offers an interactive map with the locations and histories of the 700+ publicly visible Confederate monuments in the United States and encourages individuals to share their photos and views about these

Box 3.1

University College London (UCL) Formal Apology for Its History and Legacy of Eugenics (Issued January 7, 2021)

UCL acknowledges with deep regret that it played a fundamental role in the development, propagation and legitimisation of eugenics. This dangerous ideology cemented the spurious idea that varieties of human life could be assigned different value. It provided justification for some of the most appalling crimes in human history: genocide, forced euthanasia, colonialism and other forms of mass murder and oppression based on racial and ableist hierarchy.

The legacies and consequences of eugenics still cause direct harm through the racism, antisemitism, ableism and other harmful stereotyping that they feed. These continue to impact on people's lives directly, driving discrimination and denying opportunity, access and representation.

UCL considers its history of involvement in eugenics to be in direct contradiction to its founding values of equality, openness and humanity. As a community, we reject eugenics entirely.

We apologise fully, and with humility, to all those who have suffered and to those who are still suffering because of our role in creating the conditions that enabled eugenics to become established and acted upon.

We apologise for being slow to interrogate properly the history and legacy of eugenics at UCL, and for failing to act with sufficient speed to remedy the ongoing effects on those in our community who are the targets of the eugenic mentality.

We apologise for honouring individuals who were leading eugenicists through the naming of spaces on campus.

We recognise that the legacy of eugenics is ongoing and we commit to closing down any opportunity for this legacy to continue unacknowledged and unchallenged. We operate in a climate of academic freedom, but we recognise that the right to freedom of expression is not unfettered.

UCL pledges to continue to confront its history of eugenics and ongoing legacies openly and critically, and to ensure that UCL staff and students are enabled to do the same.

Source: University College London. UCL makes formal public apology for its history and legacy of eugenics. January 7, 2021. https://www.ucl.ac.uk/news/2021/jan/ucl-makes-formal-public-apology-its-history-and-legacy-eugenics; accessed February 14, 2021.

sites on social media, plus provides advice on how to lobby representatives to take them down.[169]

What might it look like to conduct such population health research? Here again an integrative ecosocial approach to analyzing the embodiment of (in)justice can be of use, in part to point to the complexities. At issue is not only people's appraisal of these symbols (conceptualized as psychosocial stressors if oppressive, or buffers if inclusive) but also their relationship to the very events, organizing, and shifts in power dynamics propelling or blocking these shifts in public symbols. Conceptualizing the impact(s) requires considering whom, in what timeframe, and what outcomes. For example, did being involved as an organizer or protestor for change positively affect people's mental health, health behaviors, and physical health? And might it also have positively affected their expectations for respect and dignity from, and interactions with, diverse institutions, including in relation to health care? What about persons not directly involved in the organizing? Might their health impacts differ by historical generation: Elders who have seen these symbols for decades? Younger people still new to the histories of what these symbols represent? Those in between? Future generations? How might news of these events impact the health and well-being of people who live in other locales? Any and all of these impacts may be relevant, requiring studies to pay heed to what they are and are not measuring; in whom; where; in relation to what exposures and potential confounders, mediators, and modifiers; and using what study designs.

One likely useful initial strategy, given the dearth of evidence, would be to employ, at the population level, the classic difference-in-difference design.[170] Such an approach would compare changes in the specified health outcomes, before versus after the change in symbols, across otherwise comparable sites that did versus did not remove the symbols. Specification of the appropriate pre- versus postchange time period would need to take into account the etiologic period for the selected outcomes. Moreover, to avoid or minimize selection bias, the health outcomes data would ideally come from public health monitoring systems, for example, vital statistics (birth or death records) or representative population surveys.

Suggesting such a design is both feasible and warranted is provocative evidence from a 2019 study led by Heidi A. Vuletich and B. Keith Payne. They extended a prior study's longitudinal investigation of the impact of interventions on students' implicit association test (IAT) scores for racial bias across 18 US college campuses[171] by additionally testing for associations with the presence of Confederate monuments, conceptualized as being measures of structural inequality.[172] A range of other covariates were included, both institutional (e.g., measures of faculty diversity, selectivity in admission rates) and individual (e.g., SAT scores). In addition to showing, in nested models, that individual students' IAT scores were unstable and fluctuated over time, but not systematically, they found that average campus scores both were effectively stable over time and were strongly correlated with presence of Confederate monuments.[173] While unable to rule out selection effects (i.e., which students seek to enroll at which campuses), the investigators did a secondary analysis that randomly assigned students to 18 campus-size groups, so as to delink individuals from their campus context while still maintaining "group" size. The results indicated that individual-level scores remained unstable over time and the group-level effects shrank, suggesting that campus-based characteristics, not just aggregation or individual-level scores, are at issue. Inferring that IAT scores are not a stable property of individuals, as conventionally portrayed, but instead "may more properly be considered a property of social context,"[174] the investigators said the results implied that "certain contexts encourage discrimination more than others," and that "changing the social environment may be more effective than changing individual attitudes."[175]

Substitute concurrent measures of psychological distress for the IAT, or, with relevant etiologic lags, other mental or physical health outcomes or health behaviors, include relevant covariates, and the potential should be clear for estimating the health impacts of structural change that brings about the removal and replacement of oppressive public symbols with more inclusive ones. Similar designs could be used to estimate the need for and likely impact of long-overdue material reparations to mitigate historically forced structural chasms in power, sovereignty, property, wealth, income, and education between the groups at issue—but for now, these can only be predictions, because securing such reparations has yet to be won in the United States or elsewhere.[176]

Example 4: Light, Vision, and the Health of People and Other Species

My last set of empirical examples focuses on light and vision, since light from the sun is crucial for life on Earth, and the capacity to sense and respond to light is ancient and present in all branches of life: from single-celled organisms both with and without nuclei to multicellular organisms of all types, whether plants, fungi, or animals.[177] They serve to illustrate the utility of thinking through and raising new questions about the realities, for every life form, of living as emergent embodied phenotypes,[178] ceaselessly expressing and integrating social and biological phenomena within and across levels, from macro to nano—and with drivers of disease rates and inequities increasingly residing in the uniquely human body politic. The point, again, is to use the ecosocial theory of disease distribution both to illuminate presently neglected issues and to make causal predictions about thus far untested hypotheses regarding drivers of population health distributions.

The first two examples stem from the mounting global scale of political economies that alter light and temperature, two fundamental cues and enabling conditions for biological existence on Earth.[179] Although estimates vary as to when people first gained the history-transforming capacity to control and manipulate fire for both light and heat (likely dating back some 400,000 years, and for cooking, perhaps 50,000 years), for millennia the use and impacts were predominantly local, with wood, vegetation, animal fats, and tar pitch being the primary fuels.[180] Jump ahead to the 19th century CE, with the ratcheting up of scale via industrial capitalism's ravenous burning of coal and oil for power and electricity, conjoined with the invention of electrical lighting and its global spread in the 20th century CE, and alteration of light and heat have become global.[181]

One potential consequence, increasingly a focus of research, is the rising risk of a potential mismatch between the temperature and light cues that inform seasonal cycles of pollination that link plants and their animal pollinators.[182] The timing of when plants produce buds and flowers depends on both light (i.e., photoperiod), which is unaffected by climate change (since the earth just keeps on rotating as it does, on its axis as it circles the sun), and temperature, which is affected by climate change. Amplifying concerns about climate change, if light and temperature start to uncouple from current configurations that have been extant for millennia, fundamental cycles of reproduction for entire ecosystems, as well as human agriculture, could be adversely affected, likely magnifying health inequities.[183] Change environs, change emergent expressed phenotypes, and changes in population patterns of health will follow.

Even switching to sustainable sources of energy that do not disrupt the earth's climate while also mitigating the health-damaging energy insecurity of energy-poor communities globally[184] will not, however, address the second problem gaining growing attention: massively increasing light pollution.[185] Research has already demonstrated the adverse effects on the reproductive, migratory, hunting, and foraging behaviors of animals of all types.[186] Among people, studies increasingly implicate outdoor light pollution in disrupting sleep (above and beyond people's light exposure when they are awake), thereby harming daily well-being while also potentially increasing risk of cardiometabolic diseases and cancer.[187] As per the issue of climate change, however, only recently has scholarship and advocacy to reduce light pollution started to render visible its linked impacts on ecosystem and human health and health inequities.[188] Deeper and more widespread understanding of how people necessarily embody, biologically, our societal and ecological context, within and across historical generations, is essential for confronting and changing political and economic drivers of unsustainable practices that disrupt ecosystems and both harm people's health and magnify health inequities.

My third and last example concerns vision and the exquisite sensitivity of eye development to context and use.[189] When nationally representative US data on the prevalence of myopia (nearsightedness) first became available in the early 1970s, shortly after the legal abolishment of racial segregation of public schools, the prevalence was two times higher among US white compared to

Black children ages 12 to 17 (25.8% vs. 12.0%), an embodied sign of their disparate educational and reading opportunities.[190] By 2000, this racialized gap had disappeared (34.5% vs. 31.2%), driven by the faster rise in myopia among the Black versus white children (200% vs. 30%), in parallel with improved reading.[191] Magnifying problems with extant inequities in access to and insurance for visual care, higher prevalence of myopia in turn raises needs for adequate visual screening and provision of eyeglasses, which if not provided can then further compound educational inequities, since children who cannot see well are at greater risk of not being able to do well at school.[192] Other countries, such as China, whose policies have dramatically increased reading, computer use, and other screen time among children, have also experienced quickly rising rates of childhood myopia and contingent concerns about adequate detection and treatment.[193] Supporting the hypothesis that societal changes underlie these trends, two recent Mendelian randomization studies found strong evidence that factors affecting educational attainment increase risk of myopia.[194] Moreover, a 2020 Spanish study found that heritability of myopia among twins declined from 79% to 25% among those born in the mid- versus late 20th century CE, a rapid huge drop explicable only by social changes.[195]

Do these emerging trends mean, however, that myopia is a necessary "price" for "progress"? No. Spurred by concerns about potential links between rising myopia rates and children's increasing screen time (whether for educational or recreational purposes), new research has begun to yield evidence, including from both longitudinal studies and randomized clinical trials, that school-based activities that increase children's time outdoors can reduce risk and progression of myopia.[196] Two of the more prominent hypotheses as to why, both requiring more rigorous research, concern the likely beneficial effects on eye development of more exposure to sunlight and seeing more things at a greater distance.[197] These hypotheses notably are consonant with growing public health interest in the health benefits of people connecting with life outdoors,[198] as well as long-standing critical Indigenous knowledge and theorizing about the centrality of people's relationships with their environs to health—that is, reationships between people, their communities, and the ecosystems of which they are a part.[199]

The larger embodied truth is that it literally matters to be outside and able to see the forest, from afar, not just individual trees, and to see the real horizon, not just an image on a phone or computer screensaver. Additionally, issues of social justice will drive which children have access to quality educational programs that simultaneously encourage reading and computer use and structure and ensure the children's ability and access to see the bigger world, literally and figuratively.

Embodiment is real, socially structured, and ingrained in our very marrow – and our capacity for sight, and also insight.

Ecosocial Theory, Vision, and Embodying (In)justice—the Work "Embodied Truths" Can Do

My ending on examples concerned with vision is of course deliberate, because vision is what animates theory—and, vice versa, theory affects what one even sees.[200] Theory is thus fundamental to conceptualization and understanding of, and action to address, the integral connections between embodied truths and health justice. As I have noted both before and earlier in this book,[201] the etymology of the word *theory* traces back to the Greek word *theorin*, referring to inner vision, and is itself derived from a word pertaining to *theatre* and, by inference, viewing spectacles.[202] Theory is indeed inner spectacle that coherently structures the way people see and understand their world, as well as who and what they ignore, and thus shapes approaches to conceptualizing causal processes and who and what count as causal agents.[203] Theory as spectacle, and spectacles as eyeglasses, together conjoin in metaphor how seeing and processing what the world affords—as phenomena relevant to whatever are the tasks at hand, including living, surviving, and thriving[204]—is integrally bound with being alive on our planet. And, in the case of science, theory rests on and spurs the identification of robust phenomena and causal processes that can and must be publicly tested and replicated by independent researchers.[205] It is not a matter of belief.

Reflecting on the notion of critical science, scientific evidence, and embodied truths is warranted in this present moment of heightened falsehoods about scientific evidence (especially regarding climate change and COVID-19) fueled and funded by self-interested agendas of power and economics,[206] coupled with public mistrust rooted in awareness of injustices perpetrated in the name of science (including scientific racism).[207] Here it is worth quoting again the trenchant reply provided in 2019 by Donna Haraway, a well-known critic and practitioner of science,[208] in an interview asking about scientific knowledge in the face of anti-science campaigns and mistrust:[209]

> Our view was *never* that truth is just a question of which perspective you see it from. . . . The idea that reality is a question of belief is a barely secularized legacy of the religious wars. In fact, reality is a matter of worlding and inhabiting. It is a matter of testing the holdingness of things. Do things *hold* or not? Take evolution. The notion that you would or would not "believe" in evolution already gives away the game. If you say, "Of course I believe in evolution," you have lost, because you have entered the semiotics of representationalism—and post-truth, frankly. You have entered an arena where these are all just matters of internal conviction and have nothing to do with the world. You have left the domain of worlding.

A corollary is that if the world we live in is complex, integrated, and historically contingent, so too must be our theories to understand and engage with worldly processes.

For this, we need the kind of vision that theory affords. Of course, this does not mean literal vision. After all, in the anti-slavery hymn "Amazing Grace," written and published in 1779 by John Newton, a reformed ex-slaver, he framed his realization of the wrongs of enslavement as: "Was blind, but now I see."[210] And, as observed by Dr. H. Jack Geiger (1925–2020), a lifelong civil rights activist who cofounded the community health center movement in the United States, Physicians for Social Responsibility and Physicians for Human Rights,[211] when he started losing his eyesight toward the end of his life: "[I'm] living proof that you can lose your sight, but not your vision."[212]

The notions of "embodied truths" and "embodying (in)justice" that I have developed in this brief book are firmly grounded in the ecosocial theory of disease distribution. This theory's premise of integrated embodied interconnection, in historical and ecological context as tied to political economy, political ecology, and social justice, of course has its roots, as described, in the fruitful theorizing that many have done, in and outside of public health, and in and outside of academia, as discussed in Chapter 1. The work that the ecosocial constructs of *embodying (in)justice* and *embodied truths* can do is to provide clarity, concisely, about who and what give rise to the phenomena that define the focus of public health and population health sciences—that is, population patterns of health, disease, and health injustice. As I have argued throughout this book (and elsewhere), contrary to the premises of reductionist and individualist biomedical and lifestyle theories of disease distribution, the origin of these patterns lies not in innate biology and individual characteristics, even as the processes of health and disease necessarily manifest in individual bodies. Needed instead is an approach that comprehends individuals as part of populations structured by the existence of, resistance to, and transformations of societal economies and their structures of power and property and their impacts on people and the ecosystems in which we live.

The constructs of embodying (in)justice and embodied truths likewise give leverage to translating the abstract analyses offered by theory into concrete hypotheses that can be tested. I have offered several such examples in this book. Of course, no one study tests all ideas in a theory. Rather, theory helps to frame not only the specific hypotheses to be tested but also analysis of how these tests may be flawed, whether through selection bias of study populations, measurement bias, omitted variable bias, or additional biases imposed by the limitations of apt statistical methods.[213] Theory frames the specific questions scientists ask, how they interpret the answers, and which questions and possible answers they ignore.[214] In public health, a reflexive field in which the purpose of research is to generate knowledge that can change the phenomena studied (i.e., improve population health and promote health equity),[215] a central question for any study is: what is the knowledge gap you are addressing, and whom do you envision doing what, with whom, based on the knowledge produced? An expansive view is what provides the capacity to focus wisely, so as to generate sound, specific, and actionable knowledge. It is worth employing a quote I have excerpted before,[216] from Edgar Sydenstricker's classic 1933 monograph *Health and Environment*, but this time in full, with my flagging a critical admonition in italics:[217]

> Environment, according to the concept that has been emphasized in the foregoing pages, includes all of the external circumstances, both physical and social, which come into relation to the lives of human beings.
>
> This is a broad concept, so broad as to seem to preclude any possibility of determining accurately the reaction of individuals, differentiated as they are by heredity, to environmental conditions. Actually, however, there is no need for so complete an array of facts as the concept appears to imply. Data are relatively important in proportion to their relevancy. The search for knowledge cannot be carried out in all directions simultaneously by any one student. The most practicable procedure obviously is to arrive at a reasonably accurate understanding of one specific environmental factor at a time. If the student is in a position to collect the data

> that appear pertinent as the logic of the particular situation unfolds, he [*sic*] will be able to suggest conclusions that, if confirmed by other inquiries, will possess scientific value and ultimately social utility. All experience of an army of painstaking workers in scientific fields attests to the soundness and the practicability of this method. *Yet it is a dangerous procedure because the too cocksure are likely to overlook important factors other than the one which is the especial object of study for the moment* [italics added]. It is for this reason that so broad a concept of environment has been insisted upon and the perils of too narrow an outlook upon the complexities of the subject of environment and health have been so persistently emphasized in this monograph.

Arguing against the simplistic yet powerful eugenic credos of his time, Sydenstricker concluded:[218]

> As new discoveries are made in the relations of environment and health, social efforts will be further developed to combat the effects of deleterious conditions. To an increasing extent we shall strive through social action to determine what our environment and our heredity will be. What is needed is more knowledge, dispassionately collected and scientifically analyzed with a wholesome respect for the complexities of human society and of the individuals who compose it, to form a sound basis for the conscious control of our destinies.

In the nearly century since Sydenstricker penned these words, the dangers of decontextualized analyses and genetic reductionism have become clear, and as I have reviewed, empirical discoveries and conceptual developments in social, biological, and population health sciences underscore the urgent need for contextualized integrated theorizing and action. Yes, this is complex, but so too is our world—and the causal narratives can, I posit, be stated clearly.

It is in this spirit that I offer this *ecosocial theory's* concise, clear, and vivid constructs of *embodying (in)justice* and *embodied truths*, together at play in the lived realities of *emergent embodied phenotypes*. May these big ideas, presented in this small book, be of use—to researchers, practitioners, advocates, and activists, in and outside of public health—to collective cross-cutting efforts to understand and alter the structures of injustice that give rise to health inequities and planetary peril. The goal is clear: to rectify current harms and provide sound knowledge to inform the actions needed to make good on the idea that another world is possible,[219] premised on principles of social and health justice broadly writ, so that all can thrive.

—February 15, 2021

NOTES

EPIGRAPHS

1. Morrison, T. *The Source of Self- Regard: Selected Essays, Speeches, and Meditations* (New York: Alfred A. Knopf, 2019, p. 239).
2. Pullman, P. *Dæmon Voices: On Stories and Story Telling* (New York: Alfred A. Knopf, 2017, pp. 72–73).
3. Haraway, D. Interview: "A Giant Bumptious Litter: Donna Haraway on Truth, Technology, and Resisting Extinction". *Logic* 2019; December 9. https://logicmag.io/nature/a-giant-bumptious-litter/; accessed December 30, 2020.

PREFACE

1. See, for selected examples, Krieger N. Epidemiology and the web of causation: has anyone seen the spider? *Soc Sci Med* 1994;39(7):887–903. doi:10.1016/0277-9536(94)90202-x; Krieger N. Embodying inequality: a review of concepts, measures, and methods for studying health consequences of discrimination. *Int J Health Serv* 1999;29(2):295–352. doi:10.2190/M11W-VWXE-KQM9-G97Q; Krieger N. Refiguring "race": epidemiology, racialized biology, and biological expressions of race relations. *Int J Health Serv* 2000;30(1):211–216. doi:10.2190/672J-1PPF-K6QT-9N7U; Krieger N. Epidemiology and social sciences: towards a critical reengagement in the 21st century. *Epidemiol Rev* 2000;22(1):155–163. doi:10.1093/oxfordjournals.epirev.a018014; Krieger N. Theories for social epidemiology in the 21st century: an ecosocial perspective. *Int J Epidemiol* 2001;30(4):668–677. doi:10.1093/ije/30.4.668; Krieger N. A glossary for social epidemiology. *J Epidemiol Community Health* 2001;55(10):693–700. doi:10.1136/jech.55.10.693; Krieger N (ed). *Embodying Inequality: Epidemiologic Perspectives.* Amityville, NY: Baywood Publishing Co., 2004; Krieger N. Embodiment: a conceptual glossary for epidemiology. *J Epidemiol Community Health* 2005;59(5):350–355. doi:10.1136/jech.2004.024562; Krieger N. Proximal, distal, and the politics of causation: what's level got to do with it? [published online ahead of print January 2, 2008]. *Am J Public Health* 2008;98(2):221–230. doi:10.2105/AJPH.2007.111278.; Krieger N. *Epidemiology and the People's Health: Theory and Context.* New York: Oxford

University Press, 2011; Krieger N. Who and what is a "population"? Historical debates, current controversies, and implications for understanding "population health" and rectifying health inequities. *Milbank Q* 2012;90(4):634–681. doi:10.1111/j.1468-0009.2012.00678.x; Krieger N. History, biology, and health inequities: emergent embodied phenotypes and the illustrative case of the breast cancer estrogen receptor [published online ahead of print November 15, 2012]. *Am J Public Health* 2013;103(1):22–27. doi:10.2105/AJPH.2012.300967; Krieger N. Got theory? On the 21st c. CE rise of explicit use of epidemiologic theories of disease distribution: a review and ecosocial analysis. *Curr Epidemiol Rep* 2014;1(1):45–56; Krieger N. Living and dying at the crossroads: racism, embodiment, and why theory is essential for a public health of consequence. *Am J Public Health* 2016;106(5):832–833. doi:10.2105/AJPH.2016.303100; Krieger N. Theoretical frameworks and cancer inequities. In: Vaccarella S, Lortet-Vieulent J, Saracci R, Conway DI, Straif K, Wild CP (eds). *Reducing Social Inequalities in Cancer: Evidence and Priorities for Research* (IARC Scientific Publication No. 168). Lyon, France: International Agency for Research on Cancer, 2019; 189–204. http://publications.iarc.fr/580; Krieger N. Measures of racism, sexism, heterosexism, and gender binarism for health equity research: from structural injustice to embodied harm-an ecosocial analysis [published online ahead of print November 25, 2019]. *Annu Rev Public Health* 2020;41:37–62. doi:10.1146/annurev-publhealth-040119-094017.

2. Oxford University Press: *Small Books Big Ideas in Population Health*. https://global.oup.com/academic/content/series/s/small-books-big-ideas-in-population-health-sbbi/?cc=us&lang=en&; accessed August 16, 2020.
3. Beckfield J, with series edited by Nancy Krieger. *Political Sociology and the People's Health*. New York: Oxford University Press, 2018. https://global.oup.com/academic/product/political-sociology-and-the-peoples-health-9780190492472?lang=en&cc=us.
4. Friel S, with series edited by Nancy Krieger. *Climate Change and the People's Health*. New York: Oxford University Press, 2019. https://global.oup.com/academic/product/climate-change-and-the-peoples-health-9780190492731?lang=en&cc=us.
5. Breilh J, with series edited by Nancy Krieger. *Critical Epidemiology and the People's Health*. New York: Oxford University Press, 2021. https://global.oup.com/academic/product/critical-epidemiology-and-the-peoples-health-9780190492786?lang=en&cc=us#.
6. Krieger N. ENOUGH: COVID-19, structural racism, police brutality, plutocracy, climate change-and time for health justice, democratic governance, and an equitable, sustainable future [published online ahead of print August 20, 2020]. *Am J Public Health* 2020;110(11):1620–1623. doi:10.2105/AJPH.2020.305886; Buchanan L, Bui Q, Patel JK. Black Lives Matter may be

the largest movement in U.S. history. *New York Times*, July 3, 2020. https://www.nytimes.com/interactive/2020/07/03/us/george-floyd-protests-crowd-size.html; accessed February 15, 2021; Cave D, Albeck-Ripka L, Magra I. Huge crowds around the globe march in solidarity against police brutality. *New York Times*, June 6, 2020; updated June 9, 2020. https://www.nytimes.com/2020/06/06/world/george-floyd-global-protests.html; accessed February 15, 2021; Asmelash L. How Black Lives Matter went from a hashtag to a global rallying cry. *CNN*, July 26, 2020. https://www.cnn.com/2020/07/26/us/black-lives-matter-explainer-trnd/index.html; accessed February 15, 2021.

7. Krieger N. ENOUGH: COVID-19, structural racism, police brutality, plutocracy, climate change-and time for health justice, democratic governance, and an equitable, sustainable future [published online ahead of print August 20, 2020]. *Am J Public Health* 2020;110(11):1620–1623. doi:10.2105/AJPH.2020.305886; Chotiner I. The interwoven threads of inequality and health. The coronavirus crisis is revealing the inequities inherent in public health due to societal factors, Nancy Krieger, a professor of social epidemiology, says. (Interview with Nancy Krieger). *New Yorker*, April 14, 2020. https://www.newyorker.com/news/q-and-a/the-coronavirus-and-the-interwoven-threads-of-inequality-and-health; accessed February 14, 2021; for a full listing of the numerous COVID-19 empirical and conceptual publications I have authored or coauthored, see: https://www.hsph.harvard.edu/nancy-krieger/covid-19-publications/; more generally, see https://www.hsph.harvard.edu/nancy-krieger/.
8. See, for example, Krieger N. ENOUGH: COVID-19, structural racism, police brutality, plutocracy, climate change-and time for health justice, democratic governance, and an equitable, sustainable future [published online ahead of print August 20, 2020]. *Am J Public Health* 2020;110(11):1620–1623. doi:10.2105/AJPH.2020.305886; Boyd RW, Krieger N, Jones CP. In the 2020 US election, we can choose a just future [published online ahead of print October 19, 2020]. *Lancet* 2020;396(10260):1377–1380. doi:10.1016/S0140-6736(20)32140-1; Hanage WP, Testa C, Chen JT, Davis L, Pechter E, Seminario P, Santillana M, Krieger N. COVID-19: US federal accountability for entry, spread, and inequities-lessons for the future [published online ahead of print November 2, 2020]. *Eur J Epidemiol* 2020;35(11):995–1006. doi:10.1007/s10654-020-00689-2; for a full listing of the numerous COVID-19 empirical and conceptual publications I have authored or coauthored, see https://www.hsph.harvard.edu/nancy-krieger/covid-19-publications/.
9. Haberman M. Trump told crowd "You will never take back the country with weakness." *New York Times*, January 6, 2021. https://www.nytimes.com/2021/01/06/us/politics/trump-speech-capitol.html; accessed February 14, 2021; Fandos F. Trump impeached for insurrection. *New York Times*, January 13,

2021; updated February 11, 2021. https://www.nytimes.com/2021/01/13/us/politics/trump-impeached.html; accessed February 14, 2021; Leatherby L, Ray A, Singhvi A, Triebart C, Watkins D, Willis H. How a presidential rally turned into a Capitol rampage. *New York Times*, January 12, 2021. https://www.nytimes.com/interactive/2021/01/12/us/capitol-mob-timeline.html; accessed February 14, 2021; Fandos F. Trump acquitted of inciting insurrection, even as bipartisan majority votes "Guilty." *New York Times*, February 13, 2021. https://www.nytimes.com/2021/02/13/us/politics/trump-impeachment.html; accessed February 14, 2021; Holpuch A. "White supremacy won today": critics condemn Trump acquittal as racist vote. *The Guardian*, February 14, 2021. https://www.theguardian.com/us-news/2021/feb/14/trump-acquittal-white-supremacy-racist-vote; accessed February 14, 2021.

10. Krieger N. The censorship of seven words by Trump's CDC could well cost American lives [Op-Ed]. *New York Daily News*, December 18, 2017. http://www.nydailynews.com/opinion/censorship-words-trump-cdc-cost-lives-article-1.3707447; accessed December 30, 2020; Krieger N. Climate crisis, health equity, and democratic governance: the need to act together. *J Public Health Policy* 2020;41(1):4–10. doi:10.1057/s41271-019-00209-x; Krieger N. ENOUGH: COVID-19, structural racism, police brutality, plutocracy, climate change-and time for health justice, democratic governance, and an equitable, sustainable future [published online ahead of print August 20, 2020]. *Am J Public Health* 2020;110(11):1620–1623. doi:10.2105/AJPH.2020.305886; Frickel S, Rea CM. Drought, hurricane, or wildfire? Assessing the Trump Administration's anti-science disaster. *Engaging Sci Tech Society* 2020 Jan 8;6:66–75; Rutledge PE. Trump, COVID-19, and the war on expertise. *Am Rev Public Admin* 2020;50(6-7):505–511; Oreskes N, Conway EM. *Merchants of Doubt: How a Handful of Scientists Obscured the Truth on Issues from Tobacco Smoke to Global Warming*. New York: Bloomsbury Publishing USA, 2010; Michaels D. *Doubt Is Their Product: How Industry's Assault on Science Threatens Your Health*. Oxford, New York: Oxford University Press, 2008; Michaels D. *The Triumph of Doubt: Dark Money and the Science of Deception*. Oxford, New York: Oxford University Press, 2020; Markowitz G, Rosner, D. *Deceit and Denial: The Deadly Politics of Industrial Pollution* (California/Milbank Books on Health and the Public; 6). Berkeley: University of California Press, 2007; Brandt A. *The Cigarette Century: The Rise, Fall, and Deadly Persistence of the Product That Defined America*. New York: Basic Books, 2007; Proctor R. *Golden Holocaust: Origins of the Cigarette Catastrophe and the Case for Abolition*. Berkeley: University of California Press, 2011; Proctor R, Schiebinger L. *Agnotology: The Making and Unmaking of Ignorance*. Stanford, CA: Stanford University Press, 2008; Mooney C. *The Republican War on Science*. New York: Basic Books, 2005.

CHAPTER 1

1. Pan American Health Organization (PAHO). *Just Societies: Health Equity and Dignified Lives. Report of the Commission of the Pan American Health Organization on Equity and Health Inequalities in the Americas.* Washington, DC: PAHO, 2019. https://iris.paho.org/handle/10665.2/51571; accessed December 30, 2020; Birn AE, Pillay Y, Holtz T. *Textbook of Global Health.* New York: Oxford University Press, 2017; Krieger N. *Epidemiology and the People's Health: Theory and Context.* New York: Oxford University Press, 2011; Beckfield J. *Political Sociology and the People's Health.* New York: Oxford University Press, 2018.
2. Sigerist HE. *A History of Medicine, Volume I: Primitive and Archaic Medicine.* New York: Oxford University Press, 1951; Sigerist HE. *A History of Medicine, Volume II: Early Greek, Hindu, and Persian Medicine.* New York: Oxford University Press, 1951; Rosen G. *A History of Public Health* (1958). Revised expanded version, with introduction by Elizabeth Fee and biographical essay and new bibliography by Edward T. Morman. Baltimore, MD: Johns Hopkins University Press, 2015; Hutchinson DL. *Disease and Discrimination: Poverty and Pestilence in Colonial Atlantic America.* FL: University Press of Florida, 2016; Birn AE, Pillay Y, Holtz T. *Textbook of Global Health.* New York: Oxford University Press, 2017; Cuetos M. *Medicine and Public Health in Latin America: A History.* New York: Cambridge University Press, 2015; Krieger N. *Epidemiology and the People's Health: Theory and Context.* New York: Oxford University Press, 2011.
3. Rosen G. *A History of Public Health* (1958). Revised expanded version, with introduction by Elizabeth Fee and biographical essay and new bibliography by Edward T. Morman. Baltimore, MD: Johns Hopkins University Press, 2015; Rosenkrantz B. *Public Health and the State: Changing Views in Massachusetts, 1842–1936.* Cambridge, MA: Harvard University Press, 1972; Hamlin C. *Public Health and Social Justice in the Age of Chadwick: Britain, 1800–1854.* Cambridge, UK: Cambridge University Press, 1998; Birn AE, Pillay Y, Holtz T. *Textbook of Global Health.* New York: Oxford University Press, 2017; Krieger N. *Epidemiology and the People's Health: Theory and Context.* New York: Oxford University Press, 2011.
4. Birn AE, Pillay Y, Holtz T. *Textbook of Global Health.* New York: Oxford University Press, 2017; Krieger N. *Epidemiology and the People's Health: Theory and Context.* New York: Oxford University Press, 2011; Krieger N. Who and what is a "population"? Historical debates, current controversies, and implications for understanding "population health" and rectifying health inequities. *Milbank Q* 2012;90(4):634–681; Beckfield J. *Political Sociology and the People's Health.* New York: Oxford University Press, 2018.
5. Wade N. *A Troublesome Inheritance: Genes, Race and Human History.* New York: Penguin Press, 2014; Plomin R. *Blueprint: How DNA Makes Us Who We Are.* Cambridge, MA: MIT Press, 2018; Murray C. *Human Diversity: The Biology of Gender, Race, and Class.* New York: Twelve, 2020.

6. Whitehead M. The concepts and principles of equity and health. *Int J Health Serv* 1992;22(3):429–445; Macinko JA, Starfield B. Annotated bibliography on equity in health, 1980–2001. *Int J Equity Health* 2002;1(1):1. doi:10.1186/1475-9276-1-1; Braveman P, Gruskin S. Defining equity in health. *J Epidemiol Community Health* 2003;57(4):254–258; Pan American Health Organization (PAHO). *Just Societies: Health Equity and Dignified Lives—Report of the Commission of the Pan American Health Organization on Equity and Health Inequalities in the Americas.* Washington, DC: Pan American Health Organization, 2019. https://iris.paho.org/handle/10665.2/51571; accessed December 30, 2020.
7. Krieger N. *Epidemiology and the People's Health: Theory and Context.* New York: Oxford University Press, 2011; Krieger N. Who and what is a "population"? Historical debates, current controversies, and implications for understanding "population health" and rectifying health inequities. *Milbank Q* 2012;90(4):634–681; Gilbert SF, Sapp J, Tauber AI. A symbiotic view of life: we have never been individuals. *Q Rev Biol* 2012;87(4):325–341; Gilbert SF, Tauber AI. Rethinking individuality: the dialectics of the holobiont. *Biol Philos* 2016;31(6):839–853; Clarke AE, Haraway D. *Making Kin Not Population.* Chicago: Prickly Paradigm Press, 2018; Eldredge N, Grene M. *Interactions: The Biological Context of Social Systems.* New York: Columbia University Press, 1992; Latour B, Schaffer S, Gagliardi P (eds). *A Book of the Body Politic: Connecting Biology, Politics, and Social Theory.* San Giorgio Dialogue 2017. Venezia, Italy: Fondazione Giorgio Cini, 2020.
8. Krieger N. Embodying inequality: a review of concepts, measures, and methods for studying health consequences of discrimination. *Int J Health Serv* 1999;29(2):295–352; Krieger N. Embodiment: a conceptual glossary for epidemiology. *J Epidemiol Community Health* 2005;59(5):350–355; Krieger N. Theories for social epidemiology in the 21st century: an ecosocial perspective. *Int J Epidemiol* 2001;30(4):668–677; Krieger N, Davey Smith G. "Bodies count," and body counts: social epidemiology and embodying inequality. *Epidemiol Rev* 2004;26:92–103.
9. Krieger N. History, biology, and health inequities: emergent embodied phenotypes and the illustrative case of the breast cancer estrogen receptor. *Am J Public Health* 2013;103(1):22–27; Piersma T, van Gils JA. *The Flexible Phenotype: A Body-Centered Integration of Ecology, Physiology, and Behavior.* Oxford: Oxford University Press, 2011; Gilbert SF, Epel D. *Ecological Developmental Biology: The Environmental Regulation of Development, Health, and Evolution.* 2nd ed. Sunderland, MA: Sinauer Associates, 2015; Niewohner J, Lock M. Situating local biologies: anthropological perspectives on environment/human entanglements. *Biosocieties* 2018;13(4):681–697; Lock M. Recovering the body. *Annu Rev Anthropol* 2017;46:1–14.
10. Krieger N. History, biology, and health inequities: emergent embodied phenotypes and the illustrative case of the breast cancer estrogen receptor [published online ahead

of print November 15, 2012]. *Am J Public Health* 2013;103(1):22–27. doi:10.2105/AJPH.2012.300967; Krieger N. Inheritance and health: what really matters? *Am J Public Health* 2018;108(5):606–607. doi:10.2105/AJPH.2018.304353.

11. Krieger N. Discrimination and health. In: Berkman L, Kawachi I (eds). *Social Epidemiology*. Oxford: Oxford University Press, 2000; 36–75; Krieger N. Discrimination and health inequities. In: Berkman LF, Kawachi I, Glymour M (eds). *Social Epidemiology*. 2nd ed. New York: Oxford University Press, 2014; 63–125; Krieger N. Measures of racism, sexism, heterosexism, and gender binarism for health equity research: from structural injustice to embodied harm—an ecosocial analysis [published online ahead of print November 25, 2019]. *Annu Rev Public Health* 2020 April 2;41:37–62. doi:10.1146/annurev-publhealth-040119-094017; Schramme T. *Theories of Health Justice*. London: Rowman & Littlefield, 2018; Capeheart L, Milovanovic D. *Social Justice* (revised and expanded edition). New Brunswick, NJ: Rutgers University Press, 2020.
12. Krieger N. Discrimination and health. In: Berkman L, Kawachi I (eds). *Social Epidemiology*. Oxford: Oxford University Press, 2000; 36–75; Krieger N. Discrimination and health inequities. In: Berkman LF, Kawachi I, Glymour M (eds). *Social Epidemiology*. 2nd ed. New York: Oxford University Press, 2014; 63–125; Krieger N. Measures of racism, sexism, heterosexism, and gender binarism for health equity research: from structural injustice to embodied harm—an ecosocial analysis [published online ahead of print November 25, 2019]. *Annu Rev Public Health* 2020 April 2;41:37–62. doi:10.1146/annurev-publhealth-040119-094017.
13. See, for selected examples: Krieger N. Epidemiology and the web of causation: has anyone seen the spider? *Soc Sci Med* 1994;39(7):887–903. doi:10.1016/0277-9536(94)90202-x; Krieger N. Embodying inequality: a review of concepts, measures, and methods for studying health consequences of discrimination. *Int J Health Serv* 1999;29(2):295–352. doi:10.2190/M11W-VWXE-KQM9-G97Q; Krieger N. Refiguring "race": epidemiology, racialized biology, and biological expressions of race relations. *Int J Health Serv* 2000;30(1):211–216. doi:10.2190/672J-1PPF-K6QT-9N7U; Krieger N. Epidemiology and social sciences: towards a critical reengagement in the 21st century. *Epidemiol Rev* 2000;22(1):155–163. doi:10.1093/oxfordjournals.epirev.a018014; Krieger N. Theories for social epidemiology in the 21st century: an ecosocial perspective. *Int J Epidemiol* 2001;30(4):668–677. doi:10.1093/ije/30.4.668; Krieger N. A glossary for social epidemiology. *J Epidemiol Community Health* 2001;55(10):693–700. doi:10.1136/jech.55.10.693; Krieger N (ed). *Embodying Inequality: Epidemiologic Perspectives*. Amityville, NY: Baywood Publishing Co., 2004; Krieger N. Embodiment: a conceptual glossary for epidemiology. *J Epidemiol Community Health* 2005;59(5):350–355. doi:10.1136/jech.2004.024562; Krieger N. Proximal, distal, and the politics of causation: what's level got to do with it? [published

online ahead of print January 2, 2008]. *Am J Public Health* 2008;98(2):221–230. doi:10.2105/AJPH.2007.111278; Krieger N. *Epidemiology and the People's Health: Theory and Context*. New York: Oxford University Press, 2011;Krieger N. Who and what is a "population"? Historical debates, current controversies, and implications for understanding "population health" and rectifying health inequities. *Milbank Q* 2012;90(4):634–681. doi:10.1111/j.1468-0009.2012.00678.x; Krieger N. History, biology, and health inequities: emergent embodied phenotypes and the illustrative case of the breast cancer estrogen receptor [published online ahead of print November 15, 2012]. *Am J Public Health* 2013;103(1):22–27. doi:10.2105/AJPH.2012.300967; Krieger N. Got theory? On the 21 st c. CE rise of explicit use of epidemiologic theories of disease distribution: a review and ecosocial analysis. *Curr Epidemiol* Rep 2014;1(1):45–56; Krieger N. Living and dying at the crossroads: racism, embodiment, and why theory is essential for a public health of consequence. *Am J Public Health* 2016;106(5):832–833. doi:10.2105/AJPH.2016.303100; Krieger N. Theoretical frameworks and cancer inequities. In: Vaccarella S, Lortet-Vieulent J, Saracci R, Conway DI, Straif K, Wild CP (eds). *Reducing Social Inequalities in Cancer: Evidence and Priorities for Research* (IARC Scientific Publication No. 168). Lyon, France: International Agency for Research on Cancer, 2019; 189–204. http://publications.iarc.fr/580.

14. Quote investigator. It's déjà vu all over again. https://quoteinvestigator.com/2013/10/08/deja-vu-again/; accessed December 30, 2020.
15. Rosen G. *A History of Public Health* (1958). Revised expanded version, with introduction by Elizabeth Fee and biographical essay and new bibliography by Edward T. Morman. Baltimore, MD: Johns Hopkins University Press, 2015; Rosenkrantz B. *Public Health and the State: Changing Views in Massachusetts, 1842–1936*. Cambridge, MA: Harvard University Press, 1972; Hamlin C. *Public Health and Social Justice in the Age of Chadwick: Britain, 1800–1854*. Cambridge, UK: Cambridge University Press, 1998; Birn AE, Pillay Y, Holtz T. *Textbook of Global Health*. New York: Oxford University Press, 2017; Krieger N. *Epidemiology and the People's Health: Theory and Context*. New York: Oxford University Press, 2011.
16. Bashford A, Levine P (eds). *Oxford Handbook of the History of Eugenics*. New York: Oxford University Press, 2010; Chase A. *The Legacy of Malthus: The Social Costs of the New Scientific Racism*. 1st ed. New York: Knopf, 1977; Leonard TC. *Illiberal Reformers: Race, Eugenics, and American Economics in the Progressive Era*. Princeton, NJ: Princeton University Press, 2016; Whitman JQ. *Hitler's American Model: The United States and the Making of Nazi Race Law*. Princeton, NJ: Princeton University Press, 2017; Hogben L. *Nature and Nurture*. New York: W. W. Norton & Company, 1933; Sydenstricker E. *Health and Environment*. New York: McGraw Hill, 1933; Allen GE. Eugenics and modern biology: critiques of eugenics, 1910–1945. *Ann Human*

Genet 2011;75(3):314–325; Comfort N. *The Science of Human Perfection: How Genes Became the Heart of American Medicine.* New Haven, CT: Yale University Press, 2012; Stern A. *Eugenic Nation: Faults and Frontiers of Better Breeding in Modern America.* 2nd ed. Berkeley: University of California Press, 2016.

17. Wade N. *A Troublesome Inheritance: Genes, Race and Human History.* New York: Penguin Press, 2014; Murray C. *Human Diversity: The Biology of Gender, Race, and Class.* New York: Twelve, 2020; Ford CL, Griffith D, Bruce M, Gilbert K (eds). *Racism: Science and Tools for the Public Health Professional.* Washington, DC: American Public Health Association Press, 2019; Benjamin R. Cultura obscura: race, power, and "culture talk" in the health sciences. *Am J Law Med* 2017;43(2–3):225–238. doi:10.1177/0098858817723661; Bailey ZD, Krieger N, Agénor M, Graves J, Linos N, Bassett MT. Structural racism and health inequities in the USA: evidence and interventions. *Lancet* 2017;389(10077):1453–1463. doi:10.1016/S0140-6736(17)30569-X.
18. Beckfield J. *Political Sociology and the People's Health.* New York: Oxford University Press, 2018; Shostak S. *Exposed Science: Genes, the Environment, and the Politics of Population Health.* Berkeley: University of California Press, 2017; Landecker H, Panofsky A. From social structure to gene regulation, and back: a critical introduction to environmental epigenetics for sociology. *Annu Rev Sociol* 2013;39:333–357; Latour B, Schaffer S, Gagliardi P (eds). *A Book of the Body Politic: Connecting Biology, Politics, and Social Theory.* San Giorgio Dialogue 2017. Venezia, Italy: Fondazione Giorgio Cini, 2020; Gilbert SF, Epel D. *Ecological Developmental Biology: The Environmental Regulation of Development, Health, and Evolution.* 2nd ed. Sunderland, MA: Sinauer Associates, 2015; Lock M, Pálsson G. *Can Science Resolve the Nature/Nurture Debate?* Cambridge: Polity Press, 2016; Crooks A, Andrews GJ, Pearce J (eds). *Routledge Handbook of Health Geography.* 1st ed. New York: Routledge, 2018; Raworth K. *Doughnut Economics: Seven Ways to Think Like a 21st Century Economist.* White River Junction, VT: Chelsea Green Publishing, 2017; Felt U, Fouché R, Miller CA, Smith-Doerr L (eds). *The Handbook of Science and Technology Studies.* 4th ed. Cambridge, MA: MIT Press, 2017; Jasonoff S (ed). *States of Knowledge: The Co-Production of Science and Social Order.* London: Routledge, 2004.
19. Krieger N. *Epidemiology and the People's Health: Theory and Context.* New York: Oxford University Press, 2011; Hammonds EM, Reverby SM. Toward a historically informed analysis of racial health disparities since 1619. *Am J Public Health* 2019;109(10):1348–1349. doi:10.2105/AJPH.2019.305262; Benjamin R. Innovating inequity: if race is a technology, postracialism is the genius bar. *Ethn Racial Stud* 2016;39(13):2227–2234. doi:10.1080/01419870.2016.1202423; Felt U, Fouché R, Miller CA, Smith-Doerr L (eds). *The Handbook of Science and Technology Studies.* 4th ed. Cambridge, MA: MIT Press, 2017.

20. Wade N. *A Troublesome Inheritance: Genes, Race and Human History*. New York: Penguin Press, 2014; Plomin R. *Blueprint: How DNA Makes Us Who We Are*. Cambridge, MA: MIT Press, 2018; Murray C. *Human Diversity: The Biology of Gender, Race, and Class*. New York: Twelve, 2020.
21. Krieger N. *Epidemiology and the People's Health: Theory and Context*. New York: Oxford University Press, 2011; Beckfield J. *Political Sociology and the People's Health*. New York: Oxford University Press, 2018; Latour B, Schaffer S, Gagliardi P (eds). *A Book of the Body Politic: Connecting Biology, Politics, and Social Theory*. San Giorgio Dialogue 2017. Venezia, Italy: Fondazione Giorgio Cini, 2020; Landecker H, Panofsky A. From social structure to gene regulation, and back: a critical introduction to environmental epigenetics for sociology. *Annu Rev Sociol* 2013;39:333–357; Crooks A, Andrews GJ, Pearce J (eds). *Routledge Handbook of Health Geography*. 1st ed. New York: Routledge, 2018; Birn AE, Pillay Y, Holtz T. *Textbook of Global Health*. New York: Oxford University Press, 2017.
22. Smith GD. Epidemiology, epigenetics and the "Gloomy Prospect": embracing randomness in population health research and practice. *Int J Epidemiol* 2011;40(3):537–562. doi:10.1093/ije/dyr117; Davey Smith G. Post-modern epidemiology: when methods meet matter. *Am J Epidemiol* 2019;188(8):1410–1419. doi:10.1093/aje/kwz064; Krieger N. Who and what is a "population"? Historical debates, current controversies, and implications for understanding "population health" and rectifying health inequities. *Milbank Q* 2012;90(4):634–681. doi:10.1111/j.1468-0009.2012.00678.x; Krieger N. Health equity and the fallacy of treating causes of population health as if they sum to 100%. *Am J Public Health* 2017;107(4):541–549. doi:10.2105/AJPH.2017.303655. Erratum in: *Am J Public Health* 2017;107(9):e16; Gould SJ, Lewontin RC. The Spandrels of San Marco and the Panglossian paradigm: a critique of the adaptationist programme. *Proc R Soc Lond B* 1979;205:581–598; Coggon DIW, Martyn CN. Time and chance: the stochastic nature of disease causation. *Lancet* 2005;365:1434–1437.
23. Hogben, L. *Nature and Nurture*. New York: W. W. Norton & Company, 1933; Sydenstricker E. *Health and Environment*. New York: McGraw Hill, 1933; Keller EF. *The Mirage of a Space Between Nature and Nurture*. Durham, NC: Duke University Press, 2010; Tabery J. *Beyond Versus: The Struggle to Understand the Interaction of Nature and Nurture*. Cambridge, MA: MIT Press, 2014; Eldredge N, Grene M. *Interactions: The Biological Context of Social Systems*. New York: Columbia University Press, 1992; Shostak S. *Exposed Science: Genes, the Environment, and the Politics of Population Health*. Berkeley: University of California Press, 2017; Landecker H, Panofsky A. From social structure to gene regulation, and back: a critical introduction to environmental epigenetics for sociology. *Annu Rev Sociol* 2013;39:333–357; Latour B, Schaffer S, Gagliardi P (eds). *A Book of the Body Politic: Connecting Biology, Politics, and Social Theory*. San Giorgio Dialogue 2017.

Venezia, Italy: Fondazione Giorgio Cini, 2020; Gilbert SF, Epel D. *Ecological Developmental Biology: The Environmental Regulation of Development, Health, and Evolution*. 2nd ed. Sunderland, MA: Sinauer Associates, 2015; Lock M, Pálsson G. *Can Science Resolve the Nature/Nurture Debate?* Cambridge: Polity Press, 2016.

24. Institute of Medicine (US) Committee on Assessing Interactions Among Social, Behavioral, and Genetic Factors in Health. *Genes, Behavior, and the Social Environment: Moving Beyond the Nature/Nurture Debate*. Hernandez LM, Blazer DG (eds). Washington, DC: National Academies Press, 2006; Bookman EB, McAllister K, Gillanders E, Wanke K, Balshaw D, Rutter J, Reedy J, Shaughnessy D, Agurs-Collins T, Paltoo D, Atienza A, Bierut L, Kraft P, Fallin MD, Perera F, Turkheimer E, Boardman J, Marazita ML, Rappaport SM, Boerwinkle E, Suomi SJ, Caporaso NE, Hertz-Picciotto I, Jacobson KC, Lowe WL, Goldman LR, Duggal P, Gunnar MR, Manolio TA, Green ED, Olster DH, Birnbaum LS; NIH GxE Interplay Workshop participants. Gene-environment interplay in common complex diseases: forging an integrative model—recommendations from an NIH workshop. *Genet Epidemiol* 2011;35(4):217–225. doi:10.1002/gepi.20571; Shostak S. *Exposed Science: Genes, the Environment, and the Politics of Population Health*. Berkeley: University of California Press, 2017; Darling KW, Ackerman SL, Hiatt RH, Lee SS, Shim JK. Enacting the molecular imperative: how gene-environment interaction research links bodies and environments in the post-genomic age [published online ahead of print March 9, 2016]. *Soc Sci Med* 2016;155:51–60. doi:10.1016/j.socscimed.2016.03.007.
25. Keller EF. *The Mirage of a Space Between Nature and Nurture*. Durham, NC: Duke University Press, 2010; Tabery J. *Beyond Versus: The Struggle to Understand the Interaction of Nature and Nurture*. Cambridge, MA: MIT Press, 2014; Eldredge N, Grene M. *Interactions: The Biological Context of Social Systems*. New York: Columbia University Press, 1992; Shostak S. *Exposed Science: Genes, the Environment, and the Politics of Population Health*. Berkeley: University of California Press, 2017; Shostak S, Beckfield J. Making a case for genetics: interdisciplinary visions and practices in the contemporary social sciences. In: Perry BL (ed). *Genetics, Health, and Society. Advances in Medical Sociology*. Vol. 16. Bingley, UK: Emerald Group Publishing, 2015; 97–126; Latour B, Schaffer S, Gagliardi P (eds). *A Book of the Body Politic: Connecting Biology, Politics, and Social Theory*. San Giorgio Dialogue 2017. Venezia, Italy: Fondazione Giorgio Cini, 2020; Gilbert SF, Epel D. *Ecological Developmental Biology: The Environmental Regulation of Development, Health, and Evolution*. 2nd ed. Sunderland, MA: Sinauer Associates, 2015; Lock M, Pálsson G. *Can Science Resolve the Nature/Nurture Debate?* Cambridge: Polity Press, 2016.
26. Krieger N. Embodying inequality: a review of concepts, measures, and methods for studying health consequences of discrimination. *Int J Health Serv* 1999;29(2):295–352; Krieger N. Embodiment: a conceptual glossary for

epidemiology. *J Epidemiol Community Health* 2005;59(5):350–355; Krieger N. Theories for social epidemiology in the 21st century: an ecosocial perspective. *Int J Epidemiol* 2001;30(4):668–677; Krieger N, Davey Smith G. "Bodies count," and body counts: social epidemiology and embodying inequality. *Epidemiol Rev* 2004;26:92–103.

27. Krieger N. History, biology, and health inequities: emergent embodied phenotypes and the illustrative case of the breast cancer estrogen receptor [published online ahead of print November 15, 2012]. *Am J Public Health* 2013;103(1):22–27. doi:10.2105/AJPH.2012.300967; Niewohner J, Lock M. Situating local biologies: anthropological perspectives on environment/human entanglements. *Biosocieties* 2018;13(4):681–697.
28. Institute of Medicine (US) Committee on Assessing Interactions Among Social, Behavioral, and Genetic Factors in Health. *Genes, Behavior, and the Social Environment: Moving Beyond the Nature/Nurture Debate.* Hernandez LM, Blazer DG (eds). Washington, DC: National Academies Press, 2006; Bookman EB, McAllister K, Gillanders E, Wanke K, Balshaw D, Rutter J, Reedy J, Shaughnessy D, Agurs-Collins T, Paltoo D, Atienza A, Bierut L, Kraft P, Fallin MD, Perera F, Turkheimer E, Boardman J, Marazita ML, Rappaport SM, Boerwinkle E, Suomi SJ, Caporaso NE, Hertz-Picciotto I, Jacobson KC, Lowe WL, Goldman LR, Duggal P, Gunnar MR, Manolio TA, Green ED, Olster DH, Birnbaum LS; NIH GxE Interplay Workshop participants. Gene-environment interplay in common complex diseases: forging an integrative model—recommendations from an NIH workshop. *Genet Epidemiol* 2011;35(4):217–225. doi:10.1002/gepi.20571; Shostak S. *Exposed Science: Genes, the Environment, and the Politics of Population Health.* Berkeley: University of California Press, 2017; Darling KW, Ackerman SL, Hiatt RH, Lee SS, Shim JK. Enacting the molecular imperative: how gene-environment interaction research links bodies and environments in the post-genomic age [published online ahead of print March 9, 2016]. *Soc Sci Med* 2016;155:51–60. doi:10.1016/j.socscimed.2016.03.007.
29. Galton F. *Inquiries into Human Faculty and Its Development.* New York: MacMillan, 1883; 24–25; Galton F. Eugenics: its definition, scope and aims. *Am J Sociol* 1904;X(1):1–25; Keller EF. *The Mirage of a Space Between Nature and Nurture.* Durham, NC: Duke University Press, 2010.
30. Hogben L. *Nature and Nurture.* New York: W. W. Norton & Company, 1933; Sydenstricker E. *Health and Environment.* New York: McGraw Hill, 1933; Keller EF. *The Mirage of a Space Between Nature and Nurture.* Durham, NC: Duke University Press, 2010; Tabery J. R. A. Fisher, Lancelot Hogben, and the origin(s) of genotype–phenotype interaction. *J Hist Biol* 2008;41(4):717–761; Tabery J. *Beyond Versus: The Struggle to Understand the Interaction of Nature and Nurture.* Cambridge, MA: MIT Press, 2014; Shostak S. *Exposed Science: Genes,*

the Environment, and the Politics of Population Health. Berkeley: University of California Press, 2017; Latour B, Schaffer S, Gagliardi P (eds). *A Book of the Body Politic: Connecting Biology, Politics, and Social Theory*. San Giorgio Dialogue 2017. Venezia, Italy: Fondazione Giorgio Cini, 2020; Gilbert SF, Epel D. *Ecological Developmental Biology: The Environmental Regulation of Development, Health, and Evolution*. 2nd ed. Sunderland, MA: Sinauer Associates, 2015; Lock M, Pálsson G. *Can Science Resolve the Nature/Nurture Debate?* Cambridge: Polity Press, 2016; Institute of Medicine (US) Committee on Assessing Interactions Among Social, Behavioral, and Genetic Factors in Health. *Genes, Behavior, and the Social Environment: Moving Beyond the Nature/Nurture Debate*. Hernandez LM, Blazer DG (eds). Washington, DC: National Academies Press, 2006; Bookman EB, McAllister K, Gillanders E, Wanke K, Balshaw D, Rutter J, Reedy J, Shaughnessy D, Agurs-Collins T, Paltoo D, Atienza A, Bierut L, Kraft P, Fallin MD, Perera F, Turkheimer E, Boardman J, Marazita ML, Rappaport SM, Boerwinkle E, Suomi SJ, Caporaso NE, Hertz-Picciotto I, Jacobson KC, Lowe WL, Goldman LR, Duggal P, Gunnar MR, Manolio TA, Green ED, Olster DH, Birnbaum LS; NIH GxE Interplay Workshop participants. Gene-environment interplay in common complex diseases: forging an integrative model—recommendations from an NIH workshop. *Genet Epidemiol* 2011;35(4):217–225. doi:10.1002/gepi.20571.

31. Krieger N. Health equity and the fallacy of treating causes of population health as if they sum to 100%. *Am J Public Health* 2017;107(4):541–549. doi:10.2105/AJPH.2017.303655. Erratum in: *Am J Public Health* 2017;107(9):e16; Doll R, Peto R. The causes of cancer: quantitative estimates of avoidable risks of cancer in the United States today. *J Natl Cancer Inst* 1981;66(6):1191–1308; Peto R. The preventability of cancer. In: Vessey MP, Gray M (eds). *Cancer Risks and Prevention*. Oxford: Oxford University Press, 1985; 1–14; Tomasetti C, Vogelstein B. Variation in cancer risk among tissues can be explained by the number of stem cell divisions. *Science* 2015;347(6217):78–81; County Health Rankings and Roadmap. County Health Rankings Model. A Robert Wood Johnson Foundation Program/University of Wisconsin Population Health Institute, School of Medicine and Public Health. https://www.countyhealthrankings.org/explore-health-rankings/measures-data-sources/county-health-rankings-model; accessed December 30, 2020; Remington PL, Catlin BB, Gennuso KP. The County Health Rankings: rationale and methods. *Popul Health Metr* 2015;13:11. doi:10.1186/s12963-015-0044-2; America's Health Rankings. Measures, weights, and direction. https://www.americashealthrankings.org/about/methodology/measures-weights-and-direction; accessed December 30, 2020.

32. Krieger N. Health equity and the fallacy of treating causes of population health as if they sum to 100%. *Am J Public Health* 2017;107(4):541–549. doi:10.2105/

AJPH.2017.303655. Erratum in: *Am J Public Health* 2017;107(9):e16; Hogben L. *Nature and Nurture*. London: Williams & Norgate, 1933; Weinberg CR, Zaykin D. Is bad luck the main cause of cancer? *J Natl Cancer Inst* 2015;107(7):djv125.

33. Hogben L. *Nature and Nurture*. New York: W. W. Norton & Company, 1933; Tabery J. R. A. Fisher, Lancelot Hogben, and the origin(s) of genotype–phenotype interaction. *J Hist Biol* 2008;41(4):717–761; Tabery J. *Beyond Versus: The Struggle to Understand the Interaction of Nature and Nurture*. Cambridge, MA: MIT Press, 2014.
34. Hogben L. *Nature and Nurture*. London: Williams & Norgate, 1933; Krieger N. Health equity and the fallacy of treating causes of population health as if they sum to 100%. *Am J Public Health* 2017;107(4):541–549. doi:10.2105/AJPH.2017.303655. Erratum in: *Am J Public Health* 2017;107(9):e16; Weinberg CR, Zaykin D. Is bad luck the main cause of cancer? *J Natl Cancer Inst* 2015;107(7):djv125; Tabery J. R. A. Fisher, Lancelot Hogben, and the origin(s) of genotype–phenotype interaction. *J Hist Biol* 2008;41(4):717–761; Tabery J. *Beyond Versus: The Struggle to Understand the Interaction of Nature and Nurture*. Cambridge, MA: MIT Press, 2014.
35. Hogben L. *Nature and Nurture*. London: Williams & Norgate, 1933; Keller EF. *The Mirage of a Space Between Nature and Nurture*. Durham, NC: Duke University Press, 2010; Tabery J. R. A. Fisher, Lancelot Hogben, and the origin(s) of genotype–phenotype interaction. *J Hist Biol* 2008;41(4):717–761; Tabery J. *Beyond Versus: The Struggle to Understand the Interaction of Nature and Nurture*. Cambridge, MA: MIT Press, 2014; Shostak S. *Exposed Science: Genes, the Environment, and the Politics of Population Health*. Berkeley: University of California Press, 2017; Latour B, Schaffer S, Gagliardi P (eds). *A Book of the Body Politic: Connecting Biology, Politics, and Social Theory*. San Giorgio Dialogue 2017. Venezia, Italy: Fondazione Giorgio Cini, 2020; Gilbert SF, Epel D. *Ecological Developmental Biology: The Environmental Regulation of Development, Health, and Evolution*. 2nd ed. Sunderland, MA: Sinauer Associates, 2015; Niewohner J, Lock M. Situating local biologies: anthropological perspectives on environment/human entanglements. *Biosocieties* 2018;13(4):681–697; Landecker H, Panofsky A. From social structure to gene regulation, and back: a critical introduction to environmental epigenetics for sociology. *Annu Rev Sociol* 2013;39:333–357.
36. Gilbert SF, Sapp J, Tauber AI. A symbiotic view of life: we have never been individuals. *Q Rev Biol* 2012;87(4):325–341; Gilbert SF, Tauber AI. Rethinking individuality: the dialectics of the holobiont. *Biol Philos* 2016;31(6):839–853; Gilbert SF, Epel D. *Ecological Developmental Biology: The Environmental Regulation of Development, Health, and Evolution*. 2nd ed. Sunderland, MA: Sinauer Associates, 2015; Eldredge N, Grene M. *Interactions: The Biological Context of Social Systems*. New York: Columbia University Press, 1992; West-Eberhard MJ. *Developmental Plasticity and Evolution*. New York: Oxford University Press, 2003; Latour B,

Schaffer S, Gagliardi P (eds). *A Book of the Body Politic: Connecting Biology, Politics, and Social Theory*. San Giorgio Dialogue 2017. Venezia, Italy: Fondazione Giorgio Cini, 2020; Landecker H, Kelty C. Outside in | Microbiomes, epigenomes, visceral sensing, and metabolic ethics In: The Laboratory: Anthropology of Environment | Human Relations (ed). *After Practice: Thinking Through Matter(s) and Meaning Relationally*. Vol. I. Berlin: Humboldt University, 2019; 53–65.

37. Oreskes N. *Why Trust Science?* Princeton, NJ: Princeton University Press, 2019; Frickel S, Rea CM. Drought, hurricane, or wildfire? Assessing the Trump administration's anti-science disaster. *Engaging Sci Tech Society* 2020 Jan 8;6:66–75; Rutledge PE. Trump, COVID-19, and the war on expertise. *Am Rev Public Admin* 2020;50(6–7):505–511; Oreskes N, Conway EM. *Merchants of Doubt: How a Handful of Scientists Obscured the Truth on Issues from Tobacco Smoke to Global Warming*. New York: Bloomsbury Publishing USA, 2010; Michaels D. *Doubt Is Their Product: How Industry's Assault on Science Threatens Your Health*. Oxford; New York: Oxford University Press, 2008; Michaels D. *The Triumph of Doubt: Dark Money and the Science of Deception*. Oxford; New York: Oxford University Press, 2020; Markowitz G, Rosner D. *Deceit and Denial: The Deadly Politics of Industrial Pollution*. Berkeley: University of California Press, 2007; Brandt A. *The Cigarette Century: The Rise, Fall, and Deadly Persistence of the Product That Defined America*. New York: Basic Books, 2007; Proctor R. *Golden Holocaust: Origins of the Cigarette Catastrophe and the Case for Abolition*. Berkeley: University of California Press, 2011; Proctor R, Schiebinger L. *Agnotology: The Making and Unmaking of Ignorance*. Stanford: Stanford University Press, 2008; Mooney C. *The Republican War on Science*. New York: Basic Books, 2005.
38. Krieger N. Climate crisis, health equity, and democratic governance: the need to act together. *J Public Health Policy* 2020;41(1):4–10. doi:10.1057/s41271-019-00209-x; Krieger N. ENOUGH: COVID-19, structural racism, police brutality, plutocracy, climate change-and time for health justice, democratic governance, and an equitable, sustainable future [published online ahead of print August 20, 2020]. *Am J Public Health* 2020;110(11):1620–1623. doi:10.2105/AJPH.2020.305886; Michaels D. *The Triumph of Doubt: Dark Money and the Science of Deception*. Oxford; New York: Oxford University Press, 2020; Leonard C. *Kochland: The Secret History of Koch Industries and Corporate Power in America*. New York: Simon & Schuster, 2019; Mayer J. *Dark Money: The Hidden History of the Billionaires Behind the Rise of the Radical Right*. New York: Doubleday, 2016; Hertel-Fernandez A. *State Capture: How Conservative Activists, Big Businesses, and Wealthy Donors Reshaped the American States—And the Nation*. New York: Oxford University Press, 2019; Freudenberg N. *Lethal but Legal: Corporations, Consumption, and Protecting Public Health*. New York: Oxford University Press, 2014; Birn AE, Pillay Y, Holtz T. *Textbook of Global Health*. New York: Oxford University Press,

2017; Hanna-Attisha M, Olson ED. Preexisting conditions that kill us [published online ahead of print December 4, 2020]. *Fam Community Health*. doi:10.1097/FCH.0000000000000288.

39. Michaels D. *The Triumph of Doubt: Dark Money and the Science of Deception*. Oxford; New York: Oxford University Press, 2020; Leonard C. *Kochland: The Secret History of Koch Industries and Corporate Power in America*. New York: Simon & Schuster, 2019; Mayer J. *Dark Money: The Hidden History of the Billionaires Behind the Rise of the Radical Right*. New York: Doubleday, 2016; Hertel-Fernandez A. *State Capture: How Conservative Activists, Big Businesses, and Wealthy Donors Reshaped the American States—And the Nation*. New York: Oxford University Press, 2019; Freudenberg N. *Lethal but Legal: Corporations, Consumption, and Protecting Public Health*. New York: Oxford University Press, 2014; Nestle M. *Unsavory Truth: How Food Companies Skew the Science of What We Eat*. New York: Basic Books, 2018; Birn AE, Pillay Y, Holtz T. *Textbook of Global Health*. New York: Oxford University Press, 2017; Anderson C. *One Person, No Vote: How Voter Suppression Is Destroying Our Democracy*. New York: Bloomsbury Publishing, 2018.
40. Corredor ES. Unpacking "gender ideology" and the global right's antigender countermovement. *Signs* 2019;44(3):613–638; Kuhar R, Paternotte D (eds). *Anti-Gender Campaigns in Europe: Mobilizing Against Equality*. Lanham, MD: Rowman & Littlefield Intl, 2017; Stern AM. *Proud Boys and the White Ethnostate: How the Alt-Right Is Warping the American Imagination*. Boston: Beacon Press, 2019; Turshen M. *Women's Health Movements: A Global Force for Change*. 2nd ed. Singapore: Palgrave MacMillan, 2020; Marchlewska M, Chichoka A. How a gender conspiracy is spreading across the world. *The Conversation*, March 23, 2020. https://theconversation.com/how-a-gender-conspiracy-theory-is-spreading-across-the-world-133854; accessed December 30, 2020; Kane G. "Gender ideology": big, bogus, and coming to a fear campaign near you. *The Guardian*, March 30, 2018. https://www.theguardian.com/global-development/2018/mar/30/gender-ideology-big-bogus-and-coming-to-a-fear-campaign-near-you; accessed December 30, 2020; Gallo M. "Gender ideology" is a fiction that could do real harm. Voices, Open Society Foundation, August 29, 2017. https://www.opensocietyfoundations.org/voices/gender-ideology-fiction-could-do-real-harm; accessed December 30, 2020; Hosie D. What is "gender ideology"? And why are so many anti-LGBTQ activists obsessed with it? *Slate*, March 22, 2019. https://slate.com/human-interest/2019/03/gender-ideology-trump-bolsonaro-lgbtq.html; accessed December 30, 2020; United States Conference of Catholic Bishops. Created Male and Female: An Open Letter from Religious Leaders. December 15, 2017. http://www.usccb.org/issues-and-action/marriage-and-family/marriage/promotion-and-defense-of-marriage/created-male-and-female.cfm; accessed December 30, 2020; The Vatican. Male and female he created them. For a path of dialogue on the issue

of gender in education. Rome: The Vatican, June 10, 2019. https://zenit.org/2019/06/10/new-vatican-document-provides-schools-with-guidance-on-gender-issues/; accessed December 30, 2020.

41. Oreskes N. *Why Trust Science?* Princeton, NJ: Princeton University Press, 2019; Boyd RW, Lindo EG, Weeks LD, McLemore MR. On racism: a new standard for publishing on racial health inequities. *Health Affairs* blog, July 2, 2020. https://www.healthaffairs.org/do/10.1377/hblog20200630.939347/full/; accessed December 30, 2020; Ford CL, Griffith D, Bruce M, Gilbert K (eds). *Racism: Science and Tools for the Public Health Professional.* Washington, DC: American Public Health Association Press, 2019; Saini A. *Superior: The Return of Race Science.* Boston: Beacon Press, 2019; DeSalle R, Tattersall I. *Troublesome Science: The Misuse of Genetics and Genomics in Understanding Race.* New York: Columbia University Press, 2018; Currano E, Marsh L, Vance K, White D. *The Bearded Lady Project Challenging the Face of Science.* New York: Columbia University Press, 2020; Saini A. *Inferior: How Science Got Women Wrong—and the New Research That's Rewriting the Story.* Boston: Beacon Press, 2017; 500 Women Scientists. Who We Are. https://500womenscientists.org/who-we-are; accessed December 30, 2020.

42. Michaels D. *The Triumph of Doubt: Dark Money and the Science of Deception.* New York: Oxford University Press, 2020; Leonard C. *Kochland: The Secret History of Koch Industries and Corporate Power in America.* New York: Simon & Schuster, 2019; Grodin MA, Tarantola D, Annas GJ, Gruskin S (eds). *Health and Human Rights in a Changing World.* New York: Routledge, 2013; Burris S, Berman ML, Penn M, Holiday TR. *The New Public Health Law: A Transdisciplinary Approach to Practice and Advocacy.* New York: Oxford University Press, 2018; Hanna-Attisha M, Olson ED. Preexisting conditions that kill us [published online ahead of print December 4, 2020]. *Fam Community Health.* doi:10.1097/FCH.0000000000000288; Levy B (ed). *Social Injustice and Public Health.* 3rd ed. New York: Oxford University Press, 2019; Beckfield J. *Political Sociology and the People's Health.* New York: Oxford University Press, 2018; Friel S. *Climate Change and the People's Health.* New York: Oxford University Press, 2019; Krieger N. Climate crisis, health equity, and democratic governance: the need to act together [published online ahead of print January 21, 2020]. *J Public Health Policy.* https://doi.org/10.1057/s41271-019-00209-x; Latour B, Schaffer S, Gagliardi P (eds). *A Book of the Body Politic: Connecting Biology, Politics, and Social Theory.* San Giorgio Dialogue 2017. Venezia, Italy: Fondazione Giorgio Cini, 2020; Birn AE, Pillay Y, Holtz T. *Textbook of Global Health.* New York: Oxford University Press, 2017.

43. Michaels D. *The Triumph of Doubt: Dark Money and the Science of Deception.* New York: Oxford University Press, 2020; Leonard C. *Kochland: The Secret History of Koch Industries and Corporate Power in America.* New York: Simon & Schuster,

2019; Grodin MA, Tarantola D, Annas GJ, Gruskin S (eds). *Health and Human Rights in a Changing World*. New York: Routledge, 2013; Burris S, Berman ML, Penn M, Holiday TR. *The New Public Health Law: A Transdisciplinary Approach to Practice and Advocacy*. New York: Oxford University Press, 2018; Levy B (ed). *Social Injustice and Public Health*. 3rd ed. New York: Oxford University Press, 2019; Beckfield J. *Political Sociology and the People's Health*. New York: Oxford University Press, 2018; Friel S. *Climate Change and the People's Health*. New York: Oxford University Press, 2019; Krieger N. Climate crisis, health equity, and democratic governance: the need to act together [published online ahead of print January 21, 2020]. *J Public Health Policy*. https://doi.org/10.1057/s41271-019-00209-x; Latour B, Schaffer S, Gagliardi P (eds). *A Book of the Body Politic: Connecting Biology, Politics, and Social Theory*. San Giorgio Dialogue 2017. Venezia, Italy: Fondazione Giorgio Cini, 2020; Felt U, Fouché R, Miller CA, Smith-Doerr L (eds). *The Handbook of Science and Technology Studies*. 4th ed. Cambridge, MA: MIT Press, 2017; Jasonoff S (ed). *States of Knowledge: The Co-Production of Science and Social Order*. London: Routledge, 2004; Birn AE, Pillay Y, Holtz T. *Textbook of Global Health*. New York: Oxford University Press, 2017.

44. Pan American Health Organization (PAHO). *Just Societies: Health Equity and Dignified Lives. Report of the Commission of the Pan American Health Organization on Equity and Health Inequalities in the Americas*. Washington, DC: PAHO, 2019. https://iris.paho.org/handle/10665.2/51571; accessed December 30, 2020; Levy B (ed). *Social Injustice and Public Health*. 3rd ed. New York: Oxford University Press, 2019; Beckfield J. *Political Sociology and the People's Health*. New York: Oxford University Press, 2018; Friel S. *Climate Change and the People's Health*. New York: Oxford University Press, 2019; Grodin MA, Tarantola D, Annas GJ, Gruskin S (eds). *Health and Human Rights in a Changing World*. New York: Routledge, 2013; Brown P, Morello-Frosch R, Zavestoski S. *Contested Illnesses: Citizens, Science, and Health Social Movements*. Berkeley: University of California Press, 2011; Hofrichter R, Bhatia R (eds). *Tackling Health Inequities Through Public Health Practice: Theory to Action*. 2nd ed. New York: Oxford University Press, 2010; Turshen M. *Women's Health Movements: A Global Force for Change*. 2nd ed. Singapore: Palgrave MacMillan, 2020; Mack A, Baciu A, Goel N. *Supporting a Movement for Health and Health Equity: Lessons from Social Movements: Workshop Summary*. Institute of Medicine Roundtable on Population Health Improvement. Washington, DC: National Academies Press, 2014; Birn AE, Pillay Y, Holtz T. *Textbook of Global Health*. New York: Oxford University Press, 2017.
45. Hobsbawm E. *The Age of Revolution: Europe 1789–1848*. London: Abacus, 1977; Hobsbawm E. *The Age of Capital, 1848–1875*. New York: Scribner, 1975; Beckert S. *Empire of Cotton: A Global History*. New York: Alfred A. Knopf, 2014; Rosen G. *A History of Public Health* (1958). Revised expanded version, with introduction by

Elizabeth Fee and biographical essay and new bibliography by Edward T. Morman. Baltimore, MD: Johns Hopkins University Press, 2015; Rosenkrantz B. *Public Health and the State: Changing Views in Massachusetts, 1842–1936*. Cambridge, MA: Harvard University Press, 1972; Hamlin C. *Public Health and Social Justice in the Age of Chadwick: Britain, 1800–1854*. Cambridge: Cambridge University Press, 1998; Birn AE, Pillay Y, Holtz T. *Textbook of Global Health*. New York: Oxford University Press, 2017.

46. Rosen G. *A History of Public Health* (1958). Revised expanded version, with introduction by Elizabeth Fee and biographical essay and new bibliography by Edward T. Morman. Baltimore, MD: Johns Hopkins University Press, 2015; Rosenkrantz B. *Public Health and the State: Changing Views in Massachusetts, 1842–1936*. Cambridge, MA: Harvard University Press, 1972; Hamlin C. *Public Health and Social Justice in the Age of Chadwick: Britain, 1800–1854*. Cambridge: Cambridge University Press, 1998; Porter D. *Health, Civilization and the State*. London: Routledge, 1999; Birn AE, Pillay Y, Holtz T. *Textbook of Global Health*. New York: Oxford University Press, 2017.
47. Great Britain, Poor Law Commissioners. *Report to Her Majesty's principal Secretary of State for the Home Department, from the Poor Law Commissioners on an inquiry into the sanitary condition of the labouring population of Great Britain: with appendices*. London: W. Clowes and Sons, for H.M.S.O., 1842; Hamlin C. *Public Health and Social Justice in the Age of Chadwick: Britain, 1800–1854*. Cambridge: Cambridge University Press, 1998.
48. Tristan F. *Promenades dans Londres: L'Aristocracie et les Prolétaires Anglais*. Paris: Indigo & Côte-femmes, 1842 (for translation, see Tristan F. *The London Journals of Flora Tristan, 1842, or The Aristocracy and the Working Class of England*. London: Virago Press, 1982); Engels F. *The Condition of the Working Class in England* (1845). Translated by W. O. Henderson and W. H. Chaloner. Stanford, CA: Stanford University Press, 1958.
49. Griscom J. *The Sanitary Condition of the Laboring Population of New York: With Suggestions for Its Improvement*. New York: Harper & Brothers, 1845.
50. American Statistical Association. *Collections of the American Statistical Association*. Vol. 1. Boston: T. R. Marvin, 1847.
51. Massachusetts Sanitary Commission (Lemuel Shattuck, Nathaniel Prentiss Banks, Jehiel Abbott). *Report of the Sanitary Commission of Massachusetts, 1850*. Boston: Dutton & Wentworth, 1850; Rosenkrantz BG. *Public Health and the State: Changing Views in Massachusetts, 1842–1936*. Cambridge, MA: Harvard University Press, 1972.
52. Rosenkrantz BG. *Public Health and the State: Changing Views in Massachusetts, 1842–1936*. Cambridge, MA: Harvard University Press, 1972; Rosen G. *A History of Public Health* (1958). Revised expanded version, with introduction by Elizabeth

Fee and biographical essay and new bibliography by Edward T. Morman. Baltimore, MD: Johns Hopkins University Press, 2015.

53. Krieger N. *Epidemiology and the People's Health: Theory and Context.* New York: Oxford University Press, 2011; Krieger N. Shades of difference: theoretical underpinnings of the medical controversy on black/white differences in the United States, 1830–1870. *Int J Health Serv* 1987;17:259–278; Krieger N. The US census and the people's health: public health engagement from enslavement and "Indians not taxed" to census tracts and health equity (1790–2018). *Am J Public Health* 2019;109(8):1092–1101; Falk LA. Black abolitionist doctors and healers, 1810–1855. *Bull Hist Med* 1980;54:258–272; Levesque GA. Boston's Black Brahmin: Dr. John. S. Rock. *Civil War History* 1980;54:326–346; Link EP. The civil rights activities of three great Negro physicians (1840–1940). *J Negro History* 1967;52:169–184; Smith JM. On the fourteenth query of Thomas Jefferson's Notes on Virginia. *Anglo-African Magazine* 1859;1:225–238; Eglash R. Anti-racist technoscience: a generative tradition. In: Benjamin R (ed). *Captivating Technology: Race, Carceral Technoscience, and the Liberatory Imagination in Everyday Life.* Durham, NC: Duke University Press, 2019; 227–251; Ferry G. Rebecca Lee Crumpler: first Black woman physician in the USA. *Lancet* 2021;397:572.
54. Krieger N. Shades of difference: theoretical underpinnings of the medical controversy on black/white differences in the United States, 1830–1870. *Int J Health Serv* 1987;17:259–278; Eglash R. Anti-racist technoscience: a generative tradition. In: Benjamin R (ed). *Captivating Technology: Race, Carceral Technoscience, and the Liberatory Imagination in Everyday Life.* Durham, NC: Duke University Press, 2019; 227–251; Besak J. W.E.B. Du Bois embraced science to fight racism as editor of NAACP's magazine The Crisis. *The Conversation*, December 14, 2020. https://theconversation.com/w-e-b-du-bois-embraced-science-to-fight-racism-as-editor-of-naacps-magazine-the-crisis-150825; accessed February 14, 2021; Cooper RS, Rotimi CN. The practice of anti-racist science requires an internationalist perspective. *Am J Hum Genet* 2020;107(5):793–796; Gil-Riaño S. Relocating anti-racist science: the 1950 UNESCO Statement on Race and economic development in the global South. *Br J History Sci* 2018;51(2):281–303; Rusert B. *Fugitive Science: Empiricism and Freedom in Early African American Culture.* New York: New York University Press, 2017; Ferry G. Rebecca Lee Crumpler: first Black woman physician in the USA. *Lancet* 2021;397:572.
55. Black D, Morris JN, Smith C, Townsend P. *Inequalities in Health: A Report of a Research Working Group.* London: Department of Health and Social Security, 1980. (The Black Report); original available at: https://www.sochealth.co.uk/national-health-service/public-health-and-wellbeing/poverty-and-inequality/the-black-report-1980/; accessed December 30, 2020; Townsend, P, Davidson N, Whitehead M. *Inequalities in Health: The Black Report and the Health Divide.* London: Penguin Books, 1988.

56. Declaration of Alma-Ata. International Conference on Primary Health Care, Alma-Ata, USSR, September 6–12, 1978. https://www.who.int/publications/almaata_declaration_en.pdf; accessed December 30, 2020; Birn AE. Back to Alma-Ata, from 1978 to 2018 and Beyond. *Am J Public Health* 2018;108(9):1153–1155.
57. Steger M, Roy R. *Neoliberalism: A Very Short Introduction.* New York: Oxford University Press, 2010; Stiglitz J. *Globalization and Its Discontents Revisited: Anti-Globalization in the Era of Trump.* New York: W. W. Norton & Company, 2018; Birn AE, Pillay Y, Holtz T. *Textbook of Global Health.* New York: Oxford University Press, 2017.
58. World Health Organization (WHO), Commission on the Social Determinants of Health. *Closing the Gap in a Generation: Health Equity Through Action on the Social Determinants of Health.* Geneva: WHO, 2008. https://www.who.int/social_determinants/thecommission/finalreport/en/; accessed February 28, 2020.
59. Marmot M, Allen J, Goldblatt P, Boyce T, McNeish D, Grady M. *Fair Society, Healthy Lives: The Marmot Review.* London: Institute of Health Equity, 2010. http://www.instituteofhealthequity.org/resources-reports/fair-society-healthy-lives-the-marmot-review/fair-society-healthy-lives-exec-summary-pdf.pdf; accessed December 30, 2020; Marmot M, Allen J, Boyce T, Goldblatt P, Morrison J. *Health Equity in England: The Marmot Review 10 Years On.* London: Institute of Health Equity, 2020. https://www.health.org.uk/publications/reports/the-marmot-review-10-years-on; accessed December 30, 2020.
60. Commission of the Pan American Health Organization on Equity and Health Inequalities in the Americas. *Sociedades Justas: Equidad en la Salud y Viva Digna/Just Societies: Health Equity and Dignified Lives.* PAHO/IHE. Washington, DC: PAHO, 2019. https://iris.paho.org/handle/10665.2/49505; accessed December 30, 2020.
61. Commission of the Pan American Health Organization on Equity and Health Inequalities in the Americas. *Sociedades Justas: Equidad en la Salud y Viva Digna/Just Societies: Health Equity and Dignified Lives.* PAHO/IHE. Washington, DC: PAHO, 2019. https://iris.paho.org/handle/10665.2/49505; accessed December 30, 2020.
62. Commission of the Pan American Health Organization on Equity and Health Inequalities in the Americas. *Sociedades Justas: Equidad en la Salud y Viva Digna/Just Societies: Health Equity and Dignified Lives.* PAHO/IHE. Washington, DC: PAHO, 2019. https://iris.paho.org/handle/10665.2/49505; accessed December 30, 2020.
63. Global Health Watch. Mobilising civil society around an alternative World Health Report. https://www.ghwatch.org/about.html; accessed December 30, 2020; People's Health Movement. Global Health Watch. https://phmovement.org/global-health-watch/; accessed December 30, 2020; People's Health Movement.

About the People's Health Movement. https://phmovement.org/about-3/; accessed December 30, 2020.

64. Krieger N. Embodiment: a conceptual glossary for epidemiology. *J Epidemiol Community Health* 2005;59:350–355; Krieger N. Living and dying at the crossroads: racism, embodiment, and why theory is essential for a public health of consequence. *Am J Public Health* 2016;106:832–833; Krieger N, Davey Smith G. Bodies count and body counts: social epidemiology and embodying inequality. *Epidemiol Rev* 2004;26:92–103; Krieger N. Discrimination and health. In: Berkman L, Kawachi I (eds). *Social Epidemiology*. Oxford: Oxford University Press, 2000; 36–75; Krieger N. Discrimination and health inequities. In: Berkman LF, Kawachi I, Glymour M (eds). *Social Epidemiology*. 2nd ed. New York: Oxford University Press, 2014; 63–125.
65. Rosen G. *A History of Public Health* (1958). Revised expanded version, with introduction by Elizabeth Fee and biographical essay and new bibliography by Edward T. Morman. Baltimore, MD: Johns Hopkins University Press, 2015; Porter D. *Health, Civilization and the State*. London: Routledge, 1999; Bynum WF. *History of Medicine: A Very Short Introduction*. New York: Oxford University Press, 2008; Rosenberg CE, Golden J (eds). *Framing Disease: Studies in Cultural History*. New Brunswick, NJ: Rutgers University Press, 1992.
66. Krieger N. *Epidemiology and the People's Health: Theory and Context*. New York: Oxford University Press, 2011; Krieger N. A critical research agenda for social justice and public health: an ecosocial proposal. In: Levy B (ed). *Social Injustice and Public Health*. 3rd ed. New York: Oxford University Press, 2019; 531–552.
67. Oreskes N. *Why Trust Science?* Princeton, NJ: Princeton University Press, 2019; Ziman J. *Real Science: What It Is, and What It Means*. Cambridge: Cambridge University Press, 2000; Okasha S. *Philosophy of Science: A Very Short Introduction*. 2nd ed. New York: Oxford University Press, 2016.
68. Huff C. Situating Donna Haraway in the life-narrative web. *a/b: Auto/Biography Stud* 2019;34(3):375–384. doi:10.1080/08989575.2019.1664167; Haraway D. *The Donna Haraway Reader*. New York: Routledge, 2004.
69. Weigel M. A giant bumptious litter: Donna Haraway on truth, technology, and resisting extinction. *Logic Magazine* 2019 (December 7);9. https://logicmag.io/nature/a-giant-bumptious-litter/; accessed December 30, 2020.
70. Weigel M. Feminist cyborg scholar Donna Haraway: "The disorder of our era isn't necessary." *The Guardian*, June 20, 2019; https://www.theguardian.com/world/2019/jun/20/donna-haraway-interview-cyborg-manifesto-post-truth; accessed December 30, 2020.
71. Krieger N. Epidemiology and the web of causation: has anyone seen the spider? *Soc Sci Med* 1994;39(7):887–903. doi:10.1016/0277-9536(94)90202-x.

72. Krieger N. Epidemiology and the web of causation: has anyone seen the spider? *Soc Sci Med* 1994;39(7):887–903. doi:10.1016/0277-9536(94)90202-x; Krieger N. Embodying inequality: a review of concepts, measures, and methods for studying health consequences of discrimination. *Int J Health Serv* 1999;29(2):295–352. doi:10.2190/M11W-VWXE-KQM9-G97Q; Krieger N. Refiguring "race": epidemiology, racialized biology, and biological expressions of race relations. *Int J Health Serv* 2000;30(1):211–216. doi:10.2190/672J-1PPF-K6QT-9N7U; Krieger N. Epidemiology and social sciences: towards a critical reengagement in the 21st century. *Epidemiol Rev* 2000;22(1):155–163. doi:10.1093/oxfordjournals.epirev.a018014; Krieger N. Theories for social epidemiology in the 21st century: an ecosocial perspective. *Int J Epidemiol* 2001;30(4):668–677. doi:10.1093/ije/30.4.668; Krieger N. A glossary for social epidemiology. *J Epidemiol Community Health* 2001;55(10):693–700. doi:10.1136/jech.55.10.693; Krieger N (ed). *Embodying Inequality: Epidemiologic Perspectives*. Amityville, NY: Baywood Publishing Co., 2004; Krieger N. Embodiment: a conceptual glossary for epidemiology. *J Epidemiol Community Health* 2005;59(5):350–355. doi:10.1136/jech.2004.024562; Krieger N. Proximal, distal, and the politics of causation: what's level got to do with it? [published online ahead of print January 2, 2008]. *Am J Public Health* 2008;98(2):221–230. doi:10.2105/AJPH.2007.111278; Krieger N. *Epidemiology and the People's Health: Theory and Context*. New York: Oxford University Press, 2011; Krieger N. Who and what is a "population"? Historical debates, current controversies, and implications for understanding "population health" and rectifying health inequities. *Milbank Q* 2012;90(4):634–681. doi:10.1111/j.1468-0009.2012.00678.x; Krieger N. History, biology, and health inequities: emergent embodied phenotypes and the illustrative case of the breast cancer estrogen receptor [published online ahead of print November 15, 2012]. *Am J Public Health* 2013;103(1):22–27. doi:10.2105/AJPH.2012.300967; Krieger N. Got theory? On the 21 st c. CE rise of explicit use of epidemiologic theories of disease distribution: a review and ecosocial analysis. *Curr Epidemiol Rep* 2014;1(1):45–56; Krieger N. Living and dying at the crossroads: racism, embodiment, and why theory is essential for a public health of consequence. *Am J Public Health* 2016;106(5):832–833. doi:10.2105/AJPH.2016.303100; Krieger N. Theoretical frameworks and cancer inequities. In: Vaccarella S, Lortet-Vieulent J, Saracci R, Conway DI, Straif K, Wild CP (eds). *Reducing Social Inequalities in Cancer: Evidence and Priorities for Research* (IARC Scientific Publication No. 168). Lyon, France: International Agency for Research on Cancer, 2019; 189–204. http://publications.iarc.fr/580.
73. Krieger N. Epidemiology and the web of causation: has anyone seen the spider? *Soc Sci Med* 1994;39(7):887–903. doi:10.1016/0277-9536(94)90202-x; Krieger N. Embodying inequality: a review of concepts, measures, and methods for

studying health consequences of discrimination. *Int J Health Serv* 1999;29(2):295–352. doi:10.2190/M11W-VWXE-KQM9-G97Q; Krieger N, Davey Smith G. "Bodies count," and body counts: social epidemiology and embodying inequality. *Epidemiol Rev* 2004;26:92–103; Krieger N. Embodiment: a conceptual glossary for epidemiology. *J Epidemiol Community Health* 2005;59(5):350–355. doi:10.1136/jech.2004.024562; Krieger N. *Epidemiology and the People's Health: Theory and Context*. New York: Oxford University Press, 2011; Krieger N. Who and what is a "population"? Historical debates, current controversies, and implications for understanding "population health" and rectifying health inequities. *Milbank Q* 2012;90(4):634–681. doi:10.1111/j.1468-0009.2012.00678.x; Krieger N. History, biology, and health inequities: emergent embodied phenotypes and the illustrative case of the breast cancer estrogen receptor [published online ahead of print November 15, 2012]. *Am J Public Health* 2013;103(1):22–27. doi:10.2105/AJPH.2012.300967; Gibson JJ. *The Ecological Approach to Visual Perception* (Classic Edition; 1st ed.: 1979). New York: Psychology Press, Taylor and Francis, 2014; Reed E, Jones R (eds). *Reasons for Realism: Selected Essays of James Gibson*. New York: Routledge, 2020.

74. Krieger N. Who and what is a "population"? Historical debates, current controversies, and implications for understanding "population health" and rectifying health inequities. *Milbank Q* 2012;90(4):634–681. doi:10.1111/j.1468-0009.2012.00678.x; Williams R. *Keywords: A Vocabulary of Culture and Society* (Rev. ed.). New York: Oxford University Press, 1985; "individual": pp. 161–165; "society": pp. 291–295; Haraway D. *When Species Meet*. Minneapolis: University of Minnesota Press, 2008; Haraway DJ. *Staying with the Trouble: Making Kin in Chthulucene*. Durham, NC: Duke University Press, 2016; Clarke AE, Haraway D. *Making Kin Not Population*. Chicago: Prickly Paradigm Press, 2018; Gilbert SF, Sapp J, Tauber AI. A symbiotic view of life: we have never been individuals. *Q Rev Biol* 2012;87(4):325–341; Gilbert SF, Tauber AI. Rethinking individuality: the dialectics of the holobiont. *Biol Philos* 2016;31(6):839–853; Gilbert SF, Epel D. *Ecological Developmental Biology: The Environmental Regulation of Development, Health, and Evolution*. 2nd ed. Sunderland, MA: Sinauer Associates, 2015.
75. Krieger N. Who and what is a "population"? Historical debates, current controversies, and implications for understanding "population health" and rectifying health inequities. *Milbank Q* 2012;90(4):634–681. doi:10.1111/j.1468-0009.2012.00678.x; Eldredge N, Grene M. *Interactions: The Biological Context of Social Systems*. New York: Columbia University Press, 1992; Gilbert SF, Sapp J, Tauber AI. A symbiotic view of life: we have never been individuals. *Q Rev Biol* 2012;87(4):325–341; Gilbert SF, Tauber AI. Rethinking individuality: the dialectics of the holobiont. *Biol Philos* 2016;31(6):839–853; Gilbert SF, Epel D.

Ecological Developmental Biology: The Environmental Regulation of Development, Health, and Evolution. 2nd ed. Sunderland, MA: Sinauer Associates, 2015; Shostak S. *Exposed Science: Genes, the Environment, and the Politics of Population Health*. Berkeley: University of California Press, 2017; Latour B, Schaffer S, Gagliardi P (eds). *A Book of the Body Politic: Connecting Biology, Politics, and Social Theory*. San Giorgio Dialogue 2017. Venezia, Italy: Fondazione Giorgio Cini, 2020; Landecker H, Kelty C. Outside in | Microbiomes, epigenomes, visceral sensing, and metabolic ethics. In: The Laboratory: Anthropology of Environment | Human Relations (ed). *After Practice: Thinking Through Matter(s) and Meaning Relationally*. Vol. I. Berlin: Humboldt University, 2019; 53–65.

76. Krieger N. Embodiment: a conceptual glossary for epidemiology. *J Epidemiol Community Health* 2005;59(5):350–355. doi:10.1136/jech.2004.024562; Krieger N. *Epidemiology and the People's Health: Theory and Context*. New York: Oxford University Press, 2011; Krieger N. Who and what is a "population"? Historical debates, current controversies, and implications for understanding "population health" and rectifying health inequities. *Milbank Q* 2012;90(4):634–681. doi:10.1111/j.1468-0009.2012.00678.x; Krieger N. Living and dying at the crossroads: racism, embodiment, and why theory is essential for a public health of consequence. *Am J Public Health* 2016;106(5):832–833. doi:10.2105/AJPH.2016.303100.

77. Krieger N. Epidemiology and the web of causation: has anyone seen the spider? *Soc Sci Med* 1994;39(7):887–903. doi:10.1016/0277-9536(94)90202-x; Krieger N. Embodying inequality: a review of concepts, measures, and methods for studying health consequences of discrimination. *Int J Health Serv* 1999;29(2):295–352. doi:10.2190/M11W-VWXE-KQM9-G97Q; Krieger N. Refiguring "race": epidemiology, racialized biology, and biological expressions of race relations. *Int J Health Serv* 2000;30(1):211–216. doi:10.2190/672J-1PPF-K6QT-9N7U; Krieger N. Epidemiology and social sciences: towards a critical reengagement in the 21st century. *Epidemiol Rev* 2000;22(1):155–163. doi:10.1093/oxfordjournals.epirev.a018014; Krieger N. Theories for social epidemiology in the 21st century: an ecosocial perspective. *Int J Epidemiol* 2001;30(4):668–677. doi:10.1093/ije/30.4.668; Krieger N. A glossary for social epidemiology. *J Epidemiol Community Health* 2001;55(10):693–700. doi:10.1136/jech.55.10.693; Krieger N (ed). *Embodying Inequality: Epidemiologic Perspectives*. Amityville, NY: Baywood Publishing Co., 2004; Krieger N. Embodiment: a conceptual glossary for epidemiology. *J Epidemiol Community Health* 2005;59(5):350–355. doi:10.1136/jech.2004.024562; Krieger N. Proximal, distal, and the politics of causation: what's level got to do with it? [published online ahead of print January 2, 2008]. *Am J Public Health* 2008;98(2):221–230. doi:10.2105/AJPH.2007.111278; Krieger N. *Epidemiology and the People's Health: Theory and Context*. New York: Oxford

University Press, 2011; Krieger N. Who and what is a "population"? Historical debates, current controversies, and implications for understanding "population health" and rectifying health inequities. *Milbank Q* 2012;90(4):634–681. doi:10.1111/j.1468-0009.2012.00678.x; Krieger N. History, biology, and health inequities: emergent embodied phenotypes and the illustrative case of the breast cancer estrogen receptor [published online ahead of print November 15, 2012]. *Am J Public Health* 2013;103(1):22–27. doi:10.2105/AJPH.2012.300967; Krieger N. Got theory? On the 21 st c. CE rise of explicit use of epidemiologic theories of disease distribution: a review and ecosocial analysis. *Curr Epidemiol Rep* 2014;1(1):45–56; Krieger N. Living and dying at the crossroads: racism, embodiment, and why theory is essential for a public health of consequence. *Am J Public Health* 2016;106(5):832–833. doi:10.2105/AJPH.2016.303100; Krieger N. Theoretical frameworks and cancer inequities. In: Vaccarella S, Lortet-Vieulent J, Saracci R, Conway DI, Straif K, Wild CP (eds). *Reducing Social Inequalities in Cancer: Evidence and Priorities for Research* (IARC Scientific Publication No. 168). Lyon, France: International Agency for Research on Cancer, 2019; 189–204. http://publications.iarc.fr/580.

78. Krieger N. Epidemiology and the web of causation: has anyone seen the spider? *Soc Sci Med* 1994;39(7):887–903. doi:10.1016/0277-9536(94)90202-x; Krieger N. *Epidemiology and the People's Health: Theory and Context.* New York: Oxford University Press, 2011; see especially Chapter 7: "Ecosocial theory of disease distribution: embodying societal and ecologic context" (pp. 202–235).
79. "ecology, n." *OED Online*, Oxford University Press. https://www.oed.com/view/Entry/59380; accessed December 31, 2020; Sydenstricker E. *Health and Environment.* New York: McGraw-Hill, 1933; Hogben L. *Nature and Nurture.* London: Williams & Norgate Ltd, 1933; Eldredge N, Grene M. *Interactions: The Biological Context of Social Systems.* New York: Columbia University Press, 1992; Gilbert SF, Epel D. *Ecological Developmental Biology: The Environmental Regulation of Development, Health, and Evolution.* 2nd ed. Sunderland, MA: Sinauer Associates, Inc., 2015; Rosenberg CE. Epilogue: airs, waters, places. A status report. *Bull Hist Med* 2012;86(4):661–670. doi:10.1353/bhm.2012.0082; McCarthy J, Perreault T, Bridge G (eds). *Routledge Handbook of Political Ecology.* New York: Routledge, 2015; Bryant R (ed). *The International Handbook of Political Ecology.* Cheltenham, UK: Edward Elgar Publishing, 2015; Ghazoul J. *Ecology: A Very Short Introduction.* New York: Oxford University Press, 2020.
80. McMichael AJ. *Human Frontiers, Environments and Disease: Past Patterns, Uncertain Futures.* Cambridge: Cambridge University Press, 2001; Crosby AW. *Ecological Imperialism: Biological Expansion of Europe, 900–1900.* 2nd ed. Cambridge: Cambridge University Press, 2004; McNeill JR, Roe A. *Global Environmental History: An Introductory Reader.* New York: Routledge, 2013; Penna A. *The Human Footprint: A Global Environmental History.* 2nd ed.

Malden, MA: Wiley-Blackwell, 2015; McCarthy J, Perreault T, Bridge G (eds). *Routledge Handbook of Political Ecology*. New York: Routledge, 2015; Bryant R (ed). *The International Handbook of Political Ecology*. Cheltenham, UK: Edward Elgar Publishing, 2015; Birn AE, Pillay Y, Holtz T. *Textbook of Global Health*. New York: Oxford University Press, 2017.

81. "social, adj. and n." *OED Online*, Oxford University Press. https://www.oed.com/view/Entry/183739; accessed December 31, 2020; Eldredge N, Grene M. *Interactions: The Biological Context of Social Systems*. New York: Columbia University Press, 1992; Beckwith J. *Political Sociology and The People's Health*. New York: Oxford University Press, 2018; Erikson K. *The Sociologist's Eye: Reflections on a Social Life*. New Haven, CT: Yale University Press, 2017.
82. "society, n." *OED Online*, Oxford University Press. https:// www.oed.com/view/Entry/183776; accessed December 31, 2020; Williams R. *Keywords: A Vocabulary of Culture and Society*. Revised ed. New York: Oxford University Press, 1983 ("society": pp. 291–295); Erikson K. *The Sociologist's Eye: Reflections on a Social Life*. New Haven, CT: Yale University Press, 2017.
83. Beckwith J. *Political Sociology and The People's Health*. New York: Oxford University Press, 2018; Erikson K. *The Sociologist's Eye: Reflections on a Social Life*. New Haven, CT: Yale University Press, 2017.
84. "theory, n." *OED Online*, Oxford University Press. https://www.oed.com/view/Entry/200431; accessed December 31, 2020; Krieger N. *Epidemiology and the People's Health: Theory and Context*. New York: Oxford University Press, 2011; see especially Chapter 1: "Does epidemiologic theory exist? On science, data, and explaining disease distribution" (pp. 3–41).
85. Krieger N. *Epidemiology and the People's Health: Theory and Context*. New York: Oxford University Press, 2011; quote: p. 17. See also: Ziman J. *Real Science: What It is and What It Means*. Cambridge: Cambridge University Press, 2000; Mendelsohn E, Weingart P, Whitley R (eds). *The Social Production of Scientific Knowledge*. Dordrecht, Holland: D. Reidel Publ, 1997; Curd M, Psillos S (eds). *The Routledge Companion to the Philosophy of Science*. New York: Routledge, 2014; Okasha S. *Philosophy of Science: A Very Short Introduction*. 2nd ed. New York: Oxford University Press, 2016; Felt U, Fouché R, Miller CA, Smith-Doerr L (eds). *The Handbook of Science and Technology Studies*. 4th ed. Cambridge, MA: MIT Press, 2017; Jasonoff S (ed). *States of Knowledge: The Co-Production of Science and Social Order*. London: Routledge, 2004.
86. Krieger N. Epidemiology and the web of causation: has anyone seen the spider? *Soc Sci Med* 1994;39(7):887–903. doi:10.1016/0277-9536(94)90202-x; Krieger N. *Epidemiology and the People's Health: Theory and Context*. New York: Oxford University Press, 2011; see especially Chapter 1: "Does epidemiologic theory exist? On science, data, and explaining disease distribution" (pp. 3–41); Baake

K, Bernhardt SA. *Metaphor and Knowledge: The Challenges of Writing Science.* Albany: State University of New York Press, 2003; Reynolds A. *The Third Lens: Metaphor and the Creation of Modern Cell Biology.* Chicago: University of Chicago Press, 2018; Squier SM. *Epigenetic Landscapes: Drawings as Metaphors.* Durham, NC: Duke University Press, 2017.

87. Krieger N. *Epidemiology and the People's Health: Theory and Context.* New York: Oxford University Press, 2011; see especially Chapter 1: "Does epidemiologic theory exist? On science, data, and explaining disease distribution" (pp. 3–41); Ziman J. *Real Science: What It Is and What It Means.* Cambridge: Cambridge University Press, 2000; Oreskes N. *Why Trust Science?* Princeton, NJ: Princeton University Press, 2019; Mendelsohn E, Weingart P, Whitley R (eds). *The Social Production of Scientific Knowledge.* Dordrecht, Holland: D. Reidel Publ, 1997; Felt U, Fouché R, Miller CA, Smith-Doerr L (eds). *The Handbook of Science and Technology Studies.* 4th ed. Cambridge, MA: MIT Press, 2017; Jasonoff S (ed). *States of Knowledge: The Co-Production of Science and Social Order.* London: Routledge, 2004.
88. Krieger N. Epidemiology and the web of causation: has anyone seen the spider? *Soc Sci Med* 1994;39(7):887–903. doi:10.1016/0277-9536(94)90202-x; Krieger N. *Epidemiology and the People's Health: Theory and Context.* New York: Oxford University Press, 2011; see especially Chapter 1: "Does epidemiologic theory exist? On science, data, and explaining disease distribution" (pp. 3–41); Golden J, Rosenberg C. (eds). *Framing Disease: Studies in Cultural History.* New Brunswick, NJ: Rutgers University Press, 1992; Lock MM, Nguyen V-K. *An Anthropology of Biomedicine.* 2nd ed. Hoboken, NJ: John Wiley & Sons, 2018.
89. Sigerist HE. Health. *J Public Health Policy* 1996;17(2):204–234 (chapter republished from Sigerist HE. *Medicine and Human Welfare.* New Haven, CT: Yale University Press, 1941); Golden J, Rosenberg C. (eds). *Framing Disease: Studies in Cultural History.* New Brunswick, NJ: Rutgers University Press, 1992; Lock MM, Nguyen V-K. *An Anthropology of Biomedicine.* 2nd ed. Hoboken, NJ: John Wiley & Sons, 2018; Saracci R. *Epidemiology: A Very Short Introduction.* New York: Oxford University Press, 2010.
90. "distribution, n." *OED Online*, Oxford University Press. https://www.oed.com/view/Entry/55781; accessed December 31, 2020; Krieger N. Who and what is a "population"? Historical debates, current controversies, and implications for understanding "population health" and rectifying health inequities. *Milbank Q* 2012;90(4):634–681. doi:10.1111/j.1468-0009.2012.00678.x; Thomopoulos N. *Statistical Distributions: Applications and Parameter Estimates.* Cham, Switzerland: Springer International Publishing, 2017.
91. Krieger N. Who and what is a "population"? Historical debates, current controversies, and implications for understanding "population health" and rectifying health inequities. *Milbank Q* 2012;90(4):634–681. doi:10.1111/j.1468-0009.2012.00678.x.

92. Saracci R. *Epidemiology: A Very Short Introduction*. New York: Oxford University Press, 2010; Porta M (ed). *Dictionary of Epidemiology*. 6th ed. New York: Oxford University Press, 2014.
93. Krieger N. Who and what is a "population"? Historical debates, current controversies, and implications for understanding "population health" and rectifying health inequities. *Milbank Q* 2012;90(4):634–681. doi:10.1111/j.1468-0009.2012.00678.x.
94. Krieger N, Davey Smith G. "Bodies count," and body counts: social epidemiology and embodying inequality. *Epidemiol Rev* 2004;26:92–103; Komlos J. Anthropometric history: an overview of a quarter century of research. *Anthropol Anz* 2009;67(4):341–356; Harris B. Health, height, and history: an overview of recent developments in anthropometric history. *Soc Hist Med* 1994;7(2):297–320.
95. Glanz K, Rimer BK, Viswanath K (eds). *Health Behavior: Theory, Research, and Practice*. 5th ed. San Francisco, CA: Jossey-Bass, 2015; Shelton LG. *The Bronfenbrenner Primer: A Guide to Develecology*. New York: Routledge, 2019; Bronfenbrenner U (ed). *Making Human Beings Human: Bioecological Perspectives on Human Development*. Thousand Oaks, CA: Sage, 2005; Moen P, Elder G, Lüscher K, Bronfenbrenner, U. *Examining Lives in Context: Perspectives on the Ecology of Human Development*. 1st ed., APA science volumes. Washington, DC: American Psychological Association, 1995; Eriksson M, Ghazinour M, Hammarström A. Different uses of Bronfenbrenner's ecological theory in public mental health research: what is their value for guiding public mental health policy and practice? *Social Theory Health* 2018;16(4):414–433.
96. *Eugenics Review* (1909–1968, published from 1909 to 1925 by the Eugenics Education Society and from 1926 to 1968 by the Eugenics Society), and continued as the *Journal of Biosocial Science* (1969–present; Cambridge: Cambridge University Press); Oakley A. Eugenics, social medicine, and the career of Richard Titmuss in Britain, 1935–50. *Br J Sociol* 1991;42(2):165–194; Massey DS. Brave new world of biosocial science. *Criminology* 2015;53(1):127–131; Guy R, Chomczyński PA. Bioethics and biosocial criminology: hurdling the status quo. *Ethics Med Public Health* 2018;7:95–102; Meloni M, Williams S, Martin P (eds). *Biosocial Matters: Rethinking Sociology-Biology Relations in the Twenty-First Century*. Chichester: Wiley-Blackwell, 2016; Meloni M, Müller R. Transgenerational epigenetic inheritance and social responsibility: perspectives from the social sciences. *Env Epigenetics* 2018;4(2):1–10; Singer M, Clair S. Syndemics and public health: reconceptualizing disease in bio-social context. *Medical Anthro Q* 2003;17(4):423–441; Singer M, Bulled N, Ostrach B, Mendenhall E. Syndemics and the biosocial conception of health. *The Lancet* 2017;389(10072):941–950.
97. Krieger N. Epidemiology and the web of causation: has anyone seen the spider? *Soc Sci Med* 1994;39(7):887–903.

98. Krieger N. Epidemiology and the web of causation: has anyone seen the spider? *Soc Sci Med* 1994;39(7):887–903. doi:10.1016/0277-9536(94)90202-x; Krieger N. *Epidemiology and the People's Health: Theory and Context.* New York: Oxford University Press, 2011; see especially Chapter 7: "Ecosocial theory of disease distribution: embodying societal and ecologic context" (pp. 202–235), including Figure 7.2, p. 229.
99. Krieger N. Epidemiology and the web of causation: has anyone seen the spider? *Soc Sci Med* 1994;39(7):887–903; quote: pp. 896–899. doi:10.1016/0277-9536(94)90202-x.
100. Krieger N. *Epidemiology and the People's Health: Theory and Context.* New York: Oxford University Press, 2011; see Chapter 7: "Ecosocial theory of disease distribution: embodying societal and ecologic context" (pp. 202–235, especially pp. 208–209); Krieger N. Proximal, distal, and the politics of causation: what's level got to do with it? *Am J Public Health* 2008;98(2):221–230; Turner JH, Boyns D. The return of grand theory. In: Turner JH (ed). *Handbook of Sociological Theory.* New York: Plenum Press, 2002; 353–378; Turner JH. A new approach for theoretically integrating micro and macro analyses. In: Calhoun C, Rojek C, Turner B (eds). *The Sage Handbook of Sociology.* Thousand Oaks, CA: Sage Publications, 2005; 405–422.
101. Krieger N. *Epidemiology and the People's Health: Theory and Context.* New York: Oxford University Press, 2011; see especially pp. 205–213; "ecology, n." *OED Online,* Oxford University Press. https://www.oed.com/view/Entry/59380; accessed December 31, 2020; "economic, n. and adj." *OED Online,* Oxford University Press. https://www.oed.com/view/Entry/59384; accessed December 31, 2020.
102. Krieger N. *Epidemiology and the People's Health: Theory and Context.* New York: Oxford University Press, 2011; 206; Stauffer RC. Haeckel, Darwin, and ecology. *Q Rev Biol* 1957;32:138–144; quote: pp. 140–141.
103. Porta M (ed). *A Dictionary of Epidemiology.* 6th ed. Oxford: Oxford University Press, 2014; see definitions for "etiology" (p. 100), "induction period" (p. 147), "latency period" (p. 165), and "natural history of disease" (pp. 193–194); Krieger N, Davey Smith G. "Bodies count," and body counts: social epidemiology and embodying inequality. *Epidemiol Rev* 2004;26:92–103.
104. Lessler J, Reich NG, Brookmeyer R, Perl TM, Nelson KE, Cummings DA. Incubation periods of acute respiratory viral infections: a systematic review. *Lancet Infect Dis* 2009;9(5):291–300. doi:10.1016/S1473-3099(09)70069-6.
105. Bertazzi PA. Descriptive epidemiology of malignant mesothelioma. *Med Lav* 2005;96(4):287–303.
106. Atwoli L, Stein DJ, Koenen KC, McLaughlin KA. Epidemiology of posttraumatic stress disorder: prevalence, correlates and consequences. *Curr Opin Psychiatry*

2015;28(4):307–311. doi:10.1097/YCO.0000000000000167; Scott KM, Koenen KC, Aguilar-Gaxiola S, Alonso J, Angermeyer MC, Benjet C, Bruffaerts R, Caldas-de-Almeida JM, de Girolamo G, Florescu S, Iwata N, Levinson D, Lim CC, Murphy S, Ormel J, Posada-Villa J, Kessler RC. Associations between lifetime traumatic events and subsequent chronic physical conditions: a cross-national, cross-sectional study. *PLoS One* 2013;8(11):e80573. doi:10.1371/journal.pone.0080573.

107. Vonghia L, Leggio L, Ferrulli A, Bertini M, Gasbarrini G, Addolorato G; Alcoholism Treatment Study Group. Acute alcohol intoxication. *Eur J Intern Med* 2008;19(8):561–567. doi:10.1016/j.ejim.2007.06.033; Seitz HK, Bataller R, Cortez-Pinto H, et al. Alcoholic liver disease [published correction appears in Nat Rev Dis Primers. 2018 Aug 28;4(1):18]. *Nat Rev Dis Primers* 2018;4(1):16. Published 2018 Aug 16. doi:10.1038/s41572-018-0014-7.
108. Wexler P. *Toxicology in Antiquity II (History of Toxicology and Environmental Health).* Amsterdam: Elsevier/Academic Press, 2014; Hernberg S. Lead poisoning in a historical perspective. *Am J Ind Med* 2000;38(3):244–254.
109. Levin R, Zilli Vieira CL, Rosenbaum MH, Bischoff K, Mordarski DC, Brown MJ. The urban lead (Pb) burden in humans, animals and the natural environment [published online ahead of print October 28, 2020]. *Environ Res* 2020;110377. doi:10.1016/j.envres.2020.110377; Markowitz G, Rosner D. *Deceit and Denial: The Deadly Politics of Industrial Pollution*. Berkeley: University of California Press, 2002; Markowitz G, Rosner D. *Lead Wars*. Berkeley: University of California Press, 2013; Troesken W. *The Great Lead Water Pipe Disaster.* Cambridge, MA: MIT Press, 2006; Wexler P. *Toxicology in Antiquity II (History of Toxicology and Environmental Health).* Amsterdam: Elsevier/Academic Press, 2014; Hernberg S. Lead poisoning in a historical perspective. *Am J Ind Med* 2000;38(3):244–254.
110. Levin R, Zilli Vieira CL, Rosenbaum MH, Bischoff K, Mordarski DC, Brown MJ. The urban lead (Pb) burden in humans, animals and the natural environment [published online ahead of print October 28, 2020]. *Environ Res* 2020;110377. doi:10.1016/j.envres.2020.110377; Troesken W. *The Great Lead Water Pipe Disaster.* Cambridge, MA: MIT Press, 2006.
111. Levin R, Zilli Vieira CL, Rosenbaum MH, Bischoff K, Mordarski DC, Brown MJ. The urban lead (Pb) burden in humans, animals and the natural environment [published online ahead of print October 28, 2020]. *Environ Res* 2020;110377. doi:10.1016/j.envres.2020.110377; Hanna-Attisha M, Olson ED. Preexisting conditions that kill us [published online ahead of print December 4, 2020]. *Fam Community Health*. doi:10.1097/FCH.0000000000000288; Markowitz G, Rosner D. *Deceit and Denial: The Deadly Politics of Industrial Pollution.* Berkeley: University of California Press, 2002; Markowitz G, Rosner D. *Lead Wars*. Berkeley: University of California Press, 2013.

112. Levin R, Zilli Vieira CL, Rosenbaum MH, Bischoff K, Mordarski DC, Brown MJ. The urban lead (Pb) burden in humans, animals and the natural environment [published online ahead of print October 28, 2020]. *Environ Res* 2020;110377. doi:10.1016/j.envres.2020.110377; Hanna-Attisha M, Olson ED. Preexisting conditions that kill us [published online ahead of print December 4, 2020]. *Fam Community Health* doi:10.1097/FCH.0000000000000288; Markowitz G, Rosner D. *Deceit and Denial: The Deadly Politics of Industrial Pollution*. Berkeley: University of California Press, 2002; Markowitz G, Rosner D. *Lead Wars*. Berkeley: University of California Press, 2013; Troesken W. *The Great Lead Water Pipe Disaster*. Cambridge, MA: MIT Press, 2006.
113. Levin R, Zilli Vieira CL, Rosenbaum MH, Bischoff K, Mordarski DC, Brown MJ. The urban lead (Pb) burden in humans, animals and the natural environment [published online ahead of print October 28, 2020]. *Environ Res* 2020;110377. doi:10.1016/j.envres.2020.110377; Markowitz G, Rosner D. *Lead Wars*. Berkeley: University of California Press, 2013.
114. Levin R, Zilli Vieira CL, Rosenbaum MH, Bischoff K, Mordarski DC, Brown MJ. The urban lead (Pb) burden in humans, animals and the natural environment [published online ahead of print October 28, 2020]. *Environ Res* 2020;110377. doi:10.1016/j.envres.2020.110377; Hanna-Attisha M, Olson ED. Preexisting conditions that kill us [published online ahead of print December 4, 2020]. *Fam Community Health* doi:10.1097/FCH.0000000000000288; Markowitz G, Rosner D. *Deceit and Denial: The Deadly Politics of Industrial Pollution*. Berkeley: University of California Press, 2002; Markowitz G, Rosner D. *Lead Wars*. Berkeley: University of California Press, 2013.
115. US Census Bureau. Quick Facts: Flint, Michigan. https://www.census.gov/quickfacts/flintcitymichigan; accessed July 10, 2020.
116. Hanna-Attisha M, Lanphear B, Landrigan P. Lead poisoning in the 21st century: the silent epidemic continues. *Am J Public Health* 2018;108(11):1430. doi:10.2105/AJPH.2018.304725; Hanna-Attisha M, Olson ED. Preexisting conditions that kill us [published online ahead of print December 4, 2020]. *Fam Community Health*. doi:10.1097/FCH.0000000000000288; Ruckart PZ, Ettinger AS, Hanna-Attisha M, Jones N, Davis SI, Breysse PN. The Flint water crisis: a coordinated public health emergency response and recovery initiative. *J Public Health Manag Pract* 2019;25(Suppl 1): S84–S90. doi:10.1097/PHH.0000000000000871; Levin R, Zilli Vieira CL, Rosenbaum MH, Bischoff K, Mordarski DC, Brown MJ. The urban lead (Pb) burden in humans, animals and the natural environment [published online ahead of print October 28, 2020]. *Environ Res* 2020;110377. doi:10.1016/j.envres.2020.110377.
117. Krieger N. Epidemiology and the web of causation: has anyone seen the spider? *Soc Sci Med* 1994;39(7):887–903. doi:10.1016/0277-9536(94)90202-x; Krieger

N. *Epidemiology and the People's Health: Theory and Context*. New York: Oxford University Press, 2011; Krieger N, Davey Smith G. The tale wagged by the DAG: broadening the scope of causal inference and explanation for epidemiology. *Int J Epidemiol* 2016;45(6):1787–1808. doi:10.1093/ije/dyw114; Krieger N, Davey Smith G. Response: FACEing reality: productive tensions between our epidemiological questions, methods and mission. *Int J Epidemiol* 2016;45(6):1852–1865. doi:10.1093/ije/dyw330; Ziman J. *Real Science: What It Is and What It Means*. Cambridge: Cambridge University Press, 2000; Ermakoff I. Causality and history: modes of causal investigation in historical social sciences. *Annu Rev Sociol* 2019;45:581–606; Lieberson S. Einstein, Renoir, and Greeley: some thoughts about evidence in sociology: 1991 presidential address. *Am Sociol Rev* 1992;57(1):1–15; Mayr E. *Toward a New Philosophy of Biology: Observations of an Evolutionist*. Cambridge, MA: Belknap Press of Harvard University Press, 1988.

118. Levin R, Zilli Vieira CL, Rosenbaum MH, Bischoff K, Mordarski DC, Brown MJ. The urban lead (Pb) burden in humans, animals and the natural environment [published online ahead of print October 28, 2020]. *Environ Res* 2020;110377. doi:10.1016/j.envres.2020.110377; Wexler P. *Toxicology in Antiquity II (History of Toxicology and Environmental Health)*. Amsterdam: Elsevier/Academic Press, 2014; Hernberg S. Lead poisoning in a historical perspective. *Am J Ind Med* 2000;38(3):244–254.

119. Levin R, Zilli Vieira CL, Rosenbaum MH, Bischoff K, Mordarski DC, Brown MJ. The urban lead (Pb) burden in humans, animals and the natural environment [published online ahead of print October 28, 2020]. *Environ Res* 2020;110377. doi:10.1016/j.envres.2020.110377; Markowitz G, Rosner D. *Lead Wars*. Berkeley: University of California Press, 2013; Troesken W. *The Great Lead Water Pipe Disaster*. Cambridge, MA: MIT Press, 2006; Wexler P. *Toxicology in Antiquity II (History of Toxicology and Environmental Health)*. Amsterdam: Elsevier/Academic Press, 2014; Ruckart PZ, Ettinger AS, Hanna-Attisha M, Jones N, Davis SI, Breysse PN. The Flint water crisis: a coordinated public health emergency response and recovery initiative. *J Public Health Manag Pract* 2019;25(Suppl 1): S84–S90. doi:10.1097/PHH.0000000000000871.

120. Markowitz G, Rosner D. *Deceit and Denial: The Deadly Politics of Industrial Pollution*. Berkeley: University of California Press, 2002; Markowitz G, Rosner D. *Lead Wars*. Berkeley, CA: University of California Press, 2013; Troesken W. *The Great Lead Water Pipe Disaster*. Cambridge, MA: MIT Press, 2006; Ruckart PZ, Ettinger AS, Hanna-Attisha M, Jones N, Davis SI, Breysse PN. The Flint water crisis: a coordinated public health emergency response and recovery initiative. *J Public Health Manag Pract* 2019;25(Suppl 1): S84–S90. doi:10.1097/PHH.0000000000000871; Levin R, Zilli Vieira CL, Rosenbaum MH, Bischoff K, Mordarski DC, Brown MJ. The urban lead (Pb) burden in humans, animals

and the natural environment [published online ahead of print October 28, 2020]. *Environ Res* 2020;110377. doi:10.1016/j.envres.2020.110377.

121. Krieger N. Epidemiology and the web of causation: has anyone seen the spider? *Soc Sci Med* 1994;39(7):887–903. doi:10.1016/0277-9536(94)90202-x; Krieger N. *Epidemiology and the People's Health: Theory and Context*. New York: Oxford University Press, 2011; see especially Chapter 7: "Ecosocial theory of disease distribution: embodying societal and ecologic context" (pp. 202–235); Krieger N. Measures of racism, sexism, heterosexism, and gender binarism for health equity research: from structural injustice to embodied harm-an ecosocial analysis [published online ahead of print November 25, 2019]. *Annu Rev Public Health* 2020;41:37–62. doi:10.1146/annurev-publhealth-040119-094017.
122. Krieger N. Epidemiology and the web of causation: has anyone seen the spider? *Soc Sci Med* 1994;39(7):887–903. doi:10.1016/0277-9536(94)90202-x; Krieger N. *Epidemiology and the People's Health: Theory and Context*. New York: Oxford University Press, 2011; see especially Chapter 7: "Ecosocial theory of disease distribution: embodying societal and ecologic context" (pp. 202–235), including Figure 7.2, p. 229.
123. Krieger N. Epidemiology and the web of causation: has anyone seen the spider? *Soc Sci Med* 1994;39(7):887–903. doi:10.1016/0277-9536(94)90202-x; Krieger N. *Epidemiology and the People's Health: Theory and Context*. New York: Oxford University Press, 2011; see especially Chapter 7: "Ecosocial theory of disease distribution: embodying societal and ecologic context" (pp. 202–235); Petteway R, Mujahid M, Allen A. Understanding embodiment in place-health research: approaches, limitations, and opportunities. *J Urban Health* 2019;96(2):289–299.
124. Krieger N. Epidemiology and the web of causation: has anyone seen the spider? *Soc Sci Med* 1994;39(7):887–903. doi:10.1016/0277-9536(94)90202-x; Krieger N. *Epidemiology and the People's Health: Theory and Context*. New York: Oxford University Press, 2011; see especially Chapter 7: "Ecosocial theory of disease distribution: embodying societal and ecologic context" (pp. 202–235); Krieger N. Embodiment: a conceptual glossary for epidemiology. *J Epidemiol Community Health* 2005;59(5):350–355; Krieger N. Theories for social epidemiology in the 21st century: an ecosocial perspective. *Int J Epidemiol* 2001;30(4):668–677.
125. Krieger N. Embodiment: a conceptual glossary for epidemiology. *J Epidemiol Community Health* 2005;59(5):350–355. doi:10.1136/jech.2004.024562; Krieger N. *Epidemiology and the People's Health: Theory and Context*. New York: Oxford University Press, 2011; see especially Chapter 7: "Ecosocial theory of disease distribution: embodying societal and ecologic context" (pp. 202–235); "embodiment | imbodiment, n." *OED Online*, Oxford University Press. http://www.oed.com/view/Entry/60906; accessed December 30, 2020.

126. Krieger N. History, biology, and health inequities: emergent embodied phenotypes and the illustrative case of the breast cancer estrogen receptor [published online ahead of print November 15, 2012]. *Am J Public Health* 2013;103(1):22–27. doi:10.2105/AJPH.2012.300967; Krieger N. Inheritance and health: what really matters? *Am J Public Health* 2018;108(5):606–607. doi:10.2105/AJPH.2018.304353.

127. Krieger N. Embodying inequality: a review of concepts, measures, and methods for studying health consequences of discrimination. *Int J Health Serv* 1999;29(2):295–352; Krieger N. Refiguring "race": epidemiology, racialized biology, and biological expressions of race relations. *Int J Health Serv* 2000;30(1):211–216. doi:10.2190/672J-1PPF-K6QT-9N7U; Krieger N. Genders, sexes, and health: what are the connections—and why does it matter? *Int J Epidemiol* 2003;32(4):652–657.

128. Krieger N, Zierler S. Account for the health of women. *Curr Issues Public Health* 1995;1:251–256; Krieger N. Racial discrimination and health: an epidemiologist's perspective. In: *Report of the President's Cancer Panel. The Meaning of Race in Science—Considerations for Cancer Research (April 9, 1997)*. Bethesda, MD: National Institutes of Health, National Cancer Institute, 1998; A32–A35; Krieger N. Embodying inequality: a review of concepts, measures, and methods for studying health consequences of discrimination. *Int J Health Serv* 1999;29(2):295–352; Krieger N. Refiguring "race": epidemiology, racialized biology, and biological expressions of race relations. *Int J Health Serv* 2000;30(1):211–216. doi:10.2190/672J-1PPF-K6QT-9N7U; Krieger N. Genders, sexes, and health: what are the connections—and why does it matter? *Int J Epidemiol* 2003;32(4):652–657.

129. Krieger N. Shades of difference: theoretical underpinnings of the medical controversy on black/white differences in the United States, 1830–1870. *Int J Health Serv* 1987;17(2):259–278. doi:10.2190/DBY6-VDQ8-HME8-ME3R; Krieger N, Rowley DL, Herman AA, Avery B, Phillips MT. Racism, sexism, and social class: implications for studies of health, disease, and well-being. *Am J Prev Med* 1993;9(6 Suppl):82–122; Krieger N, Fee E. Man-made medicine and women's health: the biopolitics of sex/gender and race/ethnicity. *Int J Health Serv* 1994;24(2):265–283. doi:10.2190/LWLH-NMCJ-UACL-U80Y; Krieger N. Epidemiology and the web of causation: has anyone seen the spider? *Soc Sci Med* 1994;39(7):887–903. doi:10.1016/0277-9536(94)90202-x; Krieger N. Embodying inequality: a review of concepts, measures, and methods for studying health consequences of discrimination. *Int J Health Serv* 1999;29(2):295–352; Krieger N. Refiguring "race": epidemiology, racialized biology, and biological expressions of race relations. *Int J Health Serv* 2000;30(1):211–216. doi:10.2190/672J-1PPF-K6QT-9N7U; Krieger N. Genders, sexes, and health: what are the connections—and why does it matter? *Int J Epidemiol* 2003;32(4):652–657.

130. Fredrickson GM. *Racism: A Short History.* With a new foreword by Albert M Camarillo. Princeton, NJ: Princeton University Press, 2015; Saini A. *Superior: The Return of Race Science.* Boston: Beacon Press, 2019; Kendi IX. *Stamped from the Beginning: The Definitive History of Racist Ideas in America.* New York: Nation Books, 2016; Yudell M. *Race Unmasked: Biology and Race in the Twentieth Century.* New York: Columbia University Press, 2014; Ernst W, Harris B (eds). *Race, Science, and Medicine, 1700–1960.* London: Routledge, 1999; Painter NI. *The History of White People.* New York: W. W. Norton, 2010; Krieger N, Rowley DL, Herman AA, Avery B, Phillips MT. Racism, sexism, and social class: implications for studies of health, disease, and well-being. *Am J Prev Med* 1993;9(6 Suppl):82–122; Krieger N, Fee E. Man-made medicine and women's health: the biopolitics of sex/gender and race/ethnicity. *Int J Health Serv* 1994;24(2):265–283. doi:10.2190/LWLH-NMCJ-UACL-U80Y; Krieger N. Epidemiology and the web of causation: has anyone seen the spider? *Soc Sci Med* 1994;39(7):887–903. doi:10.1016/0277-9536(94)90202-x; Krieger N. Embodying inequality: a review of concepts, measures, and methods for studying health consequences of discrimination. *Int J Health Serv* 1999;29(2):295–352; Krieger N. Refiguring "race": epidemiology, racialized biology, and biological expressions of race relations. *Int J Health Serv* 2000;30(1):211–216. doi:10.2190/672J-1PPF-K6QT-9N7U.
131. Fredrickson GM. *Racism: A Short History.* With a new foreword by Albert M Camarillo. Princeton, NJ: Princeton University Press, 2015; Saini A. *Superior: The Return of Race Science.* Boston: Beacon Press, 2019; Kendi IX. *Stamped from the Beginning: The Definitive History of Racist Ideas in America.* New York: Nation Books, 2016; Yudell M. *Race Unmasked: Biology and Race in the Twentieth Century.* New York: Columbia University Press, 2014; Ernst W, Harris B (eds). *Race, Science, and Medicine, 1700–1960.* London: Routledge, 1999; Painter NI. *The History of White People.* New York: W. W. Norton, 2010.
132. Krieger N. Epidemiology and the web of causation: has anyone seen the spider? *Soc Sci Med* 1994;39(7):887–903. doi:10.1016/0277-9536(94)90202-x; Krieger N. Embodying inequality: a review of concepts, measures, and methods for studying health consequences of discrimination. *Int J Health Serv* 1999;29(2):295–352; Krieger N. Methods for the scientific study of discrimination and health: from societal injustice to embodied inequality—an ecosocial approach. *Am J Public Health* 2012;102:936–945; Krieger N. Discrimination and health inequities. In: Berkman LF, Kawachi I, Glymour M (eds). *Social Epidemiology.* 2nd ed. New York: Oxford University Press, 2014; 63–125; Krieger N. Measures of racism, sexism, heterosexism, and gender binarism for health equity research: from structural injustice to embodied harm-an ecosocial analysis [published online ahead of print November 25, 2019]. *Annu Rev Public Health* 2020;41:37–62. doi:10.1146/annurev-publhealth-040119-094017.

133. Krieger N. Measures of racism, sexism, heterosexism, and gender binarism for health equity research: from structural injustice to embodied harm—an ecosocial analysis [published online ahead of print November 25, 2019]. *Annu Rev Public Health* 2020;41:37–62. doi:10.1146/annurev-publhealth-040119-094017; Krieger N. Methods for the scientific study of discrimination and health: from societal injustice to embodied inequality—an ecosocial approach. *Am J Public Health* 2012;102:936–945; Krieger N. Discrimination and health inequities. In: Berkman LF, Kawachi I, Glymour M (eds). *Social Epidemiology*. 2nd ed. New York: Oxford University Press, 2014; 63–125; Bailey ZD, Krieger N, Agénor M, Graves J, Linos N, Bassett MT. Structural racism and health inequities in the USA: evidence and interventions. *Lancet* 2017;389(10077):1453–1463; Ford CL, Griffith DM, Bruce MA, Gilbert KL (eds). *Racism: Science and Tools for the Public Health Professional*. Washington, DC: American Public Health Association, 2019; Chadha N, Lim B, Kane M, Rowland B. *Towards the Abolition of Biological Race in Medicine: Transforming Clinical Education, Research, and Practice*. Oakland, CA: Institute for Healing and Justice, 2020. https://www.instituteforhealingandjustice.org/download-the-report-here; accessed December 31, 2020.
134. Krieger N. Shades of difference: theoretical underpinnings of the medical controversy on black/white differences in the United States, 1830–1870. *Int J Health Serv* 1987;17(2):259–278. doi:10.2190/DBY6-VDQ8-HME8-ME3R; Krieger N, Rowley DL, Herman AA, Avery B, Phillips MT. Racism, sexism, and social class: implications for studies of health, disease, and well-being. *Am J Prev Med* 1993;9(6 Suppl):82–122; Bailey ZD, Krieger N, Agénor M, Graves J, Linos N, Bassett MT. Structural racism and health inequities in the USA: evidence and interventions. *The Lancet* 2017;389(10077):1453–1463; Vyas DA, Eisenstein LG, Jones DS. Hidden in plain sight—reconsidering the use of race correction in clinical algorithms. *New Engl J Med* 2020 June 17. doi:10.1056/NEJMms2004740; Boyd RW, Lindo EG, Weeks LD, McLemore MR. On racism: a new standard for publishing on racial health inequities. *Health Affairs* blog, July 2, 2020. https://www.healthaffairs.org/do/10.1377/hblog20200630.939347/full/; accessed December 31, 2020.
135. Krieger N. Measures of racism, sexism, heterosexism, and gender binarism for health equity research: from structural injustice to embodied harm-an ecosocial analysis [published online ahead of print November 25, 2019]. *Annu Rev Public Health* 2020;41:37–62. doi:10.1146/annurev-publhealth-040119-094017; Bailey ZD, Krieger N, Agénor M, Graves J, Linos N, Bassett MT. Structural racism and health inequities in the USA: evidence and interventions. *Lancet* 2017;389(10077):1453–1463; Ford CL, Griffith DM, Bruce MA, Gilbert KL (eds). *Racism: Science and Tools for the Public Health Professional*. Washington, DC: American Public Health

Association, 2019; Chadha N, Lim B, Kane M, Rowland B. *Towards the Abolition of Biological Race in Medicine: Transforming Clinical Education, Research, and Practice.* Oakland, CA: Institute for Healing and Justice, 2020. https://www.instituteforhealingandjustice.org/download-the-report-here; accessed December 30, 2020;Vyas DA, Eisenstein LG, Jones DS. Hidden in plain sight—reconsidering the use of race correction in clinical algorithms. *New Engl J Med* 2020 June 17. doi:10.1056/NEJMms2004740; Boyd RW, Lindo EG, Weeks LD, McLemore MR. On racism: a new standard for publishing on racial health inequities. *Health Affairs* blog, July 2, 2020. https://www.healthaffairs.org/do/10.1377/hblog20200630.939347/full/; accessed December 30, 2020.

136. Krieger N, Rowley DL, Herman AA, Avery B, Phillips MT. Racism, sexism, and social class: implications for studies of health, disease, and well-being. *Am J Prev Med* 1993;9(6 Suppl):82–122; Krieger N, Fee E. Man-made medicine and women's health: the biopolitics of sex/gender and race/ethnicity. *Int J Health Serv* 1994;24(2):265–283. doi:10.2190/LWLH-NMCJ-UACL-U80Y; Fee E, Krieger N (eds). *Women's Health, Politics, and Power: Essays on Sex/Gender, Medicine, and Public Health.* Amityville, NY: Baywood Publishing Co., 1994; Krieger N. Embodying inequality: a review of concepts, measures, and methods for studying health consequences of discrimination. *Int J Health Serv* 1999;29(2):295–352.
137. Schiebinger L. *Nature's Body: Gender in the Making of Modern Science.* New Brunswick, NJ: Rutgers University Press, 2004; Oudshoorn N. *Beyond the Natural Body: An Archaeology of Sex Hormones.* New York: Routledge, 1994; Apple RD (ed). *Women, Health and Medicine in America: A Historical Handbook.* New Brunswick, NJ: Rutgers University Press, 1992; King H. *Hippocrates' Woman: Reading the Female Body in Ancient Greece.* London: Routledge, 1988; King H. *The One-Sex Body on Trial: The Classical and Early Modern Evidence.* Farnham, Surrey: Ashgate Publishing Ltd., 2013; Hubbard R. *The Politics of Women's Biology.* New Brunswick, NJ: Rutgers University Press, 1990; Saini A. *Inferior: How Science Got Women Wrong—and the New Research That's Rewriting the Story.* Boston: Beacon Press, 2017.
138. Krieger N. Genders, sexes, and health: what are the connections—and why does it matter? *Int J Epidemiol* 2003;32:652–657; Richardson SR. Sexes, species, and genomes: why males and females are not like humans and chimpanzees. *Biol Philos* 2010;25:832–841; Ah-King M, Nylin S. Sex in evolutionary perspective: just another reaction norm. *Evol Biol* 2010;37:234–246; Fausto-Sterling A. *Sex/Gender: Biology in a Social World.* New York: Routledge, 2012; Spring KW, Stellman JM, Jordan-Young RM. Beyond a catalogue of difference: a theoretical frame and good practice guidelines for researching sex/gender in human health. *Soc Sci Med* 2012;74:1817–1824; Oudshoorn N. *Beyond the Natural Body: An Archaeology of Sex Hormones.* New York: Routledge, 1994.

139. Richardson SR. Sexes, species, and genomes: why males and females are not like humans and chimpanzees. *Biol Philos* 2010;25:832–841.

140. Krieger N. Genders, sexes, and health: what are the connections—and why does it matter? *Int J Epidemiol* 2003;32:652–657; Krieger N. Measures of racism, sexism, heterosexism, and gender binarism for health equity research: from structural injustice to embodied harm-an ecosocial analysis [published online ahead of print November 25, 2019]. *Annu Rev Public Health* 2020;41:37–62. doi:10.1146/annurev-publhealth-040119-094017.

141. Gilbert SF, Epel D. *Ecological Developmental Biology: The Environmental Regulation of Development, Health, and Evolution*. 2nd ed. Sunderland, MA: Sinauer Associates, Inc., 2015; Marli M, Karasawa G. *Reproductive Diversity of Plants: An Evolutionary Perspective and Genetic Basis.* Cham, Switzerland: Springer International Publishing, 2015; Ramawat KG, Shivanna KR, Merillon J-M (eds). *Reproductive Biology of Plants*. London: Taylor & Francis Group, 2014.

142. Ah-King M, Nylin S. Sex in evolutionary perspective: just another reaction norm. *Evol Biol* 2010;37:234–246; Gemmell NJ, Todd EV, Goikoetxea A, Ortega-Recalde O, Hore TA. Natural sex change in fish. *Curr Top Dev Biol* 2019;134:71–117. https://doi.org/10.1016/bs.ctdb.2018.12.014.

143. Krieger N. Genders, sexes, and health: what are the connections—and why does it matter? *Int J Epidemiol* 2003;32:652–657; Krieger N. Measures of racism, sexism, heterosexism, and gender binarism for health equity research: from structural injustice to embodied harm-an ecosocial analysis [published online ahead of print November 25, 2019]. *Annu Rev Public Health* 2020;41:37–62. doi:10.1146/annurev-publhealth-040119-094017; Spring KW, Stellman JM, Jordan-Young RM. Beyond a catalogue of difference: a theoretical frame and good practice guidelines for researching sex/gender in human health. *Soc Sci Med* 2012;74:1817–1824; Homan P. Structural sexism and health in the United States: a new perspective on health inequality and the gender system. *Am Sociol Rev* 2019;84(3):486–516; Heise L, Greene ME, Opper N, et al. Gender inequality and restrictive gender norms: framing the challenges to health. *Lancet* 2019;393:2440–2454.

144. Krieger N. Genders, sexes, and health: what are the connections—and why does it matter? *Int J Epidemiol* 2003;32:652–657; Krieger N. Measures of racism, sexism, heterosexism, and gender binarism for health equity research: from structural injustice to embodied harm-an ecosocial analysis [published online ahead of print November 25, 2019]. *Annu Rev Public Health* 2020;41:37–62. doi:10.1146/annurev-publhealth-040119-094017; Ah-King M, Nylin S. Sex in evolutionary perspective: just another reaction norm. *Evol Biol* 2010;37:234–246; Gemmell NJ, Todd EV, Goikoetxea A, Ortega-Recalde O, Hore TA. Natural sex change in fish. *Curr Top Dev Biol* 2019;134:71–117. https://doi.org/10.1016/bs.ctdb.2018.12.014; Fausto-Sterling A. *Sex/Gender: Biology in a Social World.*

New York: Routledge, 2012; Saini A. *Inferior: How Science Got Women Wrong—and the New Research That's Rewriting the Story*. Boston: Beacon Press, 2017; Meredith SL, Schmitt CA. The outliers are in: queer perspectives on investigating variation in biological anthropology. *Am Anthropol* 2019;121(2):487–489.

145. Krieger N. History, biology, and health inequities: emergent embodied phenotypes and the illustrative case of the breast cancer estrogen receptor. *Am J Public Health* 2013;103:22–27; Darbre PD. *Endocrine Disruption and Human Health*. Amsterdam: Academic Press, 2015; Eyster KM. The estrogen receptors: an overview from different perspectives. *Methods Mol Biol* 2016;1366:1–10.
146. Krieger N. Genders, sexes, and health: what are the connections—and why does it matter? *Int J Epidemiol* 2003;32(4):652–657.
147. Krieger N. History, biology, and health inequities: emergent embodied phenotypes and the illustrative case of the breast cancer estrogen receptor [published online ahead of print November 12, 2012]. *Am J Public Health* 2013;103(1):22–27. doi:10.2105/AJPH.2012.300967; Krieger N. Inheritance and health: what really matters? *Am J Public Health* 2018;108(5):606–607. doi:10.2105/AJPH.2018.304353.
148. Krieger N. Measures of racism, sexism, heterosexism, and gender binarism for health equity research: from structural injustice to embodied harm-an ecosocial analysis [published online ahead of print November 25, 2019]. *Annu Rev Public Health* 2020;41:37–62. doi:10.1146/annurev-publhealth-040119-094017.
149. Krieger N. Embodying inequality: a review of concepts, measures, and methods for studying health consequences of discrimination. *Int J Health Serv* 1999;29(2):295–352; Krieger N. *Epidemiology and the People's Health: Theory and Context*. New York: Oxford University Press, 2011; Krieger N. A critical research agenda for social justice and public health: an ecosocial proposal. In: Levy B (ed). *Social Injustice and Public Health*. 3rd ed. New York: Oxford University Press, 2019; 531–552.
150. Krieger N. Epidemiology and the web of causation: has anyone seen the spider? *Soc Sci Med* 1994;39(7):887–903. doi:10.1016/0277-9536(94)90202-x; Krieger N. Embodying inequality: a review of concepts, measures, and methods for studying health consequences of discrimination. *Int J Health Serv* 1999;29(2):295–352. doi:10.2190/M11W-VWXE-KQM9-G97Q; Krieger N. Epidemiology and social sciences: towards a critical reengagement in the 21st century. *Epidemiol Rev* 2000;22(1):155–163. doi:10.1093/oxfordjournals.epirev.a018014; Krieger N. Theories for social epidemiology in the 21st century: an ecosocial perspective. *Int J Epidemiol* 2001;30(4):668–677. doi:10.1093/ije/30.4.668; Krieger N. Embodiment: a conceptual glossary for epidemiology. *J Epidemiol Community Health* 2005;59(5):350–355. doi:10.1136/jech.2004.024562; Krieger N. *Epidemiology and the People's Health: Theory and Context*. New York: Oxford

University Press, 2011; Krieger N. Who and what is a "population"? Historical debates, current controversies, and implications for understanding "population health" and rectifying health inequities. *Milbank Q* 2012;90(4):634–681. doi:10.1111/j.1468-0009.2012.00678.x; Krieger N. Got theory? On the 21 st c. CE rise of explicit use of epidemiologic theories of disease distribution: a review and ecosocial analysis. *Curr Epidemiol Rep* 2014;1(1):45–56; Krieger N. Discrimination and health inequities. In: Berkman LF, Kawachi I, Glymour M (eds). *Social Epidemiology*. 2nd ed. New York: Oxford University Press, 2014; 63–125; Krieger N. A critical research agenda for social justice and public health: an ecosocial proposal. In: Levy B (ed). *Social Injustice and Public Health*. 3rd ed. New York: Oxford University Press, 2019; 531–552; Krieger N. Measures of racism, sexism, heterosexism, and gender binarism for health equity research: from structural injustice to embodied harm-an ecosocial analysis [published online ahead of print November 25, 2019]. *Annu Rev Public Health* 2020;41:37–62. doi:10.1146/annurev-publhealth-040119-094017.

151. Crenshaw K. Mapping the margins: intersectionality, identity politics, and violence against women of color. *Stanford Law Rev* 1991;43:1241–1299; Hill Collins P, Bilge S. *Intersectionality*. 2nd ed. Cambridge: Polity Press, 2020; Hancock A-M. *Intersectionality: An Intellectual History*. New York: Oxford University Press, 2016; Grusky DB, Hill J (eds). *Inequality in the Twenty-First Century*. Boulder, CO: Westview Press, 2018.
152. Crenshaw K. Mapping the margins: intersectionality, identity politics, and violence against women of color. *Stanford Law Rev* 1991;43:1241–1299; Hill Collins P, Bilge S. *Intersectionality*. 2nd ed. Cambridge: Polity Press, 2020; Hancock A-M. *Intersectionality: An Intellectual History*. New York: Oxford University Press, 2016.
153. Burnham L. The activist roots of Black feminist theory. *Organizing Upgrade*, December 15, 2020. https://organizingupgrade.com/the-activist-roots-of-black-feminist-theory/; accessed December 29, 2020; Garza A. Embracing both/and: a response to Linda Burnham. *Organizing Upgrade*, December 17, 2020. https://organizingupgrade.com/embracing-both-and-a-response-to-linda-burnham/; accessed December 29, 2020; Third World Women's Alliance. Equal to what? (1969). Reprinted in: Baxandall R, Gordon L (eds). *Dispatches from the Women's Liberation Movement*. New York: Basic Books, 2000; 65–66; Baxandall R. Re-visioning the women's liberation movement's narrative: early second wave African American feminists. *Fem Stud* 2001;27(1):225–245. doi:10.2307/3178460; Murray P. The liberation of Black women (1970). In: Guy-Sheftall B (ed). *Words of Fire: An Anthology of African-American Feminist Thought*. New York: New Press, 2011; 145–156; Beal FM. Double jeopardy: to be Black and female (1970). In: Guy-Sheftall B (ed). *Words of Fire: An Anthology of African-American Feminist Thought*.

New York: New Press, 2011; 145–155; Smith B, Smith B, Frazier D. Combahee River Collective: a Black feminist statement (1977). In: Guy-Sheftall B (ed). *Words of Fire: An Anthology of African-American Feminist Thought*. New York: New Press, 2011; 231–240; Lorde A. Learning from the '60s. Talk delivered at the Malcom X Weekend, Harvard University, Cambridge, MA, February 1982. In: Lorde A (ed). *Sister Outsider: Essays and Speeches*. Freedom, CA: Cross Press, 1984; 134–144.

154. Krieger N, Rowley DL, Herman AA, Avery B, Phillips MT. Racism, sexism, and social class: implications for studies of health, disease, and well-being. *Am J Prev Med* 1993;9(6 Suppl):82–122; Krieger N, Fee E. Man-made medicine and women's health: the biopolitics of sex/gender and race/ethnicity. *Int J Health Serv* 1994;24(2):265–283. doi:10.2190/LWLH-NMCJ-UACL-U80Y; Krieger N. Epidemiology and the web of causation: has anyone seen the spider? *Soc Sci Med* 1994;39(7):887–903. doi:10.1016/0277-9536(94)90202-x; Krieger N. Embodying inequality: a review of concepts, measures, and methods for studying health consequences of discrimination. *Int J Health Serv* 1999;29(2):295–352. doi:10.2190/M11W-VWXE-KQM9-G97Q; Krieger N. Epidemiology and social sciences: towards a critical reengagement in the 21st century. *Epidemiol Rev* 2000;22(1):155–163. doi:10.1093/oxfordjournals.epirev.a018014; Krieger N. Theories for social epidemiology in the 21st century: an ecosocial perspective. *Int J Epidemiol* 2001;30(4):668–677. doi:10.1093/ije/30.4.668.
155. Schulz AJ, Mullings L (eds). *Gender, Race, Class, and Health: Intersectional Approaches*. San Francisco: Jossey-Bass, 2006; Hill SE. Axes of health inequalities and intersectionality (Chapter 7). In: Smith KE, Bambra C, Hill SE (eds). *Health Inequalities: Critical Perspectives*. Oxford: Oxford University Press, 2015. Oxford Scholarship Online, 2016. doi:10.1093/acprof:oso/9780198703358.003.0007; Gkiouleka A, Huijts T, Beckfield J, Bambra C. Understanding the micro and macro politics of health: inequalities, intersectionality and institutions—a research agenda. *Soc Sci Med* 2018;200:92–98; Bowleg L. The problem with the phrase women and minorities: intersectionality-an important theoretical framework for public health. *Am J Public Health* 2012;102(7):1267–1273; Bowleg L. We're not all in this together: on COVID-19, intersectionality, and structural inequality. *Am J Public Health* 2020;110(7):917; Santos CE, Toomey RB. Integrating an intersectionality lens in theory and research in developmental science. *New Dir Child Adolesc Dev* 2018;161:7–15; Abrams JA, Tabaac A, Jung S, Else-Quest NM. Considerations for employing intersectionality in qualitative health research [published online ahead of print June 16, 2020]. *Soc Sci Med* 2020;258:113138. doi:10.1016/j.socscimed.2020.113138; Bauer GR. Incorporating intersectionality theory into population health research methodology: challenges and the potential to advance health equity. *Soc Sci Med* 2014;110:10–17; Agénor M. Future directions for incorporating intersectionality into quantitative population health

research. *Am J Public Health* 2020;110(6):803–806; Richman LS, Zucker AN. Quantifying intersectionality: an important advancement for health inequality research. *Soc Sci Med* 2019;226:246–248; Hankivsky O. Women's health, men's health, and gender and health: implications of intersectionality. *Soc Sci Med* 2012;74(11):1712–1720; Hankivsky O, Doyal L, Einstein G, Kelly U, Shim J, Weber L, Repta R. The odd couple: using biomedical and intersectional approaches to address health inequities. *Glob Health Action* 2017;10(Suppl 2):1326686.

156. Hill Collins P, Bilge S. *Intersectionality*. 2nd ed. Cambridge: Polity Press, 2020; quote: p. 14.
157. Rankine C. *Citizen: An American Lyric*. Minneapolis, MN: Greywolf Press, 2014; quote: p. 63.
158. Hill Collins P, Bilge S. *Intersectionality*. 2nd ed. Cambridge: Polity Press, 2020; Hancock A-M. *Intersectionality: An Intellectual History*. New York: Oxford University Press, 2016; Grusky DB, Hill J (eds). *Inequality in the Twenty-First Century*. Boulder, CO: Westview Press, 2018.
159. Krieger N. Epidemiology and the web of causation: has anyone seen the spider? *Soc Sci Med* 1994;39(7):887–903. doi:10.1016/0277-9536(94)90202-x; Krieger N. *Epidemiology and the People's Health: Theory and Context*. New York: Oxford University Press, 2011; Krieger N. A critical research agenda for social justice and public health: an ecosocial proposal. In: Levy B (ed). *Social Injustice and Public Health*. 3rd ed. New York: Oxford University Press, 2019; 531–552; Krieger N. Measures of racism, sexism, heterosexism, and gender binarism for health equity research: from structural injustice to embodied harm-an ecosocial analysis [published online ahead of print November 25, 2019]. *Annu Rev Public Health* 2020;41:37–62. doi:10.1146/annurev-publhealth-040119-094017.
160. Schulz AJ, Mullings L (eds). *Gender, Race, Class, and Health: Intersectional Approaches*. San Francisco: Jossey-Bass, 2006; Hill SE. Axes of health inequalities and intersectionality (Chapter 7). In: Smith KE, Bambra C, Hill SE (eds). *Health Inequalities: Critical Perspectives*. Oxford: Oxford University Press, 2015. Oxford Scholarship Online, 2016. doi:10.1093/acprof:oso/9780198703358.003.0007; Gkiouleka A, Huijts T, Beckfield J, Bambra C. Understanding the micro and macro politics of health: inequalities, intersectionality and institutions—a research agenda. *Soc Sci Med* 2018;200:92–98; Bowleg L. The problem with the phrase women and minorities: intersectionality-an important theoretical framework for public health. *Am J Public Health* 2012;102(7):1267–1273; Bowleg L. We're not all in this together: on COVID-19, intersectionality, and structural inequality. *Am J Public Health* 2020;110(7):917; Santos CE, Toomey RB. Integrating an intersectionality lens in theory and research in developmental science. *New Dir Child Adolesc Dev* 2018;161:7–15; Abrams JA, Tabaac A, Jung S, Else-Quest NM. Considerations for employing intersectionality in qualitative health research

[published online ahead of print June 16, 2020]. *Soc Sci Med* 2020;258:113138. doi:10.1016/j.socscimed.2020.113138; Bauer GR. Incorporating intersectionality theory into population health research methodology: challenges and the potential to advance health equity. *Soc Sci Med* 2014;110:10–17; Agénor M. Future directions for incorporating intersectionality into quantitative population health research. *Am J Public Health* 2020;110(6):803–806; Richman LS, Zucker AN. Quantifying intersectionality: an important advancement for health inequality research. *Soc Sci Med* 2019;226:246–248; Hankivsky O. Women's health, men's health, and gender and health: implications of intersectionality. *Soc Sci Med* 2012;74(11):1712–1720; Hankivsky O, Doyal L, Einstein G, Kelly U, Shim J, Weber L, Repta R. The odd couple: using biomedical and intersectional approaches to address health inequities. *Glob Health Action* 2017;10(Suppl 2):1326686.

161. Weiss G, Haber HF (eds). *Perspectives on Embodiment: The Intersections of Nature and Culture.* New York: Routledge, 1999; Smith JEH (ed). *Embodiment: A History.* New York: Oxford Scholarship Online, 2017.
162. Zaner RM. *The Problem of Embodiment: Some Contributions to a Phenomenology of the Body.* The Hague: Martinus Nijhoff, 1964; Smith JEH (ed). *Embodiment: A History.* New York: Oxford Scholarship Online, 2017.
163. Weiss G, Haber HF (eds). *Perspectives on Embodiment: The Intersections of Nature and Culture.* New York: Routledge, 1999; Noland C. *Agency and Embodiment: Performing Gestures/Producing Culture.* Cambridge, MA: Harvard University Press, 2009; Tichi C. *Embodiment of a Nation: Human Form in American Places.* Cambridge, MA: Harvard University Press, 2001; Traub V (ed). *The Oxford Handbook of Shakespeare and Embodiment: Gender, Sexuality, and Race.* New York: Oxford Handbooks Online, 2016; Smith JEH (ed). *Embodiment: A History.* New York: Oxford Scholarship Online, 2017; Durt C, Fuchs T, Tewes C. *Embodiment, Enaction, and Culture.* Cambridge, MA: MIT Press, 2017; Chamberlen M. *Embodying Punishment: Emotions, Identities, and Lived Experiences in Women's Prisons.* New York: Oxford Scholarship Online, 2018; Malatino H. *Queer Embodiment: Monstrosity, Medical Violence, and Intersex Experience.* Lincoln: University of Nebraska Press, 2019; Sheridan TE, McGuire RH (eds). *The Border and Its Bodies: The Embodiment of Risk Along the U.S.-México Line.* Tucson: University of Arizona Press, 2019; Mascia-Lees F (ed). *A Companion to the Anthropology of the Body and Embodiment.* Malden, MA: Wiley-Blackwell, 2011; Boero N, Manson K (eds). *The Oxford Handbook of the Sociology of Body and Embodiment.* New York: Oxford Handbooks Online, 2019; Piran N, Tylka T (eds). *Handbook of Positive Body Image and Embodiment: Constructs, Protective Factors, and Interventions.* New York: Oxford University Press, 2019; Lock M, Nguyen V-K. *An Anthropology of Biomedicine.* 2nd ed. Hoboken, NJ: John S. Wiley & Sons, Inc, 2018; Lock M. The tempering of medical anthropology: troubling natural

categories. *Med Anthro Q* 2001;15(4):478–492; Lock M. Toxic environments and the embedded psyche. *Med Anthro Q* 2020;34(1/S1):21–40; Niewohner J, Lock M. Situating local biologies: anthropological perspectives on environment/human entanglements. *Biosocieties* 2018;13(4):681–697; Lock M. Recovering the body. *Annu Rev Anthropol* 2017;46:1–14; Lock M. Comprehending the body in the era of the epigenome. *Curr Anthropol* 2015;56(2):151–177; Landecker H, Panofsky A. From social structure to gene regulation, and back: a critical introduction to environmental epigenetics for sociology. *Annu Rev Sociol* 2013;39:333–357.

164. Weiss G, Haber HF (eds). *Perspectives on Embodiment: The Intersections of Nature and Culture*. New York: Routledge, 1999; Noland C. *Agency and Embodiment: Performing Gestures/Producing Culture*. Cambridge, MA: Harvard University Press, 2009; Tichi C. *Embodiment of a Nation: Human Form in American Places*. Cambridge, MA: Harvard University Press, 2001; Traub V (ed). *The Oxford Handbook of Shakespeare and Embodiment: Gender, Sexuality, and Race*. New York: Oxford Handbooks Online, 2016; Smith JEH (ed). *Embodiment: A History*. New York: Oxford Scholarship Online, 2017; Durt C, Fuchs T, Tewes C. *Embodiment, Enaction, and Culture*. Cambridge, MA: MIT Press, 2017; Chamberlen M. *Embodying Punishment: Emotions, Identities, and Lived Experiences in Women's Prisons*. New York: Oxford Scholarship Online, 2018; Malatino H. *Queer Embodiment: Monstrosity, Medical Violence, and Intersex Experience*. Lincoln: University of Nebraska Press, 2019; Sheridan TE, McGuire RH (eds). *The Border and Its Bodies: The Embodiment of Risk Along the U.S.-México Line*. Tucson: University of Arizona Press, 2019; Mascia-Lees F (ed). *A Companion to the Anthropology of the Body and Embodiment*. Malden, MA: Wiley-Blackwell, 2011; Boero N, Manson K (eds). *The Oxford Handbook of the Sociology of Body and Embodiment*. New York: Oxford Handbooks Online, 2019; Piran N, Tylka T (eds). *Handbook of Positive Body Image and Embodiment: Constructs, Protective Factors, and Interventions*. New York: Oxford University Press, 2019; Lock M, Nguyen V-K. *An Anthropology of Biomedicine*. 2nd ed. Hoboken, NJ: John S. Wiley & Sons, Inc, 2018; Lock M. The tempering of medical anthropology: troubling natural categories. *Med Anthro Q* 2001;15(4):478–492; Lock M. Toxic environments and the embedded psyche. *Med Anthro Q* 2020;34(1/S1):21–40; Niewohner J, Lock M. Situating local biologies: anthropological perspectives on environment/human entanglements. *Biosocieties* 2018;13(4):681–697; Lock M. Recovering the body. *Annu Rev Anthropol* 2017;46:1–14; Lock M. Comprehending the body in the era of the epigenome. *Curr Anthropol* 2015;56(2):151–177; Landecker H, Panofsky A. From social structure to gene regulation, and back: a critical introduction to environmental epigenetics for sociology. *Annu Rev Sociol* 2013;39:333–357.
165. Lock M, Nguyen V-K. *An Anthropology of Biomedicine*. 2nd ed. Hoboken, NJ: John S. Wiley & Sons, Inc, 2018; Lock M. The tempering of medical

anthropology: troubling natural categories. *Med Anthro Q* 2001;15(4):478–492; Lock M. Toxic environments and the embedded psyche. *Med Anthro Q* 2020;34(1/S1):21–40; Niewohner J, Lock M. Situating local biologies: anthropological perspectives on environment/human entanglements. *Biosocieties* 2018;13(4):681–697; Lock M. Recovering the body. *Annu Rev Anthropol* 2017;46:1–14; Lock M. Comprehending the body in the era of the epigenome. *Curr Anthropol* 2015;56(2):151–177.

166. Malatino H. *Queer Embodiment: Monstrosity, Medical Violence, and Intersex Experience.* Lincoln: University of Nebraska Press, 2019; Sheridan TE, McGuire RH (eds). *The Border and Its Bodies: The Embodiment of Risk Along the U.S.-México Line.* Tucson: University of Arizona Press, 2019; Mascia-Lees F (ed). *A Companion to the Anthropology of the Body and Embodiment.* Malden, MA: Wiley-Blackwell, 2011; Boero N, Manson K (eds). *The Oxford Handbook of the Sociology of Body and Embodiment.* New York: Oxford Handbooks Online, 2019; Landecker H, Panofsky A. From social structure to gene regulation, and back: a critical introduction to environmental epigenetics for sociology. *Annu Rev Sociol* 2013;39:333–357.
167. Mascia-Lees F (ed). *A Companion to the Anthropology of the Body and Embodiment.* Malden, MA: Wiley-Blackwell, 2011; Boero N, Manson K (eds). *The Oxford Handbook of the Sociology of Body and Embodiment.* New York: Oxford Handbooks Online, 2019; Niewohner J, Lock M. Situating local biologies: anthropological perspectives on environment/human entanglements. *Biosocieties* 2018;13(4):681–697; Lock M. Recovering the body. *Annu Rev Anthropol* 2017;46:1–14.
168. Medvetz T, Sallaz JJ. *The Oxford Handbook of Pierre Bourdieu.* Oxford: Oxford University Press, 2018; Reed-Danahay D. *Locating Bourdieu.* Bloomington: Indiana University Press, 2005.
169. Bourdieu P, Wacquant L. *An Invitation to Reflexive Sociology.* Chicago: University of Chicago Press, 1992; quote: p. 127.
170. Bourdieu P. *Outline of a Theory of Practice.* Richard Nice, Translator. Cambridge: Cambridge University Press, 1977; Bourdieu P, Wacquant L. *An Invitation to Reflexive Sociology.* Chicago: University of Chicago Press, 1992; Asimaki A, Koustourakis G. Habitus: an attempt at a thorough analysis of a controversial concept in Pierre Bourdieu's theory of practice. *Social Sci* 2014 Aug 19;3(4):121–131; Medvetz T, Sallaz JJ. *The Oxford Handbook of Pierre Bourdieu.* Oxford: Oxford University Press, 2018; Reed-Danahay D. *Locating Bourdieu.* Bloomington: Indiana University Press, 2005.
171. Bourdieu P. *Outline of a Theory of Practice.* Richard Nice, Translator. Cambridge: Cambridge University Press, 1977; Bourdieu P, Wacquant L. *An Invitation to Reflexive Sociology.* Chicago: University of Chicago Press, 1992; Asimaki A, Koustourakis G. Habitus: an attempt at a thorough analysis of a

controversial concept in Pierre Bourdieu's theory of practice. *Social Sci* 2014 Aug 19;3(4):121–131; Medvetz T, Sallaz JJ. *The Oxford Handbook of Pierre Bourdieu*. Oxford: Oxford University Press, 2018; Reed-Danahay D. *Locating Bourdieu*. Bloomington: Indiana University Press, 2005.

172. Bourdieu P. *Outline of a Theory of Practice*. Richard Nice, Translator. Cambridge: Cambridge University Press, 1977; Bourdieu P, Wacquant L. *An Invitation to Reflexive Sociology*. Chicago: University of Chicago Press, 1992; Asimaki A, Koustourakis G. Habitus: an attempt at a thorough analysis of a controversial concept in Pierre Bourdieu's theory of practice. *Social Sci* 2014 Aug 19;3(4):121–131; Medvetz T, Sallaz JJ. *The Oxford Handbook of Pierre Bourdieu*. Oxford: Oxford University Press, 2018; Reed-Danahay D. *Locating Bourdieu*. Bloomington: Indiana University Press, 2005.
173. Krieger N. Epidemiology and social sciences: towards a critical reengagement in the 21st century. *Epidemiol Rev* 2000;22(1):155–163. doi:10.1093/oxfordjournals.epirev.a018014; Krieger N. Theories for social epidemiology in the 21st century: an ecosocial perspective. *Int J Epidemiol* 2001;30(4):668–677. doi:10.1093/ije/30.4.668; Krieger N. History, biology, and health inequities: emergent embodied phenotypes and the illustrative case of the breast cancer estrogen receptor [published online ahead of print November 15, 2012]. *Am J Public Health* 2013;103(1):22–27. doi:10.2105/AJPH.2012.300967; Krieger N. Got theory? On the 21 st c. CE rise of explicit use of epidemiologic theories of disease distribution: a review and ecosocial analysis. *Curr Epidemiol Rep* 2014;1(1):45–56; Krieger N. Living and dying at the crossroads: racism, embodiment, and why theory is essential for a public health of consequence. *Am J Public Health* 2016;106(5):832–833. doi:10.2105/AJPH.2016.303100; Niewohner J, Lock M. Situating local biologies: anthropological perspectives on environment/human entanglements. *Biosocieties* 2018;13(4):681–697.
174. Niewohner J, Lock M. Situating local biologies: anthropological perspectives on environment/human entanglements. *Biosocieties* 2018;13(4):681–697; Mascia-Lees F (ed). *A Companion to the Anthropology of the Body and Embodiment*. Malden, MA: Wiley-Blackwell, 2011; Boero N, Manson K (eds). *The Oxford Handbook of the Sociology of Body and Embodiment*. New York: Oxford Handbooks Online, 2019.
175. Krieger N. Embodying inequality: a review of concepts, measures, and methods for studying health consequences of discrimination. *Int J Health Serv* 1999;29(2):295–352; Krieger N. Embodiment: a conceptual glossary for epidemiology. *J Epidemiol Community Health* 2005;59(5):350–355; Krieger N. Theories for social epidemiology in the 21st century: an ecosocial perspective. *Int J Epidemiol* 2001;30(4):668–677; Krieger N, Davey Smith G. "Bodies count," and body counts: social epidemiology and embodying inequality. *Epidemiol Rev*

2004;26:92–103; Walters KL, Mohammed SA, Evans-Campbell T, Beltrán RE, Chae DH, Duran B. Bodies don't just tell stories, they tell histories: embodiment of historical trauma among American Indians and Alaska Natives. *Du Bois Rev* 2011;8(1):179–189.

176. Geronimus AT. The weathering hypothesis and the health of African-American women and infants: evidence and speculations. *Ethn Dis* 1992;2(3):207–221; Gernonimus AT. Black/white differences in the relationship of maternal age to birthweight: a population-based test of the weathering hypothesis. *Soc Sci Med* 1996;42(4):589–597; Geronimus AT. Understanding and eliminating racial inequalities in women's health in the United States: the role of the weathering conceptual framework. *J Am Med Women's Assoc* 2001;56(4):133–136, 149–150; Geronimus AT, Hicken M, Keene D, Bound J. "Weathering" and age patterns of allostatic load scores among blacks and whites in the United States. *Am J Public Health* 2006;96(5):826–833; Geronimus AT, Hicken M, Pearson JA, Seashols SJ, Brown KL, Cruz TD. Do US Black women experience stress-related accelerated biological aging? *Hum Nat* 2010;21(1):19–38; Geronimus AT. Deep integration: letting the epigenome out of the bottle without losing sight of the structural origins of population health. *Am J Public Health* 2013;103(Suppl 1):56–63; Linnenbringer E, Gehlert S, Geronimus AT. Black-White disparities in breast cancer subtype: the intersection of socially patterned stress and genetic expression. *AIMS Public Health* 2017;4(5):526–556; Geronimus AT, Bound J, Waidmann TA, Rodriguez JM, Timpe B. Weathering, drugs, and whack-a-mole: fundamental and proximate causes of widening educational inequity in US life expectancy by sex and race, 1990–2015. *J Health Social Behav* 2019;60(2):222–239; Geronimus AT, Pearson JA, Linnenbringer E, Eisenberg AK, Stokes C, Hughes LD, Schulz AJ. Weathering in Detroit: place, race, ethnicity, and poverty as conceptually fluctuating social constructs shaping variation in allostatic load. *Milbank Q* 2020;98(4):1171–1218; Linnenbringer E, Geronimus AT, Davis KL, Bound J, Ellis L, Gomez SL. Associations between breast cancer subtype and neighborhood socioeconomic and racial composition among Black and White women [published online ahead of print January 30, 2020]. *Breast Cancer Res Treat* 2020;180(2):437–447. doi:10.1007/s10549-020-05545-1.

177. For examples, see Taylor SE, Repetti RL, Seeman T. Health psychology: what is an unhealthy environment and how does it get under the skin? *Annu Rev Psychol* 1997;48:411–447; Lupien SJ, King S, Meaney MJ, McEwen BS. Can poverty get under your skin? Basal cortisol levels and cognitive function in children from low and high socioeconomic status. *Dev Psychopathol* 2001;13(3):653–676; McEwen BS, Mirsky AE. How socioeconomic status may "get under the skin" and affect the heart. *Eur Heart J* 2002;23(22):1727–1728; Hatzenbuehler ML. How does sexual minority stigma "get under the skin"? A psychological mediation framework.

Psychol Bull 2009;135(5):707–730; Hatzenbuehler ML, Nolen-Hoeksema S, Dovidio J. How does stigma "get under the skin"? The mediating role of emotion regulation. *Psychol Sci* 2009;20(10):1282–1289; Blakely T, McLeod M. Will the financial crisis get under the skin and affect our health? Learning from the past to predict the future. *N Z Med J* 2009;122(1307):76–83; Hertzman C, Boyce T. How experience gets under the skin to create gradients in developmental health. *Annu Rev Public Health* 2010;31:329–347; McEwen B. Brain on stress: how the social environment gets under the skin. *Proc Natl Acad Sci* 2012;109(Suppl 2):17170–171785; Das A. How does race get "under the skin"? Inflammation, weathering, and metabolic problems in late life. *Soc Sci Med* 2013;77:75–83; Chiavegatto Filho AD, Kawachi I, Wang YP, Viana MC, Andrade LH. Does income inequality get under the skin? A multilevel analysis of depression, anxiety and mental disorders in Sao Paulo, Brazil. *J Epidemiol Community Health* 2013;67(11):966–972; Kim P, Evans GW, Chen E, Miller G, Seeman T. How socioeconomic disadvantages get under the skin and into the brain to influence health development across the lifespan. In: Halfon N, Forrest CB, Lerner RM, Faustman EM (eds). *Handbook of Life Course Health Development*. Cham: Springer; 2018; 463–497; King RB, Bures RM. How the social environment gets under the skin. *Pop Res Policy Rev* 2017;36(5):631–637; Harris KM, Schorpp KM. Integrating biomarkers in social stratification and health research. *Annu Rev Sociol* 2018;44:361–386; Lehrner A, Yehuda R. Cultural trauma and epigenetic inheritance. *Dev Psychopathol* 2018;30(5):1763–1777; Liu PZ, Nusslock R. How stress gets under the skin: early life adversity and glucocorticoid receptor epigenetic regulation. *Curr Genomics* 2018;19(8):653–664; Semenza DC, Link NW. How does reentry get under the skin? Cumulative reintegration barriers and health in a sample of recently incarcerated men [published online ahead of print October 22, 2019]. *Soc Sci Med* 2019;243:112618. doi:10.1016/j.socscimed.2019.112618; McCrory C, McLoughlin S, O'Halloran AM. Socio-economic position under the microscope: getting "under the skin" and into the cells. *Curr Epidemiol Rep* 2019;6(4):403–411; Niedzwiedz CL. How does mental health stigma get under the skin? Cross-sectional analysis using the Health Survey for England. *SSM Popul Health* 2019;8:100433. doi:10.1016/j.ssmph.2019.100433; Thyagarajan B, Shippee N, Parsons H, Vivek S, Crimmins E, Faul J, Shippee T. How does subjective age get "under the skin"? The association between biomarkers and feeling older or younger than one's age: The Health and Retirement Study. *Innov Aging* 2019;3(4):igz035. doi:10.1093/geroni/igz035; Overbeek G, Creasey N, Wesarg C, Huijzer-Engbrenghof M, Spencer H. When mummy and daddy get under your skin: a new look at how parenting affects children's DNA methylation, stress reactivity, and disruptive behavior [published online ahead of print September 10, 2020]. *New Dir Child Adolesc Dev* 2020;2020(172):25–38. doi:10.1002/

cad.20362; Oi K. Does Retirement get under the skin and into the head? Testing the pathway from retirement to cardio-metabolic risk, then to episodic memory [published online ahead of print July 15, 2020]. *Res Aging* 2021;43(1):25–36. doi:10.1177/0164027520941161.

178. Wild CP. Complementing the genome with an "exposome": the outstanding challenge of environmental exposure measurement in molecular epidemiology. *Cancer Epidemiol Biomarkers Prev* 2005;14(8):1847–1850. doi:10.1158/1055-9965.EPI-05-0456; Wild CP. The exposome: from concept to utility [published online ahead of print January 31, 2012]. *Int J Epidemiol* 2012;41(1):24–32. doi:10.1093/ije/dyr236; Wild CP, Scalbert A, Herceg Z. Measuring the exposome: a powerful basis for evaluating environmental exposures and cancer risk [published online ahead of print May 16, 2013]. *Environ Mol Mutagen* 2013;54(7):480–499. doi:10.1002/em.21777; Turner MC, Nieuwenhuijsen M, Anderson K, Balshaw D, Cui Y, Dunton G, Hoppin JA, Koutrakis P, Jerrett M. Assessing the exposome with external measures: commentary on the state of the science and research recommendations. *Annu Rev Public Health* 2017;38:215–239. doi:10.1146/annurev-publhealth-082516-012802; Miller G. *Exposome: A Primer: The ex-POZE-ohm: A Pr ĭm'-er: The Environmental Equivalent of the Genome.* Amsterdam: Elsevier/Academic Press, 2014; Miller G. *The Exposome: A New Paradigm for the Environment and Health* (2nd ed.). London: Academic Press, 2020.
179. Geronimus AT. The weathering hypothesis and the health of African-American women and infants: evidence and speculations. *Ethn Dis* 1992;2(3):207–221; quote: p. 207.
180. "weather, n." *OED Online*, Oxford University Press. https://www.oed.com/view/Entry/226640; accessed December 31, 2020; "weather, v." *OED Online*, Oxford University Press. https://www.oed.com/view/Entry/226641; accessed December 31, 2020; "weathering, n." *OED Online*, Oxford University Press. https://www.oed.com/view/Entry/226660; accessed December 31, 2020.
181. Geronimus AT. The weathering hypothesis and the health of African-American women and infants: evidence and speculations. *Ethn Dis* 1992;2(3):207–221; Gernonimus AT. Black/white differences in the relationship of maternal age to birthweight: a population-based test of the weathering hypothesis. *Soc Sci Med* 1996;42(4):589–597; Geronimus AT. Understanding and eliminating racial inequalities in women's health in the United States: the role of the weathering conceptual framework. *J Am Med Women's Assoc* 2001;56(4):133–136, 149–150; Geronimus AT, Hicken M, Keene D, Bound J. "Weathering" and age patterns of allostatic load scores among blacks and whites in the United States. *Am J Public Health* 2006;96(5):826–833; Geronimus AT, Hicken M, Pearson JA, Seashols SJ, Brown KL, Cruz TD. Do US Black women experience stress-related

accelerated biological aging? *Hum Nat* 2010;21(1):19–38; Geronimus AT. Deep integration: letting the epigenome out of the bottle without losing sight of the structural origins of population health. *Am J Public Health* 2013;103(Suppl 1):56–63; Linnenbringer E, Gehlert S, Geronimus AT. Black-White disparities in breast cancer subtype: the intersection of socially patterned stress and genetic expression. *AIMS Public Health* 2017;4(5):526–556; Geronimus AT, Bound J, Waidmann TA, Rodriguez JM, Timpe B. Weathering, drugs, and whack-a-mole: fundamental and proximate causes of widening educational inequity in US life expectancy by sex and race, 1990–2015. *J Health Social Behav* 2019;60(2):222–239; Geronimus AT, Pearson JA, Linnenbringer E, Eisenberg AK, Stokes C, Hughes LD, Schulz AJ. Weathering in Detroit: place, race, ethnicity, and poverty as conceptually fluctuating social constructs shaping variation in allostatic load. *Milbank Q* 2020;98(4):1171–1218; Linnenbringer E, Geronimus AT, Davis KL, Bound J, Ellis L, Gomez SL. Associations between breast cancer subtype and neighborhood socioeconomic and racial composition among Black and White women [published online ahead of print January 30, 2020]. *Breast Cancer Res Treat* 2020;180(2):437–447. doi:10.1007/s10549-020-05545-1.

182. MacNamara A, Durham S. Dermatobia hominis in the accident and emergency department: "I've got you under my skin." *J Accident Emergency Med* 1997;14(3):179–180.
183. Taylor SE, Repetti RL, Seeman T. Health psychology: what is an unhealthy environment and how does it get under the skin? *Annu Rev Psychol* 1997;48:411–447.
184. Cohen WB. Don't let it get under your skin: some observations of psychosomatic dermatology. *Rhode Island Med J* 1952;35(6):312–313; Bowcock AM, Krueger JG. Getting under the skin: the immunogenetics of psoriasis. *Nature Rev Immunol* 2005;5(9):699–711; Vesely MD. Getting under the skin: targeting cutaneous autoimmune disease. *Yale J Biol Med* 2020;93(1):197–206; Taylor SE, Repetti RL, Seeman T. Health psychology: what is an unhealthy environment and how does it get under the skin? *Annu Rev Psychol* 1997;48:411–447; Lupien SJ, King S, Meaney MJ, McEwen BS. Can poverty get under your skin? Basal cortisol levels and cognitive function in children from low and high socioeconomic status. *Dev Psychopathol* 2001;13(3):653–676; McEwen BS, Mirsky AE. How socioeconomic status may "get under the skin" and affect the heart. *Eur Heart J* 2002;23(22):1727–1728; Hatzenbuehler ML. How does sexual minority stigma "get under the skin"? A psychological mediation framework. *Psychol Bull* 2009;135(5):707–730; Hatzenbuehler ML, Nolen-Hoeksema S, Dovidio J. How does stigma "get under the skin"? The mediating role of emotion regulation. *Psychol Sci* 2009;20(10):1282–1289; Blakely T, McLeod M. Will the financial crisis get under the skin and affect our health? Learning from the past to predict

the future. *N Z Med J* 2009;122(1307):76–83; Hertzman C, Boyce T. How experience gets under the skin to create gradients in developmental health. *Annu Rev Public Health* 2010;31:329–347; McEwen B. Brain on stress: how the social environment gets under the skin. *Proc Natl Acad Sci* 2012;109(Suppl 2):17170–171785; Das A. How does race get "under the skin"? Inflammation, weathering, and metabolic problems in late life. *Soc Sci Med* 2013;77:75–83; Chiavegatto Filho AD, Kawachi I, Wang YP, Viana MC, Andrade LH. Does income inequality get under the skin? A multilevel analysis of depression, anxiety and mental disorders in Sao Paulo, Brazil. *J Epidemiol Community Health* 2013;67(11):966–972; Kim P, Evans GW, Chen E, Miller G, Seeman T. How socioeconomic disadvantages get under the skin and into the brain to influence health development across the lifespan. In: Halfon N, Forrest CB, Lerner RM, Faustman EM (eds). *Handbook of Life Course Health Development*. Cham: Springer, 2018; 463–497; King RB, Bures RM. How the social environment gets under the skin. *Pop Res Policy Rev* 2017;36(5):631–637; Harris KM, Schorpp KM. Integrating biomarkers in social stratification and health research. *Annu Rev Sociol* 2018;44:361–386; Lehrner A, Yehuda R. Cultural trauma and epigenetic inheritance. *Dev Psychopathol* 2018;30(5):1763–1777; Liu PZ, Nusslock R. How stress gets under the skin: early life adversity and glucocorticoid receptor epigenetic regulation. *Curr Genomics* 2018;19(8):653–664; Semenza DC, Link NW. How does reentry get under the skin? Cumulative reintegration barriers and health in a sample of recently incarcerated men [published online ahead of print October 22, 2019]. *Soc Sci Med* 2019;243:112618. doi:10.1016/j.socscimed.2019.112618; McCrory C, McLoughlin S, O'Halloran AM. Socio-economic position under the microscope: getting "under the skin" and into the cells. *Curr Epidemiol Rep* 2019;6(4):403–411; Niedzwiedz CL. How does mental health stigma get under the skin? Cross-sectional analysis using the Health Survey for England. *SSM Popul Health* 2019;8:100433. doi:10.1016/j.ssmph.2019.100433; Thyagarajan B, Shippee N, Parsons H, Vivek S, Crimmins E, Faul J, Shippee T. How does subjective age get "under the skin"? The association between biomarkers and feeling older or younger than one's age: The Health and Retirement Study. *Innov Aging* 2019;3(4):igz035. doi:10.1093/geroni/igz035; Overbeek G, Creasey N, Wesarg C, Huijzer-Engbrenghof M, Spencer H. When mummy and daddy get under your skin: a new look at how parenting affects children's DNA methylation, stress reactivity, and disruptive behavior [published online ahead of print September 10, 2020]. *New Dir Child Adolesc Dev* 2020;2020(172):25–38. doi:10.1002/cad.20362; Oi K. Does retirement get under the skin and into the head? Testing the pathway from retirement to cardio-metabolic risk, then to episodic memory [published online ahead of print July 15, 2020]. *Res Aging* 2021;43(1):25–36. doi:10.1177/0164027520941161.

185. Wild CP. Complementing the genome with an "exposome": the outstanding challenge of environmental exposure measurement in molecular epidemiology. *Cancer Epidemiol Biomarkers Prev* 2005;14(8):1847–1850; quote: p. 1848. doi:10.1158/1055-9965.EPI-05-0456.
186. Wild CP. Complementing the genome with an "exposome": the outstanding challenge of environmental exposure measurement in molecular epidemiology. *Cancer Epidemiol Biomarkers Prev* 2005;14(8):1847–1850. doi:10.1158/1055-9965.EPI-05-0456.; Wild CP. The exposome: from concept to utility [published online ahead of print January 31, 2012]. *Int J Epidemiol* 2012;41(1):24–32. doi:10.1093/ije/dyr236; Wild CP, Scalbert A, Herceg Z. Measuring the exposome: a powerful basis for evaluating environmental exposures and cancer risk [published online ahead of print May 16, 2013]. *Environ Mol Mutagen* 2013;54(7):480–499. doi:10.1002/em.21777; Turner MC, Nieuwenhuijsen M, Anderson K, Balshaw D, Cui Y, Dunton G, Hoppin JA, Koutrakis P, Jerrett M. Assessing the exposome with external measures: commentary on the state of the science and research recommendations. *Annu Rev Public Health* 2017;38:215–239. doi:10.1146/annurev-publhealth-082516-012802; Miller G. *Exposome: A Primer: The ex-POZE-ohm: A Pr ĭm'-er: The Environmental Equivalent of the Genome.* Amsterdam: Elsevier/Academic Press, 2014; Miller G. *The Exposome: A New Paradigm for the Environment and Health.* 2nd ed. London: Academic Press, 2020; Vineis P, Robinson O, Chadeau-Hyam M, Dehghan A, Mudway I, Dagnino S. What is new in the exposome? [published online ahead of print June 30, 2020]. *Environ Int* 2020;143:105887. doi:10.1016/j.envint.2020.105887; Juarez PD, Hood DB, Song MA, Ramesh A. Use of an exposome approach to understand the effects of exposures from the natural, built, and social environments on cardio-vascular disease onset, progression, and outcomes. *Front Public Health* 2020;8:379. doi:10.3389/fpubh.2020.00379.
187. Schmitz C, Moritz MB, Powers M. Preparing social workers for ecosocial work practice and community building. *J Community Pract* 2019;27(3–4):446–459; Matthies AL, Närhi K. The ecosocial approach in social work as a framework for structural social work. *Int Social Work* 2018;61(4):490–502;.Matthies AL, Närhi K, eds. *Ecosocial Transition of Societies: The Contribution of Social Work and Social Policy.* Routledge Advances in Social Work. Oxford: Routledge, 2017; Norton CL. Social work and the environment: an ecosocial approach. *Int J Social Welfare* 2012;21(3):299–308.
188. Gilbert SF, Epel D. *Ecological Developmental Biology: The Environmental Regulation of Development, Health, and Evolution.* 2nd ed. Sunderland, MA: Sinauer Associates, Inc., 2015; Spencer HG. Beyond equilibria: the neglected role of history in ecology and evolution. *Q Rev Biol* 2020;95(4):311–321; Haraway DJ. *Staying with the Trouble: Making Kin in Chthulucene.* Durham, NC: Duke University Press, 2016; Grene M. *The Philosophy of Marjorie Grene.*

Chicago: Open Court, 2002; Grene M. *The Understandings of Nature: Essays in the Philosophy of Biology*. Boston: Reidel Pub., 1974; Grene M. *The Philosophy of Biology: An Episodic History*. Cambridge: Cambridge University Press, 2004; Grene M. Biology and human nature. In: *The Philosophy of Biology: An Episodic History*. Cambridge: Cambridge University Press, 2004; 322–347; Eldredge N. *The Miner's Canary: Unraveling the Mysteries of Extinction*. Princeton, NJ: Princeton University Press, 1994; Eldrege N. *Time Frames: The Evolution of Punctuate Equilibria*. Princeton, NJ: Princeton University Press, 2014.

189. Eldredge N, Grene M. *Interactions: The Biological Context of Social Systems*. New York: Columbia University Press, 1992.
190. Krieger N. History, biology, and health inequities: emergent embodied phenotypes and the illustrative case of the breast cancer estrogen receptor. *Am J Public Health* 2013;103(1):22–27; Krieger N. Inheritance and health: what really matters? *Am J Public Health* 2018;108(5):606–607.
191. Gilbert SF, Epel D. *Ecological Developmental Biology: The Environmental Regulation of Development, Health, and Evolution*. 2nd ed. Sunderland, MA: Sinauer Associates, Inc., 2015; Eldredge N, Grene M. *Interactions: The Biological Context of Social Systems*. New York: Columbia University Press, 1992.
192. Krieger N. Inheritance and health: what really matters? *Am J Public Health* 2018;108(5):606–607; Johannsen W. The genotype conception of heredity. *Am Nat* 1911;45(531):129–159; Hogben L. *Nature and Nurture*. New York: W. W. Norton & Co, 1933; Tucker RC (ed). *The Marx-Engels Reader*. New York: W. W. Norton, 1972; Piketty T. *Capital in the 21st Century*. Cambridge, MA: Belknap Press of Harvard University, 2014.
193. Jones K. *Health, Disease, and Society: A Critical Medical Geography*. London: Routgledge & Kegan Paul, 1987; Brown T, McLafferty S, Moon G (eds). *A Companion to Health and Medical Geography*. Malden, MA: Wiley-Blackwell, 2010; Crooks VA, Andrews GJ, Pearce J (eds). *Routledge Handbook of Health Geography*. London: Routledge, 2018; Foley R, Kearns R, Kistemann T, Wheeler B (eds). *Blue Space, Health, and Well-Being: Hydrophilia Unbounded*. New York: Routledge, 2019; Dyck E, Fletcher C. *Locating Health: Historical and Anthropological Investigations of Health and Place*. London: Pickering & Chatto, 2011; Ackerknecht EH. *History and Geography of the Most Important Diseases*. New York: Hafner Pub. Co., 1965; Rosenberg CE. Epilogue: airs, waters, places. A status report. *Bull Hist Med* 2012;86(4):661–670.
194. Robins P. *Political Ecology: A Critical Introduction*. 3rd ed. Hoboken, NJ: Wiley-Blackwell, 2020; Park TK, Greenberg JB (eds). *Terrestrial Transformations: A Political Ecology Approach to Society and Nature*. Lanham, MD: Lexington Books, 2020; Baer HA, Singer M. *Global Warming and the Political Ecology of Health: Emerging Crises and Systemic Solutions*. Walnut Creek, CA: Left Coast

Press, 2009; Singer M (ed). *A Companion to the Anthropology of Environmental Health*. Malden, MA: John Wiley & Sons, 2016; King B, Crews KA (eds). *Ecologies and Politics of Health*. New York: Routledge, 2013; McCarthy J, Perreault T, Bridge G (eds). *Routledge Handbook of Political Ecology*. New York: Routledge, 2015; Bryant R (ed). *The International Handbook of Political Ecology*. Cheltenham, UK: Edward Elgar Publishing, 2015; Hayes-Conroy A, Hayes-Conroy J. Political ecology of the body: a visceral approach. In: Bryant R (ed). *The International Handbook of Political Ecology*. Cheltenham, UK: Edward Elgar Publishing, 2015; 659–672.

195. Krieger N. Who and what is a "population"? Historical debates, current controversies, and implications for understanding "population health" and rectifying health inequities. *Milbank Q* 2012;90(4):634–681; Krieger N. Health equity and the fallacy of treating causes of population health as if they sum to 100 [published correction appears in *Am J Public Health* 2017;107(9):e16]. *Am J Public Health* 2017;107(4):541–549; Krieger N. Proximal, distal, and the politics of causation: what's level got to do with it? *Am J Public Health* 2008;98(2):221–230; Krieger N. *Epidemiology and the People's Health: Theory and Context*. New York: Oxford University Press, 2011.
196. Krieger N. Who and what is a "population"? Historical debates, current controversies, and implications for understanding "population health" and rectifying health inequities. *Milbank Q* 2012;90(4):634–681; Krieger N. Health equity and the fallacy of treating causes of population health as if they sum to 100 [published correction appears in *Am J Public Health* 2017;107(9):e16]. *Am J Public Health* 2017;107(4):541–549; Krieger N. Proximal, distal, and the politics of causation: what's level got to do with it? *Am J Public Health* 2008;98(2):221–230; Krieger N. *Epidemiology and the People's Health: Theory and Context*. New York: Oxford University Press, 2011.
197. Krieger N. Who and what is a "population"? Historical debates, current controversies, and implications for understanding "population health" and rectifying health inequities. *Milbank Q* 2012;90(4):634–681; Krieger N. Health equity and the fallacy of treating causes of population health as if they sum to 100 [published correction appears in *Am J Public Health* 2017;107(9):e16]. *Am J Public Health* 2017;107(4):541–549; Krieger N. Proximal, distal, and the politics of causation: what's level got to do with it? *Am J Public Health* 2008;98(2):221–230; Krieger N. *Epidemiology and the People's Health: Theory and Context*. New York: Oxford University Press, 2011.
198. Krieger N. Who and what is a "population"? Historical debates, current controversies, and implications for understanding "population health" and rectifying health inequities. *Milbank Q* 2012;90(4):634–681; Krieger N. Health equity and the fallacy of treating causes of population health as if they sum to 100

[published correction appears in *Am J Public Health* 2017;107(9):e16]. *Am J Public Health* 2017;107(4):541–549; Krieger N. Proximal, distal, and the politics of causation: what's level got to do with it? *Am J Public Health* 2008;98(2):221–230; Krieger N. *Epidemiology and the People's Health: Theory and Context.* New York: Oxford University Press, 2011.

199. Krieger N. Who and what is a "population"? Historical debates, current controversies, and implications for understanding "population health" and rectifying health inequities. *Milbank Q* 2012;90(4):634–681; Galton F. *Natural Inheritance.* London: Macmillan, 1889.
200. Krieger N. Who and what is a "population"? Historical debates, current controversies, and implications for understanding "population health" and rectifying health inequities. *Milbank Q* 2012;90(4):634–681; Carlson EA. *The Unfit: A History of a Bad Idea.* Cold Spring Harbor, NY: Cold Spring Harbor Press, 2001; Cowan RS. Galton, Sir Francis (1822–1911). *Oxford Dictionary of National Biography.* Oxford: Oxford University Press, 2004. http://www.oxforddnb.com.ezp-prod1.hul.harvard.edu/view/article/33315, accessed June 17, 2012; Galton F. *Natural Inheritance.* London: Macmillan, 1889; Galton F. *Inquiries into Human Faculty and its Development.* New York: MacMillan, 1883; 24–25; Galton F. Eugenics: its definition, scope, and aims. *Nature* 1904;70:82; Keller EF. *The Century of the Gene.* Cambridge, MA: Harvard University Press; 2000; Keller EF. *The Mirage of a Space Between Nature and Nurture.* Durham, NC: Duke University Press; 2010; Stigler SM. Regression towards the mean, historically considered. *Stat Methods Med Res* 1997;6:103–114.
201. Galton F. *Natural Inheritance.* London: Macmillan, 1889; quote: p. 63.
202. Galton F. *Natural Inheritance.* London: Macmillan, 1889; quote: p. 66.
203. Chase A. *The Legacy of Malthus: The Social Costs of the New Scientific Racism.* 1st ed. New York: Knopf, 1976; Leonard TC. *Illiberal Reformers: Race, Eugenics, and American Economics in the Progressive Era.* Princeton, NJ: Princeton University Press, 2016; Bashford A, Levine P (eds). *The Oxford Handbook of the History of Eugenics.* New York: Oxford University Press, 2010; Carlson EA. *The Unfit: A History of a Bad Idea.* Cold Spring Harbor, NY: Cold Spring Harbor Press, 2001; Herrnstein RJ, Murray C. *The Bell Curve: Intelligence and Class Structure in American Life.* New York: Free Press, 1994; Jacoby R, Glauberman N. *The Bell Curve Debate: History, Documents, Opinions.* New York: Times Book, 1995; Fischer CS, Jankowski MS, Hout M, Lucas R, Swidler A, Voss K. *Inequality by Design: Cracking the Bell Curve Myth.* Princeton, NJ: Princeton University Press, 1996; Devlin B, Fienberg SE, Resnick DP, Roeder K (eds). *Intelligence, Genes, and Success: Scientists Respond to the Bell Curve.* New York: Springer Books, 1997; Plomin R. *Blueprint: How DNA Makes Us Who We Are.* Cambridge, MA: MIT Press, 2018; Murray CA. *Human Diversity: The Biology of Gender, Race, and*

Class. New York: Twelve, 2020; Saini A. *Superior: The Return of Race Science*. Boston: Beacon Press, 2019.

204. Galton F. *Hereditary Genius: An Inquiry into Its Laws and Consequences*. London: Macmillan, 1869; see especially the chapter titled: "The comparative worth of different races" (pp. 336–350)—and these views are retained in the 2nd edition (published in 1892); Galton F. *Inquiries into Human Faculty and its Development*. New York: MacMillan 1883; Galton F. *Natural Inheritance*. London: Macmillan, 1889.
205. Chase A. *The Legacy of Malthus: The Social Costs of the New Scientific Racism*. 1st ed. New York: Knopf, 1976; Leonard TC. *Illiberal Reformers: Race, Eugenics, and American Economics in the Progressive Era*. Princeton, NJ: Princeton University Press, 2016; Bashford A, Levine P (eds). *The Oxford Handbook of the History of Eugenics*. New York: Oxford University Press, 2010; Carlson EA. *The Unfit: A History of a Bad Idea*. Cold Spring Harbor, NY: Cold Spring Harbor Press, 2001; Hogben L. *Genetic Principles in Medicine and Social Science*. London: Williams & Norgate, Ltd, 1931; see especially Chapter V, "The concept of race" (pp. 122–144) and Chapter VI, "The nature of genetic selection in the social group" (pp. 145–172).
206. Galton F. *Hereditary Genius: An Inquiry into Its Laws and Consequences*. London: Macmillan, 1869; see especially the chapter titled: "The comparative worth of different races" (pp. 336–350)—and these views are retained in the 2nd edition (published in 1892); Galton F. *Inquiries into Human Faculty and Its Development*. New York: MacMillan, 1883.
207. Galton F. *Hereditary Genius: An Inquiry into Its Laws and Consequences*. London: Macmillan, 1869; see especially the chapter titled: "The comparative worth of different races" (pp. 336–350); quote: pp. 337–338—and these views are retained in the 2nd edition (published in 1892).
208. James CLR. *The Black Jacobins: Toussaint L'Overture and the San Domingo Revolution*. New York: Dial Press, 1938; Williams EE. *Capitalism and Slavery*. Chapel Hill: University of North Carolina Press, 1944; Fergus CK. *Revolutionary Emancipation: Slavery and Abolitionism in the British West Indies*. Baton Rouge: Louisiana State University Press, 2013.
209. Galton F. *Natural Inheritance*. London: Macmillan, 1889; see especially Chapter X, "Disease," pp. 164–186.
210. Galton F. *Natural Inheritance*. London: Macmillan, 1889; quote: p. 165.
211. Limpert E, Stahel WA, Abbt M. Log-normal distributions across the sciences: keys and clues. *BioSci* 2001;51:341–352.
212. Krieger N. Who and what is a "population"? Historical debates, current controversies, and implications for understanding "population health" and rectifying health inequities. *Milbank Q* 2012;90(4):634–681.

213. Krieger N. Who and what is a "population"? Historical debates, current controversies, and implications for understanding "population health" and rectifying health inequities. *Milbank Q* 2012;90(4):634–681; Krieger N. Health equity and the fallacy of treating causes of population health as if they sum to 100 [published correction appears in *Am J Public Health* 2017;107(9):e16]. *Am J Public Health* 2017;107(4):541–549; Coggon DIW, Martyn CN. Time and chance: the stochastic nature of disease causation. *Lancet* 2005;365:1434–1437; Smith GD. Epidemiology, epigenetics and the "Gloomy Prospect": embracing randomness in population health research and practice. *Int J Epidemiol* 2011;40(3):537–562; Davey Smith G. Post-modern epidemiology: when methods meet matter. *Am J Epidemiol* 2019;188(8):1410–1419; Davey Smith G, Relton CL, Brennan P. Chance, choice and cause in cancer aetiology: individual and population perspectives. *Int J Epidemiol* 2016;45(3):605–613; Beckfield J. *Political Sociology and the People's Health.* New York: Oxford University Press, 2018; see especially pp. 5–9.
214. United Nations. Universal Declaration of Human Rights. General Assembly Resolution 217A, December 10, 1948. https://www.un.org/en/universal-declaration-human-rights/index.html; accessed December 31, 2020; Gruskin S, Mills EJ, Tarantola D. History, principles, and practice of health and human rights. *Lancet* 2007;370(9585):449–455; Gruskin S. Reflections on 25 years of health and human rights: history, context, and the need for strategic action. *Health Hum Rights* 2020;22(1):327–329.
215. Krieger N. Proximal, distal, and the politics of causation: what's level got to do with it? *Am J Public Health* 2008;98(2):221–230; Krieger N. *Epidemiology and the People's Health: Theory and Context.* New York: Oxford University Press, 2011.
216. Garza A. *The Purpose of Power: How We Come Together When We Fall Apart.* New York: One World, Random House, 2020; Douglass F. *Two Speeches by Frederick Douglass; West India Emancipation and the Dredd Scot Decision.* Rochester, NY: C.P. Dewey, 1857. https://www.loc.gov/item/mfd.21039; accessed December 31, 2020; Hobsbawm E. *Age of Revolution: Europe, 1789–1848.* London: Phoenix, 1962; Hobsbawm E. *Age of Extremes: The Short Twentieth Century, 1914–1991.* London: Michael Joseph, 1994; Minkler M. *Community Organizing and Community Building for Health and Welfare.* Piscataway, NJ: Rutgers University Press, 2012; Miller C, Crane J (eds). *The Nature of Hope: Grassroots Organizing, Environmental Justice, and Political Change.* Longville: University Press of Colorado, 2019; Zavella P. *The Movement for Reproductive Justice: Empowering Women of Color Through Social Activism.* New York: New York University Press, 2020; Turner JH, Boyns D. The return of grand theory. In: Turner JH (ed). *Handbook of Sociological Theory.* New York: Plenum Press, 2002; 353–378; Turner

JH. A new approach for theoretically integrating micro and macro analyses. In: Calhoun C, Rojek C, Turner B (eds). *The Sage Handbook of Sociology*. Thousand Oaks, CA: Sage Publications, 2005; 405–422.

217. Diez-Roux AV. Bringing context back into epidemiology: variables and fallacies in multilevel analysis. *Am J Public Health* 1998;88(2):216–222. doi:10.2105/ajph.88.2.216; Subramanian SV, Jones K, Kaddour A, Krieger N. Revisiting Robinson: the perils of individualistic and ecologic fallacy. *Int J Epidemiol* 2009;38(2):342–360; author reply 370–373. doi:10.1093/ije/dyn359.
218. Krieger N. The real ecological fallacy: epidemiology and global climate change. *J Epidemiol Community Health* 2015;69(8):803–804; quote: p. 803.

CHAPTER 2

1. Kenneally C. Large DNA study traces violent history of American slavery. *New York Times*, July 24, 2020. https://www.nytimes.com/2020/07/23/science/23andme-african-ancestry.html; accessed January 1, 2021; BBC News. Genetic impact of colonial-era slave trade revealed in DNA study. *BBC*, July 24, 2020. https://www.bbc.com/news/world-africa-53527405; accessed January 1, 2021; Rahhal N. DNA study reveals how the slave trade's dark history of rape, disease and deadly working conditions shaped the modern-day genetics of black people in America. *Daily Mail*, July 24, 2020. https://www.dailymail.co.uk/health/article-8554635/Dark-history-transatlanic-slavery-traced-DNA-study.html; accessed January 1, 2021; AFP-JIJI. Horrors of trans-Atlantic slavery traced through DNA study. *Japan Times*, July 24, 2020. https://www.japantimes.co.jp/news/2020/07/24/world/slavery-dna-study/; accessed January 1, 2021; MSN. Black American's genes reflect the hardships of slavery. *Africa Today News*, July 24, 2020. https://africatodaynewsonline.com/2020/07/23/black-americans-genes-reflect-the-hardships-of-slavery/; accessed January 1, 2021; Marshall M. How the slave trade left its mark in the DNA of people in the Americas. *New Scientist*, July 23, 2020. https://www.newscientist.com/article/2249839-how-the-slave-trade-left-its-mark-in-the-dna-of-people-in-the-americas/#; accessed January 1, 2021; Cell Press. Using 23andMe African ancestry data to reexamine the history of slavery. *Sci Tech Daily*, July 23, 2020. https://scitechdaily.com/using-23andme-african-ancestry-data-to-reexamine-the-history-of-slavery/; accessed January 1, 2021; Marcus AD. Study sheds light on regional origins of many Black Americans' enslaved ancestors. *Wall Street Journal*, July 23, 2020. https://www.wsj.com/articles/study-finds-legacy-of-slavery-shaped-the-dna-of-black-americans-today-11595516400?mod=searchresults&page=1&pos=1; accessed January 1, 2021.
2. Bogel-Burroughs N, Mervosh S. U.S. hospitalizations for the coronavirus near April peak. *New York Times*, July 22, 2020. https://www.nytimes.com/2020/07/22/us/coronavirus-hospitalizations-near-peak.html; accessed January 1, 2021;

Leatherby L. How the U.S. compares with the world's worst coronavirus hot spots. *New York Times*, July 24, 2020. https://www.nytimes.com/interactive/2020/07/23/us/coronavirus-hotspots-countries.html; accessed January 1, 2021.

3. Micheletti SJ, Bryc K, Ancona Esselmann SG, Freyman WA, Moreno ME, Poznik GD, Shastri AJ, 23andMe Research Team, Beleza S, Mountain JL. Genetic consequences of the transatlantic slave trade in the Americas. *Am J Hum Genet* 2020. https://doi.org/10.1016/j.ajhg.2020.06.012.
4. Micheletti SJ, Bryc K, Ancona Esselmann SG, Freyman WA, Moreno ME, Poznik GD, Shastri AJ, 23andMe Research Team, Beleza S, Mountain JL. Genetic consequences of the transatlantic slave trade in the Americas. *Am J Hum Genet* 2020. https://doi.org/10.1016/j.ajhg.2020.06.012.
5. Krieger N. ENOUGH: COVID-19, structural racism, police brutality, plutocracy, climate change-and time for health justice, democratic governance, and an equitable, sustainable future [published online ahead of print August 20, 2020]. *Am J Public Health* 2020;110(11):1620–1623. doi:10.2105/AJPH.2020.305886; Buchanan L, Bui Q, Patel JK. Black Lives Matter may be the largest movement in U.S. history. *New York Times*, July 3, 2020. https://www.nytimes.com/interactive/2020/07/03/us/george-floyd-protests-crowd-size.html; accessed January 1, 2021.
6. Micheletti SJ, Bryc K, Ancona Esselmann SG, Freyman WA, Moreno ME, Poznik GD, Shastri AJ, 23andMe Research Team, Beleza S, Mountain JL. Genetic consequences of the transatlantic slave trade in the Americas. *Am J Hum Genet* 2020. https://doi.org/10.1016/j.ajhg.2020.06.012.
7. Belbin GM, Nieves-Colón MA, Kenny EE, Moreno-Estrada A, Gignoux CR. Genetic diversity in populations across Latin America: implications for population and medical genetic studies. *Curr Opin Genet Dev* 2018;53:98–104; Torres JD, Torres Colón GA. Racial experience as an alternative operationalization of race. *Hum Biol* 2015;87(4):306–312; Torres JD. Anthropological perspectives on genomic data, genetic ancestry, and race. *Yearbook Phys Anthropol* 2020;171(Suppl 70):74–86; Curet LA. *Caribbean Paleodemography: Population, Culture History, and Sociopolitical Processes in Ancient Puerto Rico*. Tuscaloosa: University of Alabama Press, 2005; Morgan K. *Slavery and the British Empire: From Africa to America*. New York: Oxford University Press, 2008; Wheat D. *Atlantic Africa and the Spanish Caribbean, 1570–1640*. Chapel Hill: University of North Carolina Press, 2016; Knight F (ed). *General History of the Caribbean. Volume III, The Slave Societies of the Caribbean*. New York; Paris: Palgrave Macmillan: UNESCO Publishing, 2007; Reséndez A. *The Other Slavery: The Uncovered Story of Indian Enslavement in America*. Boston: Houghton Mifflin Harcourt, 2016.
8. Kenneally C. Large DNA study traces violent history of American slavery. *New York Times*, July 24, 2020. https://www.nytimes.com/2020/07/23/science/23andme-african-ancestry.html; accessed January 1, 2021; BBC News. Genetic impact of

colonial-era slave trade revealed in DNA study. *BBC*, July 24, 2020. https://www.bbc.com/news/world-africa-53527405; accessed January 1, 2021; Rahhal N. DNA study reveals how the slave trade's dark history of rape, disease and deadly working conditions shaped the modern-day genetics of black people in America. *Daily Mail*, July 24, 2020. https://www.dailymail.co.uk/health/article-8554635/Dark-history-transatlanic-slavery-traced-DNA-study.html; accessed January 1, 2021; AFP-JIJI. Horrors of trans-Atlantic slavery traced through DNA study. *Japan Times*, July 24, 2020. https://www.japantimes.co.jp/news/2020/07/24/world/slavery-dna-study/; accessed January 1, 2021; MSN. Black American's genes reflect the hardships of slavery. *Africa Today News*, July 24, 2020. https://africatodaynewsonline.com/2020/07/23/black-americans-genes-reflect-the-hardships-of-slavery/; accessed January 1, 2021; Marshall M. How the slave trade left its mark in the DNA of people in the Americas. *New Scientist*, July 23, 2020. https://www.newscientist.com/article/2249839-how-the-slave-trade-left-its-mark-in-the-dna-of-people-in-the-americas/#; accessed January 1, 2021; Cell Press. Using 23andMe African ancestry data to reexamine the history of slavery. *Sci Tech Daily*, July 23, 2020. https://scitechdaily.com/using-23andme-african-ancestry-data-to-reexamine-the-history-of-slavery/; accessed January 1, 2021; Marcus AD. Study sheds light on regional origins of many Black Americans' enslaved ancestors. *Wall Street Journal*, July 23, 2020. https://www.wsj.com/articles/study-finds-legacy-of-slavery-shaped-the-dna-of-black-americans-today-11595516400?mod=searchresults&page=1&pos=1; accessed January 1, 2021.

9. Zimmer C. Biden to elevate science advisor to his cabinet. *New York Times*, January 15, 2021; updated January 26, 2021. https://www.nytimes.com/2021/01/15/science/biden-science-cabinet.html; accessed February 13, 2021; Subbaraman N. "Inspired choice": Biden appoints sociologist Alondra Nelson to top science post. *Nature*, January 21, 2021. https://www.nature.com/articles/d41586-021-00159-z; accessed February 13, 2021.
10. Kenneally C. Large DNA study traces violent history of American slavery. *New York Times*, July 24, 2020. https://www.nytimes.com/2020/07/23/science/23andme-african-ancestry.html; accessed January 1, 2021.
11. Morgan K. *Slavery and the British Empire: From Africa to America*. New York: Oxford University Press, 2008; Wheat D. *Atlantic Africa and the Spanish Caribbean, 1570–1640*. Chapel Hill: University of North Carolina Press, 2016; Knight F (ed). *General History of the Caribbean. Volume III, The Slave Societies of the Caribbean*. New York; Paris: Palgrave Macmillan: UNESCO Publishing, 2007; Reséndez A. *The Other Slavery: The Uncovered Story of Indian Enslavement in America*. Boston: Houghton Mifflin Harcourt, 2016.
12. Vitale AS. *The End of Policing*. London: Verso, 2017; Richardson AV. *Bearing Witness While Black: African Americans, Smartphones, and the New*

Protest #Journalism. New York: Oxford University Press, 2020; Krieger N. ENOUGH: COVID-19, structural racism, police brutality, plutocracy, climate change-and time for health justice, democratic governance, and an equitable, sustainable future [published online ahead of print August 20, 2020]. *Am J Public Health* 2020;110(11):1620–1623. doi:10.2105/AJPH.2020.305886; Krieger N, Chen JT, Waterman PD, Kiang MV, Feldman J. Police killings and police deaths are public health data and can be counted. *PLoS Med* 2015;12(12):e1001915. doi:10.1371/journal.pmed.1001915; Feldman JM, Gruskin S, Coull BA, Krieger N. Killed by police: validity of media-based data and misclassification of death certificates in Massachusetts, 2004–2016 [published online ahead of print August 17, 2017]. *Am J Public Health* 2017;107(10):1624–1626. doi:10.2105/AJPH.2017.303940.

13. Robert DE, Rollins D. Why sociology matters to race and biosocial science. *Annu Rev Sociol* 2020;46:195–214; Mills MC, Tropf FC. Sociology, genetics, and the coming age of sociogenomics. *Annu Rev Sociol* 2020;46:553–581; Braudt DB. Sociogenomics in the 21st century: an introduction to the history and potential of genetically informed social science. *Sociol Compass* 2018;12(10): e12626. https://doi.org/10.1111/soc4.12626.
14. Benjamin R. *Race After Technology: Abolitionist Tools for the New Jim Code*. Cambridge: Polity, 2019; Benjamin R. *Captivating Technology: Race, Carceral Technoscience, and Liberatory Imagination in Everyday Life*. Durham, NC: Duke University Press, 2019; Ferguson AG. *The Rise of Big Data Policing: Surveillance, Race, and the Future of Law Enforcement*. New York: NYU Press, 2017.
15. Fredrickson GM. *Racism: A Short History*. With a new foreword by Albert M Camarillo. Princeton, NJ: Princeton University Press, 2015; Saini A. *Superior: The Return of Race Science*. Boston: Beacon Press, 2019; Kendi IX. *Stamped from the Beginning: The Definitive History of Racist Ideas in America*. New York: Nation Books, 2016; Yudell M. *Race Unmasked: Biology and Race in the Twentieth Century*. New York: Columbia University Press, 2014; Ernst W, Harris B (eds). *Race, Science, and Medicine, 1700–1960*. London: Routledge, 1999; Painter NI. *The History of White People*. New York: W. W. Norton, 2010.
16. Fredrickson GM. *Racism: A Short History*. With a new foreword by Albert M Camarillo. Princeton, NJ: Princeton University Press, 2015; Saini A. *Superior: The Return of Race Science*. Boston: Beacon Press, 2019; Kendi IX. *Stamped from the Beginning: The Definitive History of Racist Ideas in America*. New York: Nation Books, 2016; Yudell M. Race *Unmasked: Biology and Race in the Twentieth Century*. New York: Columbia University Press, 2014; Ernst W, Harris B (eds). *Race, Science, and Medicine, 1700–1960*. London: Routledge, 1999; Painter NI. *The History of White People*. New York: W. W. Norton, 2010.

17. Pan American Health Organization (PAHO). *Just Societies: Health Equity and Dignified Lives. Report of the Commission of the Pan American Health Organization on Equity and Health Inequalities in the Americas.* Washington, DC: PAHO, 2019. https://iris.paho.org/handle/10665.2/51571; accessed December 30, 2020; Hutchinson DL. *Disease and Discrimination: Poverty and Pestilence in Colonial Atlantic America.* Gainesville, FL: University Press of Florida, 2016; Cuetos M. *Medicine and Public Health in Latin America: A History.* New York: Cambridge University Press, 2015; Birn AE, Pillay Y, Holtz T. *Textbook of Global Health.* New York: Oxford University Press, 2017; Krieger N. Shades of difference: theoretical underpinnings of the medical controversy on black/white differences in the United States, 1830–1870. *Int J Health Serv* 1987;17:259–278.
18. Seelye KQ. John Lewis: towering figure of civil rights era, dies at 80. *New York Times,* July 17, 2020; updated July 30, 2020. https://www.nytimes.com/2020/07/17/us/john-lewis-dead.html; accessed January 1, 2021; Carlson M. John Lewis obituary. *The Guardian,* July 18, 2020. https://www.theguardian.com/us-news/2020/jul/18/john-lewis-obituary; accessed January 21, 2021; Lewis J. Together, you can redeem the soul of our nation. *New York Times,* July 30, 2020. https://www.nytimes.com/2020/07/30/opinion/john-lewis-civil-rights-america.html; accessed January 21, 2021; Lewis J, Aydin A, Powell N. *March.* Marietta, GA: Top Shelf Productions, 2013.
19. Lewis J. Together, you can redeem the soul of our nation. *New York Times,* July 30, 2020. https://www.nytimes.com/2020/07/30/opinion/john-lewis-civil-rights-america.html; accessed January 21, 2021; Galofaro C. Voices of protest, crying for change, ring across the US, beyond. *Associated Press,* June 17, 2020. https://apnews.com/54e9e9117e2903e3ec99ee15c76e74a6; accessed January 1, 2021; Buchanan L, Bui Q, Patel JK. Black Lives Matter may be the largest movement in U.S. history. *New York Times,* July 3, 2020. https://www.nytimes.com/interactive/2020/07/03/us/george-floyd-protests-crowd-size.html; accessed January 1, 2021; New York Times. How statues are falling around the world: statues and monuments that have long honored racist figures are being boxed up, spray-painted—or beheaded. *New York Times,* June 24, 2020; updated September 12, 2020. https://www.nytimes.com/2020/06/24/us/confederate-statues-photos.html; accessed January 1, 2021.
20. Murray P. *States' Laws on Race and Color.* Athens, GA: Women's Division of Christian Services, 1950; Gates HL. *Stony the Road: Reconstruction, White Supremacy, and the Rise of Jim Crow.* New York, Penguin Books, 2019; Woodward CV. *The Strange Career of Jim Crow* (1955). Commemorative edition. New York: Oxford University Press, 2002; Purnell B, Theoharis J, with Woodward K (eds). *The Strange Careers of the Jim Crow North: Segregation and Struggles Outside of the South.* New York City: NYU Press, 2019; Packard J. *American*

Nightmare: The History of Jim Crow. New York: St. Martin's Press, 2002; Cole S, Ring NJ, Stein M, Perdue T. *The Folly of Jim Crow: Rethinking the Segregated South*. College Station: Texas A&M University Press, 2012; Chafe W, Gavins R, Korstad, R. *Remembering Jim Crow: African Americans Tell About Life in the Segregated South*. New York: New Press, 2011; Gilmore G. *Defying Dixie: The Radical Roots of Civil Rights, 1919–1950*. New York: W. W. Norton, 2008; Anderson C. *Eyes Off the Prize: The United Nations and the African American Struggle for Human Rights, 1944–1955*. New York: Cambridge University Press, 2003.

21. Purnell B, Theoharis J, with Woodward K (eds). *The Strange Careers of the Jim Crow North: Segregation and Struggles Outside of the South*. New York: NYU Press, 2019; Murray P. *States' Laws on Race and Color*. Athens, GA: Women's Division of Christian Services, 1950; Gates HL. *Stony the Road: Reconstruction, White Supremacy, and the Rise of Jim Crow*. New York, Penguin Books, 2019; Kendi IX. *Stamped from the Beginning: The Definitive History of Racist Ideas in America*. New York: Nation Books, 2016; Anderson C. *Eyes Off the Prize: The United Nations and the African American Struggle for Human Rights, 1944–1955*. New York: Cambridge University Press, 2003.
22. Foner E. *Reconstruction: America's Unfinished Revolution, 1863–1877* (Updated ed., New American nation series). New York: HarperPerennial, 2014; Emberton C, Baker B (eds). *Remembering Reconstruction: Struggles Over the Meaning of America's Most Turbulent Era*. Baton Rouge: Louisiana State University Press, 2017; Equal Justice Initiative. *Reconstruction in America: Racial Violence after the Civil War, 1865–1876*. Montgomery, AL: Equal Justice Initiative, 2020. https://eji.org/report/reconstruction-in-america/; accessed January 1, 2021; Gates HL. *Stony the Road: Reconstruction, White Supremacy, and the Rise of Jim Crow*. New York: Penguin Books, 2019; Kendi IX. *Stamped from the Beginning: The Definitive History of Racist Ideas in America*. New York: Nation Books, 2016; Painter NI. *The History of White People*. New York: W. W. Norton, 2010.
23. Murray P. *States' Laws on Race and Color*. Athens, GA: Women's Division of Christian Services, 1950.
24. Murray P. *Pauli Murray: The Autobiography of a Black Activist, Feminist, Lawyer, Priest, and Poet*. Knoxville: University of Tennessee Press, 1989; Saxby T. *Pauli Murray: A Personal and Political Life*. Chapel Hill: University of North Carolina Press, 2020; Bell-Scott P. *The Firebrand and the First Lady: Portrait of a Friendship: Pauli Murray, Eleanor Roosevelt, and the Struggle for Social Justice*. New York: Knopf, 2016; Rosenberg R. *Jane Crow: The Life of Pauli Murray*. New York: Oxford University Press, 2017.
25. Risen C. *The Bill of the Century: The Epic Battle for the Civil Rights Act*. New York: Bloomsbury Press, 2014; Stainback K, Tomaskovic-Devey D. *Documenting Desegregation: Racial and Gender Segregation in Private Sector*

Employment since the Civil Rights Act. New York: Russell Sage Foundation, 2012; Purnell B, Theoharis J, with Woodward K (eds). *The Strange Careers of the Jim Crow North: Segregation and Struggles Outside of the South*. New York City: NYU Press, 2019; Packard J. *American Nightmare: The History of Jim Crow*. New York: St. Martin's Press, 2002; Cole S, Ring NJ, Stein M, Perdue T. *The Folly of Jim Crow: Rethinking the Segregated South*. College Station: Texas A&M University Press, 2012.

26. Lewis J. Together, you can redeem the soul of our nation. *New York Times*, July 30, 2020. https://www.nytimes.com/2020/07/30/opinion/john-lewis-civil-rights-america.html; accessed January 21, 2021; Lewis J, Aydin A, Powell N. *March*. Marietta, GA: Top Shelf Productions, 2013.
27. Risen C. *The Bill of the Century: The Epic Battle for the Civil Rights Act*. New York: Bloomsbury Press, 2014; Stainback K, Tomaskovic-Devey D. *Documenting Desegregation: Racial and Gender Segregation in Private Sector Employment since the Civil Rights Act*. New York: Russell Sage Foundation, 2012; Lewis J, Aydin A, Powell N. *March*. Marietta, GA: Top Shelf Productions, 2013.
28. Lefkowitz B. *Community Health Centers: A Movement and the People Who Made It Happen*. New Brunswick, NJ: Rutgers University Press, 2007; Ward TJ Jr. *Out in the Rural: A Mississippi Health Center and Its War on Poverty*. With a foreword by H. Jack Geiger. New York: Oxford University Press, 2017; Geiger HJ. Community-oriented primary care: a path to community development. *Am J Public Health* 2002;92(11):1713–1716; National Association of Community Health Centers. About Health Centers. http://www.nachc.org/about/about-our-health-centers/; accessed January 1, 2021; Grady D. H. Jack Geiger, doctor who fought social ills, dies at 95. *New York Times*, December 28, 2020. https://www.nytimes.com/2020/12/28/health/h-jack-geiger-dead.html; accessed January 1, 2021; Simmons A. In Memoriam: The Passing of Civil Rights Activist, Co-Founder of Community Health Center Movement, H. Jack Geiger, MD. National Association of Community Health Centers, December 30, 2020. https://www.nachc.org/in-memoriam-the-passing-of-civil-rights-activist-co-founder-of-community-health-center-movement-h-jack-geiger-md/; accessed January 1, 2021.
29. Lefkowitz B. *Community Health Centers: A Movement and the People Who Made It Happen*. New Brunswick, NJ: Rutgers University Press, 2007; Ward TJ Jr. *Out in the Rural: A Mississippi Health Center and Its War on Poverty*. With a foreword by H. Jack Geiger. New York: Oxford University Press, 2017; Geiger HJ. Community-oriented primary care: a path to community development. *Am J Public Health* 2002;92(11):1713–1716; National Association of Community Health Centers. About Health Centers. http://www.nachc.org/about/about-our-health-centers/; accessed January 1, 2021; Grady D. H. Jack Geiger, doctor who fought social

ills, dies at 95. *New York Times*, December 28, 2020. https://www.nytimes.com/2020/12/28/health/h-jack-geiger-dead.html; accessed January 1, 2021; Simmons A. In Memoriam: The Passing of Civil Rights Activist, Co-Founder of Community Health Center Movement, H. Jack Geiger, MD. National Association of Community Health Centers, December 30, 2020. https://www.nachc.org/in-memoriam-the-passing-of-civil-rights-activist-co-founder-of-community-health-center-movement-h-jack-geiger-md/; accessed January 1, 2021.

30. Lefkowitz B. *Community Health Centers: A Movement and the People Who Made It Happen*. New Brunswick, NJ: Rutgers University Press, 2007; Ward TJ Jr. *Out in the Rural: A Mississippi Health Center and Its War on Poverty*. With a foreword by H. Jack Geiger. New York: Oxford University Press, 2017; Geiger HJ. Community-oriented primary care: a path to community development. *Am J Public Health* 2002;92(11):1713–1716; National Association of Community Health Centers. About Health Centers. http://www.nachc.org/about/about-our-health-centers/ ; accessed January 1, 2021; Grady D. H. Jack Geiger, doctor who fought social ills, dies at 95. *New York Times*, December 28, 2020. https://www.nytimes.com/2020/12/28/health/h-jack-geiger-dead.html; accessed January 1, 2021; Simmons A. In Memoriam: The Passing of Civil Rights Activist, Co-Founder of Community Health Center Movement, H. Jack Geiger, MD. National Association of Community Health Centers, December 30, 2020. https://www.nachc.org/in-memoriam-the-passing-of-civil-rights-activist-co-founder-of-community-health-center-movement-h-jack-geiger-md/; accessed January 1, 2021.
31. National Association of Community Health Centers. About Health Centers. http://www.nachc.org/about/about-our-health-centers/; accessed January 1, 2021; Simmons A. In Memoriam: The Passing of Civil Rights Activist, Co-Founder of Community Health Center Movement, H. Jack Geiger, MD. National Association of Community Health Centers, December 30, 2020. https://www.nachc.org/in-memoriam-the-passing-of-civil-rights-activist-co-founder-of-community-health-center-movement-h-jack-geiger-md/; accessed January 1, 2021.
32. Chay KY, Greenstone M. The convergence in black-white infant mortality rates during the 1960's. *Am Econ Rev* 2000;90:326–332; Almond DV, Chay KY, Greenstone M. Civil Rights, the War on Poverty, and Black-White Convergence in Infant Mortality in the Rural South and Mississippi. MIT Economics Working Paper No. 07-04, December 31, 2006. https://papers.ssrn.com/sol3/papers.cfm?abstract_id=961021; accessed January 1, 2021; Almond D, Chay KY. The Long-Run and Intergenerational Impact of Poor Infant Health: Evidence from Cohorts Born During the Civil Rights Era. Working Paper; Revision 2006. https://users.nber.org/~almond/chay_npc_paper.pdf; accessed January 1, 2021; Kaplan G, Ranjit N, Burgard S. Lifting gates, lengthening lives: did civil rights policies improve the health of African-American women in the 1960s and

1970s? In: Schoeni RF, House JS, Kaplan G, Pollack H (eds). *Making Americans Healthier: Social and Economic Policy as Health Policy*. New York: Russell Sage Foundation, 2008; 145–170; Smith DB. Racial and ethnic health disparities and the unfinished civil rights agenda. *Health Aff (Millwood)* 2005;24(2):317–324. doi:10.1377/hlthaff.24.2.317; Krieger N, Chen JT, Coull B, Waterman PD, Beckfield J. The unique impact of abolition of Jim Crow laws on reducing inequities in infant death rates and implications for choice of comparison groups in analyzing societal determinants of health [published online ahead of print October 17, 2013]. *Am J Public Health* 2013;103(12):2234–2244. doi:10.2105/AJPH.2013.301350; Krieger N, Chen JT, Coull BA, Beckfield J, Kiang MV, Waterman PD. Jim Crow and premature mortality among the US Black and White population, 1960–2009: an age-period-cohort analysis. *Epidemiology* 2014;25(4):494–504. doi:10.1097/EDE.0000000000000104; Krieger N, Jahn JL, Waterman PD, Chen JT. Breast cancer estrogen receptor status according to biological generation: US Black and White women born 1915–1979. *Am J Epidemiol* 2018;187(5):960–970. doi:10.1093/aje/kwx312; Krieger N, Jahn JL, Waterman PD. Jim Crow and estrogen-receptor-negative breast cancer: US-born black and white non-Hispanic women, 1992–2012 [published online ahead of print December 17, 2016]. *Cancer Causes Control* 2017;28(1):49–59. doi:10.1007/s10552-016-0834-2; Probst JC, Glover S, Kirksey V. Strange harvest: a cross-sectional ecological analysis of the association between historic lynching events and 2010–2014 county mortality rates. *J Racial Ethn Health Disparities* 2019;6(1):143–152. doi:10.1007/s40615-018-0509-7.

33. Krieger N, Chen JT, Coull B, Waterman PD, Beckfield J. The unique impact of abolition of Jim Crow laws on reducing inequities in infant death rates and implications for choice of comparison groups in analyzing societal determinants of health [published online ahead of print October 17, 2013]. *Am J Public Health* 2013;103(12):2234–2244. doi:10.2105/AJPH.2013.301350; Krieger N, Chen JT, Coull BA, Beckfield J, Kiang MV, Waterman PD. Jim Crow and premature mortality among the US Black and White population, 1960–2009: an age-period-cohort analysis. *Epidemiology* 2014;25(4):494–504. doi:10.1097/EDE.0000000000000104; Krieger N, Jahn JL, Waterman PD, Chen JT. Breast cancer estrogen receptor status according to biological generation: US Black and White women born 1915–1979. *Am J Epidemiol* 2018;187(5):960–970. doi:10.1093/aje/kwx312; Krieger N, Jahn JL, Waterman PD. Jim Crow and estrogen-receptor-negative breast cancer: US-born black and white non-Hispanic women, 1992–2012 [published online ahead of print December 17, 2016]. *Cancer Causes Control* 2017;28(1):49–59. doi:10.1007/s10552-016-0834-2.
34. Krieger N, Chen JT, Coull B, Waterman PD, Beckfield J. The unique impact of abolition of Jim Crow laws on reducing inequities in infant death rates and

implications for choice of comparison groups in analyzing societal determinants of health [published online ahead of print October 17, 2013]. *Am J Public Health* 2013;103(12):2234–2244. doi:10.2105/AJPH.2013.301350.

35. Chay KY, Greenstone M. The convergence in black-white infant mortality rates during the 1960's. *Am Econ Rev* 2000;90:326–332.; Almond DV, Chay KY, Greenstone M. Civil Rights, the War on Poverty, and Black-White Convergence in Infant Mortality in the Rural South and Mississippi. MIT Economics Working Paper No. 07-04, December 31, 2006. https://papers.ssrn.com/sol3/papers.cfm?abstract_id=961021; accessed January 1, 2021; Almond D, Chay KY. The Long-Run and Intergenerational Impact of Poor Infant Health: Evidence from Cohorts Born During the Civil Rights Era. Working Paper; Revision 2006. https://users.nber.org/~almond/chay_npc_paper.pdf; accessed January 1, 2021; Kaplan G, Ranjit N, Burgard S. Lifting gates, lengthening lives: did civil rights policies improve the health of African-American women in the 1960s and 1970s? In: Schoeni RF, House JS, Kaplan G, Pollack H (eds). *Making Americans Healthier: Social and Economic Policy as Health Policy*. New York: Russell Sage Foundation, 2008; 145–170; Smith DB. Racial and ethnic health disparities and the unfinished civil rights agenda. *Health Aff (Millwood)* 2005;24(2):317–324. doi:10.1377/hlthaff.24.2.317; Krieger N, Chen JT, Coull B, Waterman PD, Beckfield J. The unique impact of abolition of Jim Crow laws on reducing inequities in infant death rates and implications for choice of comparison groups in analyzing societal determinants of health [published online ahead of print October 17, 2013]. *Am J Public Health* 2013;103(12):2234–2244. doi:10.2105/AJPH.2013.301350; Krieger N, Chen JT, Coull BA, Beckfield J, Kiang MV, Waterman PD. Jim Crow and premature mortality among the US Black and White population, 1960–2009: an age-period-cohort analysis. *Epidemiology* 2014;25(4):494–504. doi:10.1097/EDE.0000000000000104; Krieger N, Jahn JL, Waterman PD, Chen JT. Breast cancer estrogen receptor status according to biological generation: US Black and White women born 1915–1979. *Am J Epidemiol* 2018;187(5):960–970. doi:10.1093/aje/kwx312; Krieger N, Jahn JL, Waterman PD. Jim Crow and estrogen-receptor-negative breast cancer: US-born black and white non-Hispanic women, 1992–2012 [published online ahead of print December 17, 2016]. *Cancer Causes Control* 2017;28(1):49–59. doi:10.1007/s10552-016-0834-2; Probst JC, Glover S, Kirksey V. Strange harvest: a cross-sectional ecological analysis of the association between historic lynching events and 2010–2014 county mortality rates. *J Racial Ethn Health Disparities* 2019;6(1):143–152. doi:10.1007/s40615-018-0509-7.

36. Krieger N, Chen JT, Coull B, Waterman PD, Beckfield J. The unique impact of abolition of Jim Crow laws on reducing inequities in infant death rates and implications for choice of comparison groups in analyzing societal determinants

of health [published online ahead of print October 17, 2013]. *Am J Public Health* 2013;103(12):2234–2244. doi:10.2105/AJPH.2013.301350; Krieger N, Chen JT, Coull BA, Beckfield J, Kiang MV, Waterman PD. Jim Crow and premature mortality among the US Black and White population, 1960–2009: an age-period-cohort analysis. *Epidemiology* 2014;25(4):494–504. doi:10.1097/EDE.0000000000000104.

37. Purnell B, Theoharis J, with Woodward K (eds). *The Strange Careers of the Jim Crow North: Segregation and Struggles Outside of the South*. New York: NYU Press, 2019; Murray P. *States' Laws on Race and Color*. Athens, GA: Women's Division of Christian Services, 1950; Gates HL. *Stony the Road: Reconstruction, White Supremacy, and the Rise of Jim Crow*. New York: Penguin Books, 2019; Kendi IX. *Stamped from the Beginning: The Definitive History of Racist Ideas in America*. New York: Nation Books, 2016; Anderson C. *Eyes Off the Prize: The United Nations and the African American Struggle for Human Rights, 1944–1955*. New York: Cambridge University Press, 2003; Woodward CV. *The Strange Career of Jim Crow* (1955). Commemorative edition. New York: Oxford University Press, 2002; Packard J. *American Nightmare: The History of Jim Crow*. New York: St. Martin's Press, 2002.
38. Lefkowitz B. *Community Health Centers: A Movement and the People Who Made It Happen*. New Brunswick, NJ: Rutgers University Press, 2007; Ward TJ Jr. *Out in the Rural: A Mississippi Health Center and Its War on Poverty*. With a foreword by H. Jack Geiger. New York: Oxford University Press, 2017; Geiger HJ. Community-oriented primary care: a path to community development. *Am J Public Health* 2002;92(11):1713–1716; National Association of Community Health Centers. About Health Centers. http://www.nachc.org/about/about-our-health-centers/; accessed July 30, 2020; Chu C. Out in the rural: a health center in Mississippi. *Soc Med* 2006;1(2):139; Rogers JS (producer and director). *Out in the Rural*. 1970. https://vimeo.com/9307557; accessed January 1, 2021; Grady D. H. Jack Geiger, doctor who fought social ills, dies at 95. *New York Times*, December 28, 2020. https://www.nytimes.com/2020/12/28/health/h-jack-geiger-dead.html; accessed January 1, 2021; Simmons A. In Memoriam: The Passing of Civil Rights Activist, Co-Founder of Community Health Center Movement, H. Jack Geiger, MD. National Association of Community Health Centers, December 30, 2020. https://www.nachc.org/in-memoriam-the-passing-of-civil-rights-activist-co-founder-of-community-health-center-movement-h-jack-geiger-md/; accessed January 1, 2021.
39. Rogers JS (producer and director). Out in the Rural. 1970. https://vimeo.com/9307557; accessed January 1, 2021.
40. Krieger N, Rehkopf DH, Chen JT, Waterman PD, Marcelli E, Kennedy M. The fall and rise of US inequities in premature mortality: 1960–2002. *PLoS*

Med 2008;5(2):e46. doi:10.1371/journal.pmed.0050046; Krieger N, Chen JT, Coull BA, Beckfield J, Kiang MV, Waterman PD. Jim Crow and premature mortality among the US Black and White population, 1960–2009: an age-period-cohort analysis. *Epidemiology* 2014;25:494–504; Kiang MV, Krieger N, Buckee CO, Onnela J-P, Chen JT. Decomposition of the US black/white inequality in premature mortality, 2010–2015: an observational study. *BMJ Open* 2019;9(11):e029373. doi:10.1136/bmjopen-2019-029373; Bassett MT, Chen JT, Krieger N. Variation in racial/ethnic disparities in COVID-19 mortality by age in the United States: a cross-sectional study. *PLoS Med* 2020;17(10):e1003402. doi:10.1371/journal.pmed.1003402; Chen Y, Freedman ND, Rodriquez EJ, Shiels MS, Napoles AM, Withrow DR, Spillane S, Sigel B, Perez-Stable EJ, Berrington de González A. Trends in premature deaths among adults in the United States and Latin America. *JAMA Netw Open* 2020;3(2):e1921085. doi:10.1001/jamanetworkopen.2019.21085; Cunningham TJ, Croft B, Liu Y, Lu H, Eke PI, Giles WH. Vital signs: racial disparities in age-specific mortality among Blacks or African-Americans—United States, 1999–2015 [published online ahead of print May 2, 2017]. *MMWR Morb Mortal Wkly Rep*. https://www.cdc.gov/mmwr/volumes/66/wr/mm6617e1.htm; accessed January 1, 2021; Shiels MS, Chernyavskiy P, Anderson WF, Best AF, Haozous EA, Hartge P, Rosenberg PS, Thomas D, Freedman ND, Berrington de Gonzalez A. Trends in premature mortality in the USA by sex, race, and ethnicity from 1999 to 2014: an analysis of death certificate data [published online ahead of print January 26, 2017]. *Lancet* 2017;389(10073):1043–1054. doi:10.1016/S0140-6736(17)30187-3; Levine RS, Foster JE, Fullilove RE, Fullilove MT, Briggs NC, Hull PC, Husaini BA, Hennekens CH. Black-white inequalities in mortality and life expectancy, 1933–1999: implications for Healthy People 2010. *Public Health Rep* 2001 Sep-Oct;116(5):474–483. doi:10.1093/phr/116.5.474; Manton KG, Patrick CH, Johnson KW. Health differentials between blacks and whites: recent trends in mortality and morbidity. *Milbank Q* 1987;65(Suppl 1):129–199.

41. Murray P. *States' Laws on Race and Color*. Athens, GA: Women's Division of Christian Services, 1950; Purnell B, Theoharis J, with Woodard K (eds). *The Strange Careers of the Jim Crow North: Segregation and Struggle Outside of the South*. New York: New York University Press, 2019; Packard J. *American Nightmare: The History of Jim Crow*. New York: St. Martin's Press, 2002; Risen C. *The Bill of the Century: The Epic Battle for the Civil Rights Act*. New York: Bloomsbury Press, 2014; Stainback K, Tomaskovic-Devey D. *Documenting Desegregation: Racial and Gender Segregation in Private Sector Employment Since the Civil Rights Act*. New York: Russell Sage Foundation, 2012.
42. Beckfield J, Krieger N. Epi + demos + cracy: linking political systems and priorities to the magnitude of health inequities—evidence, gaps, and a research agenda

[published online ahead of print May 27, 2009]. *Epidemiol Rev* 2009;31:152–177. Doi:10.1093/epirev/mxp002; Beckfield J. *Political Sociology and the People's Health*. New York: Oxford University Press, 2018; Krieger N. *Epidemiology and the People's Health: Theory and Context*. New York: Oxford University Press, 2011; Krieger N, Singh N, Chen JT, Coull BA, Beckfield J, Kiang MV, Waterman PD, Gruskin S. Why history matters for quantitative target setting: long-term trends in socioeconomic and racial/ethnic inequities in US infant death rates (1960–2010) [published online ahead of print May 14, 2015]. *J Public Health Policy* 2015;36(3):287–303. doi:10.1057/jphp.2015.12.

43. Walker AT, Smith PJ, Kolasa M; Centers for Disease Control and Prevention (CDC). Reduction of racial/ethnic disparities in vaccination coverage, 1995–2011. *MMWR Suppl* 2014 Apr 18;63(1):7–12; Rehkopf DH, Strully KW, Dow WH. The short-term impacts of Earned Income Tax Credit disbursement on health. *Int J Epidemiol* 2014;43(6):1884–1894. doi:10.1093/ije/dyu172; Bell ON, Hole MK, Johnson K, Marcil LE, Solomon BS, Schickedanz A. Medical-financial partnerships: cross-sector collaborations between medical and financial services to improve health [published online ahead of print October 13, 2019]. *Acad Pediatr* 2020 Mar;20(2):166–174. doi:10.1016/j.acap.2019.10.001; Jessel S, Sawyer S, Hernández D. Energy, poverty, and health in climate change: a comprehensive review of an emerging literature. *Front Public Health* 2019;7:357. doi:10.3389/fpubh.2019.00357; Tehranifar P, Goyal A, Phelan JC, Link BG, Liao Y, Fan X, Desai M, Terry MB. Age at cancer diagnosis, amenability to medical interventions, and racial/ethnic disparities in cancer mortality [published online ahead of print March 12, 2016]. *Cancer Causes Control* 2016;27(4):553–560. doi:10.1007/s10552-016-0729-2.
44. Krieger N. Stormy weather: race, gene expression, and the science of health disparities [published online ahead of print October 27, 2005]. *Am J Public Health* 2005;95(12):2155–2160. doi:10.2105/AJPH.2005.067108.
45. Krieger N. Inheritance and health: what really matters? *Am J Public Health* 2018;108(5):606–607. doi:10.2105/AJPH.2018.304353; Müller R. A task that remains before us: reconsidering inheritance as a biosocial phenomenon [published online ahead of print September 9, 2019]. *Semin Cell Dev Biol* 2020;97:189–194. doi:10.1016/j.semcdb.2019.07.008.
46. Halfon N, Forrest CB, Lerner RM, Faustman EM (eds). *Handbook of Life Course Health Development*. Cham, Switzerland: Springer Open, 2018; Vineis P, Avendano-Pabon M, Barros H, Bartley M, Carmeli C, Carra L, Chadeau-Hyam M, Costa G, Delpierre C, D'Errico A, Fraga S, Giles G, Goldberg M, Kelly-Irving M, Kivimaki M, Lepage B, Lang T, Layte R, MacGuire F, Mackenbach JP, Marmot M, McCrory C, Milne RL, Muennig P, Nusselder W, Petrovic D, Polidoro S, Ricceri F, Robinson O, Stringhini S, Zins M. Special report: the biology of inequalities in health: the Lifepath Consortium. *Front Public Health* 2020;8:118.

doi:10.3389/fpubh.2020.00118; Jones NL, Gilman SE, Cheng TL, Drury SS, Hill CV, Geronimus AT. Life course approaches to the causes of health disparities. *Am J Public Health* 2019;109(S1):S48–S55. doi:10.2105/AJPH.2018.304738; Godfrey KM, Sheppard A, Gluckman PD, Lillycrop KA, Burdge GC, McLean C, Rodford J, Slater-Jefferies JL, Garratt E, Crozier SR, Emerald BS, Gale CR, Inskip HM, Cooper C, Hanson MA. Epigenetic gene promoter methylation at birth is associated with child's later adiposity. *Diabetes* 2011;60(5):1528–1534; Zheng H, Tumin D. Variation in the effects of family background and birth region on adult obesity: results of a prospective cohort study of a Great Depression-era American cohort. *BMC Public Health* 2015;15:535. doi:10.1186/s12889-015-1870-7.

47. Meloni M, Müller R. Transgenerational epigenetic inheritance and social responsibility: perspectives from the social sciences. *Environ Epigenet* 2018;4(2):dvy019. doi:10.1093/eep/dvy019; Bošković A, Rando OJ. Transgenerational epigenetic inheritance [published online ahead of print August 30, 2018]. *Annu Rev Genet* 2018;52:21–41. doi:10.1146/annurev-genet-120417-031404; Lehrner A, Yehuda R. Cultural trauma and epigenetic inheritance. *Develop Psychopathol* 2018;30(5):1763–1777; Grossi É. New avenues in epigenetic research about race: online activism around reparations for slavery in the United States. *Soc Sci Info* 2020;59(1):93–116.
48. Gilbert SF, Epel D. *Ecological Developmental Biology: The Environmental Regulation of Development, Health, and Evolution*. 2nd ed. Sunderland, MA: Sinauer Associates, Inc., 2015; Relton CL, Davey Smith G. Is epidemiology ready for epigenetics? *Int J Epidemiol* 2012;41(1):5–9. doi:10.1093/ije/dys006.
49. Gilbert SF, Epel D. *Ecological Developmental Biology: The Environmental Regulation of Development, Health, and Evolution*. 2nd ed. Sunderland, MA: Sinauer Associates, Inc., 2015; Relton CL, Davey Smith G. Is epidemiology ready for epigenetics? *Int J Epidemiol* 2012;41(1):5–9. doi:10.1093/ije/dys006.
50. van Otterdijk SD, Michels KB. Transgenerational epigenetic inheritance in mammals: how good is the evidence? [published online ahead of print April 1, 2016]. *FASEB J* 2016;30(7):2457–2465. doi:10.1096/fj.201500083; Zeng Y, Chen T. DNA methylation reprogramming during mammalian development. *Genes (Basel)* 2019;10(4):257. doi:10.3390/genes10040257; Horsthemke B. A critical view on transgenerational epigenetic inheritance in humans. *Nat Commun* 2018;9(1):2973. doi:10.1038/s41467-018-05445-5.
51. van Otterdijk SD, Michels KB. Transgenerational epigenetic inheritance in mammals: how good is the evidence? [published online ahead of print April 1, 2016]. *FASEB J* 2016;30(7):2457–2465. doi:10.1096/fj.201500083; Zeng Y, Chen T. DNA methylation reprogramming during mammalian development. *Genes (Basel)* 2019;10(4):257. doi:10.3390/genes10040257; Horsthemke B. A

critical view on transgenerational epigenetic inheritance in humans. *Nat Commun* 2018;9(1):2973. doi:10.1038/s41467-018-05445-5.

52. Meloni M, Müller R. Transgenerational epigenetic inheritance and social responsibility: perspectives from the social sciences. *Environ Epigenet* 2018;4(2):dvy019. doi:10.1093/eep/dvy019; Bošković A, Rando OJ. Transgenerational epigenetic inheritance [published online ahead of print August 30, 2018]. *Annu Rev Genet* 2018;52:21–41. doi:10.1146/annurev-genet-120417-031404; Lehrner A, Yehuda R. Cultural trauma and epigenetic inheritance. *Develop Psychopathol* 2018;30(5):1763–1777; Grossi É. New avenues in epigenetic research about race: online activism around reparations for slavery in the United States. *Soc Sci Info* 2020;59(1):93–116.
53. Krieger N. Stormy weather: race, gene expression, and the science of health disparities [published online ahead of print October 27, 2005]. *Am J Public Health* 2005;95(12):2155–2160. doi:10.2105/AJPH.2005.067108.
54. Krieger N. Measures of racism, sexism, heterosexism, and gender binarism for health equity research: from structural injustice to embodied harm-an ecosocial analysis [published online ahead of print November 25, 2019]. *Annu Rev Public Health* 2020;41:37–62. doi:10.1146/annurev-publhealth-040119-094017; Krieger N. Methods for the scientific study of discrimination and health: from societal injustice to embodied inequality—an ecosocial approach. *Am J Public Health* 2012;102:936–945; Krieger N. Discrimination and health inequities. In: Berkman LF, Kawachi I, Glymour M (eds). *Social Epidemiology*. 2nd ed. New York: Oxford University Press, 2014; 63–125; Bailey ZD, Krieger N, Agénor M, Graves J, Linos N, Bassett MT. Structural racism and health inequities in the USA: evidence and interventions. *Lancet* 2017;389(10077):1453–1463.
55. Krieger N. Shades of difference: theoretical underpinnings of the medical controversy on black/white differences in the United States, 1830–1870. *Int J Health Serv* 1987;17(2):259–278; Levesque GA. Boston's Black Brahmin: Dr. John S. Rock. *Civil War Hist* 1980;54(4):326–346; Link EP. The civil rights activities of three great Negro physicians (1840–1940). *J Negro Hist* 1967;52(3):169–184; Rock JS. I will sink or swim with my race. Oration at the first annual Crispus Attucks Day Observance, Faneuil Hall, Boston, MA, March 5, 1858. https://www.blackpast.org/african-american-history/1858-john-s-rock-i-will-sink-or-swim-my-race/; accessed July 31, 2020; Middleton B. This day in Black History: Feb 1, 1865—abolitionist lawyer John S. Rock became the first African-American to practice before the United States Supreme Court. https://www.bet.com/news/national/2013/02/01/this-day-in-black-history-feb-1-1865.html; accessed July 31, 2020; Black History Now. John S. Rock. Last updated September 23, 2011. http://blackhistorynow.com/john-s-rock/; accessed July 31, 2020.

56. Krieger N. Shades of difference: theoretical underpinnings of the medical controversy on black/white differences in the United States, 1830–1870. *Int J Health Serv* 1987;17(2):259–278; Levesque GA. Boston's Black Brahmin: Dr. John S. Rock. *Civil War Hist* 1980;54(4):326–346; Rock JS. I will sink or swim with my race. Oration at the first annual Crispus Attucks Day Observance, Fanueil Hall, Boston, MA, March 5, 1858. https://www.blackpast.org/african-american-history/1858-john-s-rock-i-will-sink-or-swim-my-race/; accessed July 31, 2020; Middleton B. This day in Black History: Feb 1, 1865—abolitionist lawyer John S. Rock became the first African-American to practice before the United States Supreme Court. https://www.bet.com/news/national/2013/02/01/this-day-in-black-history-feb-1-1865.html; accessed July 31, 2020; Black History Now. John S. Rock. Last updated September 23, 2011. http://blackhistorynow.com/john-s-rock/; accessed July 31, 2020. University of Massachusetts History Club. Crispus Attucks on-line museum. http://www.crispusattucksmuseum.org/about/; accessed June 24, 2021. Mashpee Wampanoag. Timeline. https://mashpeewampanoagtribe-nsn.gov/timeline; accessed June 24, 2021.
57. Younge G. Interview–Eduardo Galeano: "My great fear is that we are all suffering from amnesia." *The Guardian*, July 23, 2013. https://www.theguardian.com/books/2013/jul/23/eduardo-galeano-children-days-interview; accessed July 31, 2020.
58. Krieger N. *Epidemiology and the People's Health: Theory and Context*. New York: Oxford University Press, 2011; Krieger N. ENOUGH: COVID-19, structural racism, police brutality, plutocracy, climate change-and time for health justice, democratic governance, and an equitable, sustainable future [published online ahead of print August 20, 2020]. *Am J Public Health* 2020;110(11):1620–1623. doi:10.2105/AJPH.2020.305886; Krieger N. Climate crisis, health equity, and democratic governance: the need to act together. *J Public Health Policy* 2020;41(1):4–10. doi:10.1057/s41271-019-00209-x.
59. Ziman J. *Real Science: What It Is, and What It Means*. Cambridge: Cambridge University Press, 2000; Curd M, Psillos S (eds). *The Routledge Companion to the Philosophy of Science*. New York: Routledge, 2014; Okasha S. *Philosophy of Science: A Very Short Introduction*. 2nd. ed. New York: Oxford University Press, 2016.
60. Some selected examples from the start of my work as a social epidemiologist include Bassett M, Krieger N. Social class and black-white differences in breast cancer survival. *Am J Public Health* 1986;76:1400–1403; Krieger N. Social class and the black/white crossover in age-specific incidence of breast cancer: a study linking census-derived data to population-based registry records. *Am J Epidemiol* 1990;131:804–814; and in the past few years (with many empirical investigations and analytic essays in between!), Krieger N, Singh N, Waterman PD. Metrics for monitoring cancer inequities: residential segregation, the Index of Concentration at the Extremes (ICE), and breast cancer estrogen receptor status (United States,

1992–2012). *Cancer Causes Control* 2016;27:1139–1151. doi:10.1007/s10552-016-0793-7; Krieger N, Jahn JL, Waterman PD. Jim Crow and estrogen-receptor negative breast cancer: US-born black and white non-Hispanic women, 1992–2012 [published online ahead of print December 17, 2016]. *Cancer Causes Control* 2017;28:49–59; Krieger N, Jahn JL, Waterman PD, Chen JT. Breast cancer estrogen receptor by biological generation: US black and white women, born 1915–1979 [published online ahead of print September 20, 2017]. *Am J Epidemiol* 2018;187:960–970; Krieger N, Wright E, Chen JT, Waterman PD, Huntley ER, Arcaya M. Cancer stage at diagnosis, historical redlining, and current neighborhood characteristics: breast, cervical, lung, and colorectal cancers, Massachusetts, 2001–2015. *Am J Epidemiol* 2020 Oct 1;189(10):1065–1075. doi:10.1093/aje/kwaa045.

61. Li C (ed). *Breast Cancer Epidemiology*. New York: Springer, 2010. https://doi.org/10.1007/978-1-4419-0685-4; Rojas K, Stuckey A. Breast cancer epidemiology and risk factors. *Clin Obstet Gynecol* 2016;59(4):651–672. doi:10.1097/GRF.0000000000000239; Tamimi R, Hankinson S, Lagiou P. Breast cancer. In: Adami H-O, Hunter DJ, Lagiou P, Mucci L (eds). *Textbook of Cancer Epidemiology*. Oxford Scholarship Online, February 2018. doi:10.1093/oso/9870190676827.003.0016; Power ML, Schulkin J. *Milk: The Biology of Lactation*. Baltimore, MD: Johns Hopkins University Press, 2016; McNally S, Stein T. Overview of mammary gland development: a comparison of mouse and human. In: Martin F, Stein T, Howlin J (eds). *Mammary Gland Development. Methods in Molecular Biology*. Vol. 1501. New York: Humana Press, 2017. https://doi-org.ezp-prod1.hul.harvard.edu/10.1007/978-1-4939-6475-8_1; Krieger N. Exposure, susceptibility, and breast cancer risk: a hypothesis regarding exogenous carcinogens, breast tissue development, and social gradients, including black/white differences, in breast cancer incidence. *Breast Cancer Res Treat* 1989;13:205–223; Krieger N, Chen JT, Waterman PD. Temporal trends in the black/white breast cancer case ratio for estrogen receptor status: disparities are historically contingent, not innate *Cancer Causes Control* 2011;22:511–514; Krieger N, Chen JT, Kosheleva A, Waterman PD. Shrinking, widening, reversing, and stagnating trends in US socioeconomic inequities in cancer mortality: 1960–2006. *Cancer Causes Control* 2012;23:297–319. doi:10.1007/s10552-011-9879-4; Krieger N. History, biology, and health inequities: emergent embodied phenotypes and the illustrative case of the breast cancer estrogen receptor. *Am J Public Health* 2013;103:22–27; Davey Smith G, Relton CL, Brennan P. Chance, choice and cause in cancer aetiology: individual and population perspectives. *Int J Epidemiol* 2016;45(3):605–613; Doll R. *The Prevention of Cancer: Pointers from Epidemiology*. London: Nuffield Provincial Hospitals Trust, 1967; 108.

62. Bassett M, Krieger N. Social class and black-white differences in breast cancer survival. *Am J Public Health* 1986;76:1400–1403; Krieger N. Exposure, susceptibility, and breast cancer risk: a hypothesis regarding exogenous carcinogens, breast tissue development, and social gradients, including black/white differences, in breast cancer incidence. *Breast Cancer Res Treat* 1989;13:205–223.
63. Wailoo K. *How Cancer Crossed the Color Line*. New York: Oxford University Press, 2001; Krieger N. Exposure, susceptibility, and breast cancer risk: a hypothesis regarding exogenous carcinogens, breast tissue development, and social gradients, including black/white differences, in breast cancer incidence. *Breast Cancer Res Treat* 1989;13:205–223; Krieger N. Racial discrimination and health: an epidemiologist's perspective. In: *Report of the President's Cancer Panel. The Meaning of Race in Science—Considerations for Cancer Research (April 9, 1997)*. Bethesda, MD: National Institutes of Health, National Cancer Institute, 1998; A-32–A35; Krieger N. Defining and investigating social disparities in cancer: critical issues. *Cancer Causes Control* 2005;16:5–14.
64. Bassett M, Krieger N. Social class and black-white differences in breast cancer survival. *Am J Public Health* 1986;76:1400–1403.
65. Bassett M, Krieger N. Social class and black-white differences in breast cancer survival. *Am J Public Health* 1986;76:1400–1403; Krieger N. Exposure, susceptibility, and breast cancer risk: a hypothesis regarding exogenous carcinogens, breast tissue development, and social gradients, including black/white differences, in breast cancer incidence. *Breast Cancer Res Treat* 1989;13:205–223.
66. Krieger N. A century of census tracts: health and the body politic (1906–2006). *J Urban Health* 2006;83(3):355–361. doi:10.1007/s11524-006-9040-y.
67. Krieger N, Chen JT, Waterman PD, Rehkopf DH, Subramanian SV. Painting a truer picture of US socioeconomic and racial/ethnic health inequalities: the Public Health Disparities Geocoding Project. *Am J Public Health* 2005;95(2):312–323. doi:10.2105/AJPH.2003.032482..
68. Bassett M, Krieger N. Social class and black-white differences in breast cancer survival. *Am J Public Health* 1986;76:1400–1403; Krieger N. Exposure, susceptibility, and breast cancer risk: a hypothesis regarding exogenous carcinogens, breast tissue development, and social gradients, including black/white differences, in breast cancer incidence. *Breast Cancer Res Treat* 1989;13:205–223.
69. Krieger N, Chen JT, Waterman PD, Rehkopf DH, Subramanian SV. Painting a truer picture of US socioeconomic and racial/ethnic health inequalities: the Public Health Disparities Geocoding Project. *Am J Public Health* 2005;95(2):312–323. doi:10.2105/AJPH.2003.032482; for the full description of the Public Health Disparities Geocoding Project, the affiliated publications, and the resources it provides, see https://www.hsph.harvard.edu/thegeocodingproject/; accessed January 1, 2021.

70. Krieger N, Chen JT, Waterman PD, Rehkopf DH, Subramanian SV. Painting a truer picture of US socioeconomic and racial/ethnic health inequalities: the Public Health Disparities Geocoding Project. *Am J Public Health* 2005;95(2):312–323. doi:10.2105/AJPH.2003.032482; for the full description of the Public Health Disparities Geocoding Project, the affiliated publications, and the resources it provides, see https://www.hsph.harvard.edu/thegeocodingproject/; accessed January 1, 2021.
71. Li C (ed). *Breast Cancer Epidemiology*. New York: Springer, 2010. https://doi.org/10.1007/978-1-4419-0685-4; Rojas K, Stuckey A. Breast cancer epidemiology and risk factors. *Clin Obstet Gynecol* 2016;59(4):651–672. doi:10.1097/GRF.0000000000000239; Tamimi R, Hankinson S, Lagiou P. Breast cancer. In: Adami H-O, Hunter DJ, Lagiou P, Mucci L (eds). *Textbook of Cancer Epidemiology*. Oxford Scholarship Online, February 2018. doi:10.1093/oso/9870190676827.003.0016.
72. Li C (ed). *Breast Cancer Epidemiology*. New York: Springer, 2010. https://doi.org/10.1007/978-1-4419-0685-4; Rojas K, Stuckey A. Breast cancer epidemiology and risk factors. *Clin Obstet Gynecol* 2016;59(4):651–672. doi:10.1097/GRF.0000000000000239; Tamimi R, Hankinson S, Lagiou P. Breast cancer. In: Adami H-O, Hunter DJ, Lagiou P, Mucci L (eds). *Textbook of Cancer Epidemiology*. Oxford Scholarship Online, February 2018. doi:10.1093/oso/9870190676827.003.0016.
73. Li C (ed). *Breast Cancer Epidemiology*. New York: Springer, 2010. https://doi.org/10.1007/978-1-4419-0685-4; Rojas K, Stuckey A. Breast cancer epidemiology and risk factors. *Clin Obstet Gynecol* 2016;59(4):651–672. doi:10.1097/GRF.0000000000000239; Tamimi R, Hankinson S, Lagiou P. Breast cancer. In: Adami H-O, Hunter DJ, Lagiou P, Mucci L (eds). *Textbook of Cancer Epidemiology*. Oxford Scholarship Online, February 2018. doi:10.1093/oso/9870190676827.003.0016.
74. Li C (ed). *Breast Cancer Epidemiology*. New York: Springer, 2010. https://doi.org/10.1007/978-1-4419-0685-4; Rojas K, Stuckey A. Breast cancer epidemiology and risk factors. *Clin Obstet Gynecol* 2016;59(4):651–672. doi:10.1097/GRF.0000000000000239; Tamimi R, Hankinson S, Lagiou P. Breast cancer. In: Adami H-O, Hunter DJ, Lagiou P, Mucci L (eds). *Textbook of Cancer Epidemiology*. Oxford Scholarship Online, February 2018. doi:10.1093/oso/9870190676827.003.0016.
75. Li C (ed). *Breast Cancer Epidemiology*. New York: Springer, 2010. https://doi.org/10.1007/978-1-4419-0685-4; Rojas K, Stuckey A. Breast cancer epidemiology and risk factors. *Clin Obstet Gynecol* 2016;59(4):651–672. doi:10.1097/GRF.0000000000000239; Tamimi R, Hankinson S, Lagiou P. Breast cancer. In: Adami H-O, Hunter DJ, Lagiou P, Mucci L (eds). *Textbook of Cancer*

Epidemiology. Oxford Scholarship Online, February 2018. doi:10.1093/oso/9870190676827.003.0016.

76. Perou CM, Sørlie T, Eisen MB, van de Rijn M, Jeffrey SS, Rees CA, Pollack JR, Ross DT, Johnsen H, Akslen LA, Fluge O, Pergamenschikov A, Williams C, Zhu SX, Lønning PE, Børresen-Dale AL, Brown PO, Botstein D. Molecular portraits of human breast tumours. *Nature* 2000;406(6797):747–752. doi:10.1038/35021093; Rojas K, Stuckey A. Breast cancer epidemiology and risk factors. *Clin Obstet Gynecol* 2016;59(4):651–672. doi:10.1097/GRF.0000000000000239.
77. Li C (ed). *Breast Cancer Epidemiology*. New York: Springer, 2010. https://doi.org/10.1007/978-1-4419-0685-4; Rojas K, Stuckey A. Breast cancer epidemiology and risk factors. *Clin Obstet Gynecol* 2016;59(4):651–672. doi:10.1097/GRF.0000000000000239; Tamimi R, Hankinson S, Lagiou P. Breast cancer. In: Adami H-O, Hunter DJ, Lagiou P, Mucci L (eds). *Textbook of Cancer Epidemiology*. Oxford Scholarship Online, February 2018. doi:10.1093/oso/9870190676827.003.0016.
78. Wailoo K. *How Cancer Crossed the Color Line*. New York: Oxford University Press, 2001; Amend K, Hicks D, Ambrosone CB. Breast cancer in African-American women: differences in tumor biology from European-American women. *Cancer Res* 2006;66(17):8327–8230. doi:10.1158/0008-5472.CAN-06-1927; Newman LA, Griffith KA, Jatoi I, Simon MS, Crowe JP, Colditz GA. Meta-analysis of survival in African American and white American patients with breast cancer: ethnicity compared with socioeconomic status. *J Clin Oncol* 2006;24(9):1342–1349. doi:10.1200/JCO.2005.03.3472; Hayanga AJ, Newman LA. Investigating the phenotypes and genotypes of breast cancer in women with African ancestry: the need for more genetic epidemiology. *Surg Clin North Am* 2007;87(2):551–568, xii. doi:10.1016/j.suc.2007.01.003
79. Krieger N, Chen JT, Waterman PD, Soobader MJ, Subramanian SV, Carson R. Geocoding and monitoring of US socioeconomic inequalities in mortality and cancer incidence: does the choice of area-based measure and geographic level matter? The Public Health Disparities Geocoding Project. *Am J Epidemiol* 2002;156(5):471–482. doi:10.1093/aje/kwf068.
80. Krieger N, Van Den Eeden SK, Zava D, Okamoto A. Race/ethnicity, social class, and prevalence of breast cancer molecular prognostic biomarkers: a study of white, black, and Asian women in the San Francisco Bay Area. *Ethn Dis* 1997;7:137–149; Vona-Davis L, Rose DP. The influence of socioeconomic disparities on breast cancer tumor biology and prognosis: a review. *J Womens Health (Larchmt)* 2009;18(6):883–893. doi:10.1089/jwh.2008.1127.
81. Krieger N, Van Den Eeden SK, Zava D, Okamoto A. Race/ethnicity, social class, and prevalence of breast cancer molecular prognostic biomarkers: a study of white, black, and Asian women in the San Francisco Bay Area. *Ethn Dis* 1997;7:137–149;

Vona-Davis L, Rose DP. The influence of socioeconomic disparities on breast cancer tumor biology and prognosis: a review. *J Womens Health (Larchmt)* 2009;18(6):883–893. doi:10.1089/jwh.2008.1127.

82. Krieger N, Van Den Eeden SK, Zava D, Okamoto A. Race/ethnicity, social class, and prevalence of breast cancer molecular prognostic biomarkers: a study of white, black, and Asian women in the San Francisco Bay Area. *Ethn Dis* 1997;7:137–149; Vona-Davis L, Rose DP. The influence of socioeconomic disparities on breast cancer tumor biology and prognosis: a review. *J Womens Health (Larchmt)* 2009;18(6):883–893. doi:10.1089/jwh.2008.1127.
83. Landrine H, Corral I, Lee JGL, Efird JT, Hall MB, Bess JJ. Residential segregation and racial cancer disparities: a systematic review [published online ahead of print December 30, 2016]. *J Racial Ethn Health Disparities* 2017;4(6):1195–1205. doi:10.1007/s40615-016-0326-9.
84. Amend K, Hicks D, Ambrosone CB. Breast cancer in African-American women: differences in tumor biology from European-American women. *Cancer Res* 2006;66(17):8327–8330. doi:10.1158/0008-5472.CAN-06-1927; Newman LA, Griffith KA, Jatoi I, Simon MS, Crowe JP, Colditz GA. Meta-analysis of survival in African American and white American patients with breast cancer: ethnicity compared with socioeconomic status. *J Clin Oncol* 2006;24(9):1342–1349. doi:10.1200/JCO.2005.03.3472; Hayanga AJ, Newman LA. Investigating the phenotypes and genotypes of breast cancer in women with African ancestry: the need for more genetic epidemiology. *Surg Clin North Am* 2007;87(2):551–568, xii. doi:10.1016/j.suc.2007.01.003; Agurs-Collins T, Dunn BK, Browne D, Johnson KA, Lubet R. Epidemiology of health disparities in relation to the biology of estrogen receptor-negative breast cancer. *Semin Oncol* 2010;37(4):384–401. doi:10.1053/j.seminoncol.2010.05.002.
85. Pellom ST Jr, Arnold T, Williams M, Brown VL, Samuels AD. Examining breast cancer disparities in African Americans with suggestions for policy [published online ahead of print June 10, 2020]. *Cancer Causes Control* 2020;31(9):795–800. doi:10.1007/s10552-020-01322-z; Newman LA. Disparities in breast cancer and African ancestry: a global perspective [published online ahead of print January 9, 2015]. *Breast J* 2015;21(2):133–139. doi:10.1111/tbj.12369.
86. Eng A, McCormack V, dos-Santos-Silva I. Receptor-defined subtypes of breast cancer in indigenous populations in Africa: a systematic review and meta-analysis. *PLoS Med* 2014 Sep 9;11(9):e1001720. doi:10.1371/journal.pmed.1001720; Campbell MC, Tishkoff SA. African genetic diversity: implications for human demographic history, modern human origins, and complex disease mapping. *Annu Rev Genomics Hum Genet* 2008;9:403–433. doi:10.1146/annurev.genom.9.081307; Parker J. *African History: A Very Short Introduction*. New York: Oxford University Press, 2007.

87. Krieger N, Chen JT, Waterman PD. Temporal trends in the black/white breast cancer case ratio for estrogen receptor status: disparities are historically contingent, not innate *Cancer Causes Control* 2011;22:511–514.
88. Krieger N. History, biology, and health inequities: emergent embodied phenotypes and the illustrative case of the breast cancer estrogen receptor. *Am J Public Health* 2013;103:22–27.
89. Krieger N. History, biology, and health inequities: emergent embodied phenotypes and the illustrative case of the breast cancer estrogen receptor. *Am J Public Health* 2013;103:22–27.
90. Krieger N. History, biology, and health inequities: emergent embodied phenotypes and the illustrative case of the breast cancer estrogen receptor. *Am J Public Health* 2013;103:22–27; quote: p. 23.
91. Krieger N. History, biology, and health inequities: emergent embodied phenotypes and the illustrative case of the breast cancer estrogen receptor. *Am J Public Health* 2013;103:22–27; quote: p. 22.
92. Krieger N. History, biology, and health inequities: emergent embodied phenotypes and the illustrative case of the breast cancer estrogen receptor. *Am J Public Health* 2013;103:22–27.
93. Krieger N. History, biology, and health inequities: emergent embodied phenotypes and the illustrative case of the breast cancer estrogen receptor. *Am J Public Health* 2013;103:22–27; quote p. 22.
94. Krieger N, Rehkopf DH, Chen JT, Waterman PD, Marcelli E, Kennedy M. The fall and rise of US inequities in premature mortality: 1960–2002. *PLoS Med* 2008;5(2):e46. doi:10.1371/journal.pmed.0050046; Krieger N, Chen JT, Coull B, Waterman PD, Beckfield J. The unique impact of abolition of Jim Crow laws on reducing inequities in infant death rates and implications for choice of comparison groups in analyzing societal determinants of health [published online ahead of print October 17, 2013]. *Am J Public Health* 2013;103(12):2234–2244. doi:10.2105/AJPH.2013.301350; Krieger N, Chen JT, Coull BA, Beckfield J, Kiang MV, Waterman PD. Jim Crow and premature mortality among the US Black and White population, 1960–2009: an age-period-cohort analysis. *Epidemiology* 2014;25(4):494–504. doi:10.1097/EDE.0000000000000104.
95. Newman LA. Disparities in breast cancer and African ancestry: a global perspective. *Breast J* 2015;21(2):133–139; Vona-Davis L, Rose DP. The influence of socioeconomic disparities on breast cancer tumor biology and prognosis: a review. *J Womens Health (Larchmt)* 2009;18(6):883–893; Dietze EC, Sistrunk C, Miranda-Carboni G, O'Regan R, Seewaldt VL. Triple-negative breast cancer in African-American women: disparities versus biology. *Nat Rev Cancer* 2015;15(4):248–254. doi:10.1038/nrc3896.

96. Krieger N, Jahn JL, Waterman PD. Jim Crow and estrogen-receptor negative breast cancer: US-born black and white non-Hispanic women, 1992–2012 [published online ahead of print December 17, 2016]. *Cancer Causes Control* 2017;28:49–59. PMID:27988896.
97. Krieger N, Jahn JL, Waterman PD, Chen JT. Breast cancer estrogen receptor by biological generation: US black and white women, born 1915–1979 [published online ahead of print September 20, 2017]. *Am J Epidemiol* 2018;187:960–970. PMID:29036268.
98. Krieger N, Singh N, Waterman PD. Metrics for monitoring cancer inequities: residential segregation, the Index of Concentration at the Extremes (ICE), and breast cancer estrogen receptor status (USA, 1992–2012). *Cancer Causes Control* 2016;27(9):1139–1151. doi:10.1007/s10552-016-0793-7; Linnenbringer E, Geronimus AT, Davis KL, Bound J, Ellis L, Gomez SL. Associations between breast cancer subtype and neighborhood socioeconomic and racial composition among Black and White women [published online ahead of print January 30, 2020]. *Breast Cancer Res Treat* 2020;180(2):437–447. doi:10.1007/s10549-020-05545-1.
99. Krieger N, Jahn JL, Waterman PD. Jim Crow and estrogen-receptor negative breast cancer: US-born black and white non-Hispanic women, 1992–2012 [published online ahead of print December 17, 2016]. *Cancer Causes Control* 2017;28:49–59.
100. Gingerich PD. Quantification and comparison of evolutionary rates. *Am J Sci* 1993;293(A):453–478; Gingerich PD. Rates of evolution. *Annu Rev Ecol Evol Syst* 2009;40:657–675; Hendry AP, Kinnison MT. The pace of modern life: measuring rates of contemporary microevolution. *Evolution* 1999;53(6):1637–1653; Hendry AP, Farrugia TJ, Kinnison MT. Human influences on rates of phenotypic change in wild animal populations. *Mol Ecol* 2008;17(1):20–29; DeLong JP, Forbes VE, Galic N, Gibert JP, Laport RG, Phillips JS, Vavra JM. How fast is fast? Eco-evolutionary dynamics and rates of change in populations and phenotypes. *Ecol Evol* 2016;6(2):573–581; Byars SG, Ewbank D, Govindaraju DR, Stearns SC. Natural selection in a contemporary human population. *PNAS* 2010;107(Suppl 1):1787–1792.
101. Krieger N, Chen JT, Waterman PD, Koshelva A, Beckfield J. History, haldanes and health inequities: exploring phenotypic changes in body size by generation and income level among the US-born white and black non-Hispanic populations, 1959–1962 to 2005–2008. *Int J Epidemiol* 2013;42(1):281–295; Byars SG. Commentary: haldanes and trends in phenotypic change in humans. *Int J Epidemiol* 2013;42(1):295–297.
102. Krieger N, Kiang MV, Kosheleva A, Waterman PD, Chen JT, Beckfield J. Age at menarche: 50-year socioeconomic trends among US-born black and white women. *Am J Public Health* 2015;105(2):388–397.
103. Hendry AP, Kinnison MT. The pace of modern life: measuring rates of contemporary microevolution. *Evolution* 1999;53(6):1637–1653; Hendry AP,

Farrugia TJ, Kinnison MT. Human influences on rates of phenotypic change in wild animal populations. *Mol Ecol* 2008;17(1):20–29; DeLong JP, Forbes VE, Galic N, Gibert JP, Laport RG, Phillips JS, Vavra JM. How fast is fast? Eco-evolutionary dynamics and rates of change in populations and phenotypes. *Ecol Evol* 2016;6(2):573–581; Krieger N, Chen JT, Waterman PD, Koshelva A, Beckfield J. History, haldanes and health inequities: exploring phenotypic changes in body size by generation and income level among the US-born white and black non-Hispanic populations, 1959–1962 to 2005–2008. *Int J Epidemiol* 2013;42(1):281–295; Byars SG. Commentary: haldanes and trends in phenotypic change in humans. *Int J Epidemiol* 2013;42(1):295–297.

104. Dowsett T, Verghese E, Pollock S, Pollard J, Heads J, Hanby A, Speirs V. The value of archival tissue blocks in understanding breast cancer biology [published online ahead of print October 29, 2013]. *J Clin Pathol* 2014;67(3):272–275. doi:10.1136/jclinpath-2013-201854.
105. Alimujiang A, Mo M, Liu Y, Huang NS, Liu G, Xu W, Wu J, Shen ZZ, Shao Z, Colditz GA. The association between China's Great famine and risk of breast cancer according to hormone receptor status: a hospital-based study [published online ahead of print October 1, 2016]. *Breast Cancer Res Treat* 2016;160(2):361–369. doi:10.1007/s10549-016-3994-6.
106. Elands RJJ, Offermans NSM, Simons CCJM, Schouten LJ, Verhage BA, van den Brandt PA, Weijenberg MP. Associations of adult-attained height and early life energy restriction with postmenopausal breast cancer risk according to estrogen and progesterone receptor status [published online ahead of print November 9, 2018]. *Int J Cancer* 2019 Apr 15;144(8):1844–1857. doi:10.1002/ijc.31890; Xiang Y, Zhou W, Duan X, Fan Z, Wang S, Liu S, Liu L, Wang F, Yu L, Zhou F, Huang S, Li L, Zhang Q, Fu Q, Ma Z, Gao D, Cui S, Geng C, Cao X, Yang Z, Wang X, Liang H, Jiang H, Wang H, Li G, Wang Q, Zhang J, Jin F, Tang J, Tian F, Ye C, Yu Z. Metabolic syndrome, and particularly the hypertriglyceridemic-waist phenotype, increases breast cancer risk, and adiponectin is a potential mechanism: a case-control study in Chinese women. *Front Endocrinol (Lausanne)* 2020 Jan 21;10:905. doi:10.3389/fendo.2019.00905. Erratum in: *Front Endocrinol (Lausanne)* 2020 Jun 04;11:227; Fang Z, Chen C, Wang H, Tang K. Association between fetal exposure to famine and anthropometric measures in adulthood: a regression discontinuity approach [published online ahead of print March 10, 2020]. *Obesity (Silver Spring)* 2020;28(5):962–969. doi:10.1002/oby.22760; Arage G, Belachew T, Hassen H, Abera M, Abdulhay F, Abdulahi M, Hassen Abate K. Effects of prenatal exposure to the 1983-1985 Ethiopian great famine on the metabolic syndrome in adults: a historical cohort study [published online ahead of print June 10, 2020]. *Br J Nutr* 2020;124(10):1052–1060. doi:10.1017/S0007114520002123; Dibaba DT, Braithwaite D, Akinyemiju T.

Metabolic syndrome and the risk of breast cancer and subtypes by race, menopause and BMI. *Cancers (Basel)* 2018;10(9):299. doi:10.3390/cancers10090299.

107. Popkin BM. The challenge in improving the diets of supplemental nutrition assistance program recipients: a historical commentary. *Am J Prev Med* 2017;52(2S2):S106–S114. doi:10.1016/j.amepre.2016.08.018. Erratum in: *Am J Prev Med* 2017;52(4):554; United States National Advisory Commission on Rural Poverty. *Rural Poverty in the United States: A Report by the President's National Advisory Commission on Rural Poverty.* Washington, DC: US Government Printing Office, 1968; Hinton E. *From the War on Poverty to the War on Crime.* Cambridge, MA: Harvard University Press, 2016; Ward TJ Jr. *Out in the Rural: A Mississippi Health Center and its War on Poverty.* New York: Oxford University Press, 2016.
108. United States National Advisory Commission on Rural Poverty. *Rural Poverty in the United States: A Report by the President's National Advisory Commission on Rural Poverty.* Washington, DC: US Government Printing Office, 1968; 288.
109. Krieger N. Workers are people too: societal aspects of occupational health disparities—an ecosocial perspective [commentary]. *Am J Indust Med* 2010;53:104–115; Barbeau EM, Hartman C, Quinn MM, Stoddard AM, Krieger N. Methods for recruiting white, black, and Hispanic working class women and men to a study of physical and social hazards at work: the United for Health Study. *Int J Health Serv* 2007;37:127–144.
110. Barbeau EM, Hartman C, Quinn MM, Stoddard AM, Krieger N. Methods for recruiting white, black, and Hispanic working class women and men to a study of physical and social hazards at work: the United for Health Study. *Int J Health Serv* 2007;37:127–144. PMID:17436989; Krieger N, Smith K, Naishadham D, Hartman C, Barbeau EM. Experiences of discrimination: validity and reliability of a self-report measure for population health research on racism and health. *Soc Sci Med* 2005;61:1576–1596. PMID:16005789; Krieger N, Waterman PD, Hartman C, Bates LM, Stoddard AM, Quinn MM, Sorensen G, Barbeau EM. Social hazards on the job: workplace abuse, sexual harassment, and racial discrimination—a study of black, Latino, and white low-income women and men workers (US). *Int J Health Serv* 2006;36:51–85. PMID:16524165; Quinn MM, Sembajwe G, Stoddard AM, Kriebel D, Krieger N, Sorensen G, Hartman C, Naishadham D, Barbeau EM. Social disparities in the burden of occupational exposures: results of a cross-sectional study. *Am J Indust Med* 2007;50:861–875. PMID:17979135; Krieger N, Chen JT, Waterman PD, Hartman C, Stoddard AM, Quinn MM, Sorensen G, Barbeau E. The inverse hazard law: blood pressure, sexual harassment, racial discrimination, workplace abuse and occupational exposures in the *United for Health* study of US low-income black, white, and Latino workers (Greater Boston Area, Massachusetts, United States, 2003–2004).

Soc Sci Med 2008;67:1970–1981. PMID:18950922; Sembajwe G, Quinn M, Kriebel D, Stoddard A, Krieger N, Barbeau E. The influence of sociodemographic characteristics on agreement between self-reports and expert exposure assessments. *Am J Ind Med* 2010;53:1019–1031. PMID:20306494; Krieger N, Kaddour A, Koenen K, Kosheleva A, Chen JT, Waterman PD, Barbeau EM. Occupational, social, and relationship hazards and psychological distress among low-income workers: implications of the "inverse hazard law." *J Epidemiol Community Health* 2011;65:260–272. PMID:20713372.

111. Krieger N, Waterman PD, Hartman C, Bates LM, Stoddard AM, Quinn MM, Sorensen G, Barbeau EM. Social hazards on the job: workplace abuse, sexual harassment, and racial discrimination—a study of black, Latino, and white low-income women and men workers (US). *Int J Health Serv* 2006;36:51–85.
112. Barbeau EM, Hartman C, Quinn MM, Stoddard AM, Krieger N. Methods for recruiting white, black, and Hispanic working class women and men to a study of physical and social hazards at work: the United for Health Study. *Int J Health Serv* 2007;37:127–144.
113. Krieger N, Waterman PD, Hartman C, Bates LM, Stoddard AM, Quinn MM, Sorensen G, Barbeau EM. Social hazards on the job: workplace abuse, sexual harassment, and racial discrimination—a study of black, Latino, and white low-income women and men workers (US). *Int J Health Serv* 2006;36:51–85; Wright EO. *Classes Count: Comparative Studies in Class Analysis.* Cambridge: Cambridge University Press, 1997; National Statistics (UK). *The National Statistics Socio-economic Classification (NS-SEC): Introduction.* 2002. At the time we did the study, the version that we accessed (in 2005) was at http://www.statistics.gov.uk/methods_quality/ns_sec/default.asp; the current version is available at https://www.ons.gov.uk/methodology/classificationsandstandards/otherclassifications/thenationalstatisticssocioeconomicclassificationnssecrebasedonsoc2010; accessed January 2, 2021; Krieger N, Barbeau EM, Soobader M-J. Class matters: US vs UK measures of occupational gradients in access to health services and health status in the 2000 US National Health Interview Survey. *Int J Health Serv* 2005;35:216–236.
114. Krieger N, Chen JT, Waterman PD, Hartman C, Stoddard AM, Quinn MM, Sorensen G, Barbeau E. The inverse hazard law: blood pressure, sexual harassment, racial discrimination, workplace abuse and occupational exposures in the *United for Health* study of US low-income black, white, and Latino workers (Greater Boston Area, Massachusetts, United States, 2003–2004). *Soc Sci Med* 2008;67:1970–1981; Krieger N, Kaddour A, Koenen K, Kosheleva A, Chen JT, Waterman PD, Barbeau EM. Occupational, social, and relationship hazards and psychological distress among low-income workers: implications of the "inverse hazard law." *J Epidemiol Community Health* 2011;65:260–272.

115. Krieger N, Chen JT, Waterman PD, Hartman C, Stoddard AM, Quinn MM, Sorensen G, Barbeau E. The inverse hazard law: blood pressure, sexual harassment, racial discrimination, workplace abuse and occupational exposures in the *United for Health* study of US low-income black, white, and Latino workers (Greater Boston Area, Massachusetts, United States, 2003–2004). *Soc Sci Med* 2008;67:1970–1981.
116. Chae DH, Krieger N, Bennett GG, Lindsey JC, Stoddard AM, Barbeau EM. Implications of discrimination based on sexuality, gender, and race for psychological distress among working class sexual minorities: the United for Health Study, 2003–2004. *Int J Health Serv* 2010;40:589–608.
117. Quinn MM, Sembajwe G, Stoddard AM, Kriebel D, Krieger N, Sorensen G, Hartman C, Naishadham D, Barbeau EM. Social disparities in the burden of occupational exposures: results of a cross-sectional study. *Am J Indust Med* 2007;50:861–875.
118. Barbeau EM, Hartman C, Quinn MM, Stoddard AM, Krieger N. Methods for recruiting white, black, and Hispanic working class women and men to a study of physical and social hazards at work: the United for Health Study. *Int J Health Serv* 2007;37:127–144; Krieger N, Smith K, Naishadham D, Hartman C, Barbeau EM. Experiences of discrimination: validity and reliability of a self-report measure for population health research on racism and health. *Soc Sci Med* 2005;61:1576–1596; Krieger N, Waterman PD, Hartman C, Bates LM, Stoddard AM, Quinn MM, Sorensen G, Barbeau EM. Social hazards on the job: workplace abuse, sexual harassment, and racial discrimination—a study of black, Latino, and white low-income women and men workers (US). *Int J Health Serv* 2006;36:51–85; Quinn MM, Sembajwe G, Stoddard AM, Kriebel D, Krieger N, Sorensen G, Hartman C, Naishadham D, Barbeau EM. Social disparities in the burden of occupational exposures: results of a cross-sectional study. *Am J Indust Med* 2007;50:861–875; Krieger N, Chen JT, Waterman PD, Hartman C, Stoddard AM, Quinn MM, Sorensen G, Barbeau E. The inverse hazard law: blood pressure, sexual harassment, racial discrimination, workplace abuse and occupational exposures in the *United for Health* study of US low-income black, white, and Latino workers (Greater Boston Area, Massachusetts, United States, 2003–2004). *Soc Sci Med* 2008;67:1970–1981; Sembajwe G, Quinn M, Kriebel D, Stoddard A, Krieger N, Barbeau E. The influence of sociodemographic characteristics on agreement between self-reports and expert exposure assessments. *Am J Ind Med* 2010;53:1019–1031; Krieger N, Kaddour A, Koenen K, Kosheleva A, Chen JT, Waterman PD, Barbeau EM. Occupational, social, and relationship hazards and psychological distress among low-income workers: implications of the "inverse hazard law." *J Epidemiol Community Health* 2011;65:260–272; Krieger N, Kosheleva A, Waterman PD, Chen JT, Koenen K. Racial discrimination, psychological distress, and self-rated

health among US-born and foreign-born black Americans. *Am J Public Health* 2011;101:1704–1713; Chae DH, Krieger N, Bennett GG, Lindsey JC, Stoddard AM, Barbeau EM. Implications of discrimination based on sexuality, gender, and race for psychological distress among working class sexual minorities: the United for Health Study, 2003–2004. *Int J Health Serv* 2010;40:589–608.

119. Krieger N. Embodying inequality: a review of concepts, measures, and methods for studying health consequences of discrimination. *Int J Health Serv* 1999;29(2):295–352; Krieger N. Does racism harm health? Did child abuse exist before 1962? On explicit questions, critical science, and current controversies: an ecosocial perspective. *Am J Public Health* 2003;93(2):194–199; Krieger N. Methods for the scientific study of discrimination and health: an ecosocial approach. *Am J Public Health* 2012;102(5):936–944; Krieger N. Discrimination and health inequities. *Int J Health Serv* 2014;44(4):643–710; Krieger N. Measures of racism, sexism, heterosexism, and gender binarism for health equity research: from structural injustice to embodied harm-an ecosocial analysis. *Annu Rev Public Health* 2020;41:37–62; Lenderink AF, Zoer I, van der Molen HF, Spreeuwers D, Frings-Dresen MH, van Dijk FJ. Review on the validity of self-report to assess work-related diseases. *Int Arch Occup Environ Health* 2012;85(3):229–251; Gibbs A, Pretorius L, Jewkes R. Test-retest stability of self-reported violence against women measures: results from the stepping stones and creating futures pilot. *Glob Health Action* 2019;12(1):1671663. doi:10.1080/16549716.2019.1671663; Midanik LT, Greenfield TK. Interactive voice response versus computer-assisted telephone interviewing (CATI) surveys and sensitive questions: the 2005 National Alcohol Survey. *J Stud Alcohol Drugs* 2008;69(4):580–588.

120. For examples of methodological and conceptual work on "intersectionality" in public health research and practice, which our United for Health study predated, including the recent flurry published since 2018, see (listed in reverse chronological order): *Methods (which, for the quantitative analyses, chiefly focus on how to model the diverse social categories to which people are assigned, as opposed to modeling measurement of the exposures associated with the categories):* Fehrenbacher AE, Patel D. Translating the theory of intersectionality into quantitative and mixed methods for empirical gender transformative research on health. *Cult Health Sex* 2020;22(Suppl 1):145–160. doi:10.1080/13691058.2019.1671494; Abrams JA, Tabaac A, Jung S, Else-Quest NM. Considerations for employing intersectionality in qualitative health research [published online ahead of print June 16, 2020]. *Soc Sci Med* 2020;258:113138. doi:10.1016/j.socscimed.2020.113138; Phillips SP, Vafaei A, Yu S, Rodrigues R, Ilinca S, Zolyomi E, Fors S. Systematic review of methods used to study the intersecting impact of sex and social locations on health outcomes. *SSM—Population Health* 2020;12:100705. doi.org/10.1016/j.ssmph.2020.100705; Lizotte DJ, Mahendran M, Churchill SM, Bauer GR. Math

versus meaning in MAIHDA: a commentary on multilevel statistical models for quantitative intersectionality. *Soc Sci Med* 2020;245:112500. doi:10.1016/j.socscimed.2019.112500; Evans CR, Leckie G, Merlo J. Multilevel versus single-level regression for the analysis of multilevel information: the case of quantitative intersectional analysis. *Soc Sci Med* 2020;245:112499. doi:10.1016/j.socscimed.2019.112499; Evans CR. Modeling the intersectionality of processes in the social production of health inequalities. *Soc Sci Med* 2019;226:249–253; Evans CR. Adding interactions to models of intersectional health inequalities: comparing multilevel and conventional methods. *Soc Sci Med* 2019;221:95–105. doi:10.1016/j.socscimed.2018.11.036; Bauer GR, Scheim AI. Advancing quantitative intersectionality research methods: intracategorical and intercategorical approaches to shared and differential constructs. *Soc Sci Med* 2019;226:260–262. doi:10.1016/j.socscimed.2019.03.018; Bauer GR, Scheim AI. Methods for analytic intercategorical intersectionality in quantitative research: discrimination as a mediator of health inequalities. *Soc Sci Med* 2019;226:236–245. doi:10.1016/j.socscimed.2018.12.015; Richman LS, Zucker AN. Quantifying intersectionality: an important advancement for health inequality research. *Soc Sci Med* 2019;226:246–248; Evans CR, Williams DR, Onnela JP, Subramanian SV. A multilevel approach to modeling health inequalities at the intersection of multiple social identities [published online ahead of print November 30, 2017]. *Soc Sci Med* 2018;203:64–73. doi:10.1016/j.socscimed.2017.11.011; Merlo J. Multilevel analysis of individual heterogeneity and discriminatory accuracy (MAIHDA) within an intersectional framework. *Soc Sci Med* 2018;203:74–80. doi:10.1016/j.socscimed.2017.12.026; Green MA, Evans CR, Subramanian SV. Can intersectionality theory enrich population health research? *Soc Sci Med* 2017;178:214–216. doi:10.1016/j.socscimed.2017.02.029; Bauer GR. Incorporating intersectionality theory into population health research methodology: challenges and the potential to advance health equity. *Soc Sci Med* 2014;110:10–17; *Conceptual (which focus mainly on the social aspects of the analytic framework and approach, and not the biological processes of embodying (in)justice):* Alvidrez J, Greenwood GL, Johnson TL, Parker KL. Intersectionality in public health research: a view from the National Institutes of Health. *Am J Public Health* 2021;111(1):95–97. doi:10.2105/AJPH.2020.305986; Bowleg L. Evolving intersectionality within public health: from analysis to action. *Am J Public Health* 2021;111(1):88–90. doi:10.2105/AJPH.2020.306031; Poteat T. Navigating the storm: how to apply intersectionality to public health in times of crisis. *Am J Public Health* 2021;111(1):91–92. doi:10.2105/AJPH.2020.305944; Aguayo-Romero RA. (Re)centering Black feminism into intersectionality research. *Am J Public Health* 2021;111(1):101–103. doi:10.2105/AJPH.2020.306005; Bauer GR, Lizotte DJ. Artificial intelligence, intersectionality, and the future of public health. *Am J Public Health* 2021;111(1):98–100. doi:10.2105/AJPH.2020.306006; Heard E, Fitzgerald L, Wigginton B, Mutch A. Applying intersectionality theory in health promotion research and practice. *Health Promot Int* 2020;35(4):866–876. doi:10.1093/heapro/daz080; Gueta K. Exploring the promise of intersectionality for promoting justice-involved women's health research and policy. *Health Justice* 2020;8(1):19. doi:10.1186/s40352-020-00120-8; Samra R, Hankivsky O. Adopting an intersectionality framework to address power and equity in medicine. *Lancet* 2020;S0140-6736(20):32513–32517. doi:10.1016/S0140-6736(20)32513-7; Agénor M. Future directions for incorporating intersectionality into quantitative population health research. *Am J Public Health* 2020;110(6):803–806; Abrams JA, Tabaac A, Jung S, Else-Quest NM. Considerations for employing intersectionality in qualitative health research. *Soc Sci Med* 2020;258:113138. doi:10.1016/j.socscimed.2020.113138; Bowleg L. We're not all in this together: on COVID-19, intersectionality, and structural inequality. *Am J Public Health* 2020;110(7):917; English D, Carter JA, Bowleg L, Malebranche DJ, Talan AJ, Rendina HJ. Intersectional social control: the roles of incarceration and police discrimination in psychological and HIV-related outcomes for Black sexual minority men. *Soc Sci Med* 2020;258:113121; Fagrell Trygg N, Gustafsson PE, Månsdotter A. Languishing in the crossroad? A scoping review of intersectional inequalities in mental health. *Int J Equity Health* 2019;18(1):115. doi:10.1186/s12939-019-1012-4; Wesp LM, Malcoe LH, Elliott A, Poteat T. Intersectionality research for transgender health justice: a theory-driven conceptual framework for structural analysis of transgender health inequities. *Transgend Health* 2019;4(1):287–296. doi:10.1089/trgh.2019.0039; Gkiouleka A, Huijts T, Beckfield J, Bambra C. Understanding the micro and macro politics of health: inequalities, intersectionality and institutions—a research agenda. *Soc Sci Med* 2018;200:92–98; Santos CE, Toomey RB. Integrating an intersectionality lens in theory and research in developmental science. *New Dir Child Adolesc Dev* 2018;161:7–15; Moradi B, Grzanka PR. Using intersectionality responsibly: toward critical epistemology, structural analysis, and social justice activism. *J Couns Psychol* 2017;64(5):500–513. doi:10.1037/cou0000203; Hill SE. Axes of health inequalities and intersectionality (Chapter 7). In: Smith KE, Bambra C, Hill SE (eds). *Health Inequalities: Critical Perspectives*. Oxford: Oxford University Press, 2015. Oxford Scholarship Online, 2016. doi:10.1093/acprof:oso/9780198703358.003.0007; Hankivsky O, Doyal L, Einstein G, Kelly U, Shim J, Weber L, Repta R. The odd couple: using biomedical and intersectional approaches to address health inequities. *Glob Health Action* 2017;10(Suppl 2):1326686; Bowleg L. The problem with the phrase women and minorities: intersectionality-an important theoretical framework for public health. *Am J Public Health* 2012;102(7):1267–1273; Hankivsky O. Women's health, men's health, and gender and health: implications of intersectionality. *Soc Sci Med* 2012;74(11):1712–1720.

121. Krieger N, Kaddour A, Koenen K, Kosheleva A, Chen JT, Waterman PD, Barbeau EM. Occupational, social, and relationship hazards and psychological distress among low-income workers: implications of the "inverse hazard law." *J Epidemiol Community Health* 2011;65:260–272.
122. Chae DH, Krieger N, Bennett GG, Lindsey JC, Stoddard AM, Barbeau EM. Implications of discrimination based on sexuality, gender, and race for psychological distress among working class sexual minorities: the United for Health Study, 2003–2004. *Int J Health Serv* 2010;40:589–608.
123. Krieger N, Kosheleva A, Waterman PD, Chen JT, Koenen K. Racial discrimination, psychological distress, and self-rated health among US-born and foreign-born black Americans. *Am J Public Health* 2011;101:1704–1713.
124. Krieger N, Chen JT, Waterman PD, Hartman C, Stoddard AM, Quinn MM, Sorensen G, Barbeau E. The inverse hazard law: blood pressure, sexual harassment, racial discrimination, workplace abuse and occupational exposures in the *United for Health* study of US low-income black, white, and Latino workers (Greater Boston Area, Massachusetts, United States, 2003–2004). *Soc Sci Med* 2008;67:1970–1981. quote: p. 1971.
125. Hart JT. The inverse care law. *Lancet* 1971;1(7696):405–412.
126. Krieger N, Chen JT, Waterman PD, Hartman C, Stoddard AM, Quinn MM, Sorensen G, Barbeau E. The inverse hazard law: blood pressure, sexual harassment, racial discrimination, workplace abuse and occupational exposures in the *United for Health* study of US low-income black, white, and Latino workers (Greater Boston Area, Massachusetts, United States, 2003–2004). *Soc Sci Med* 2008;67:1970–1981; Krieger N, Kaddour A, Koenen K, Kosheleva A, Chen JT, Waterman PD, Barbeau EM. Occupational, social, and relationship hazards and psychological distress among low-income workers: implications of the "inverse hazard law." *J Epidemiol Community Health* 2011;65:260–272.
127. Krieger N, Chen JT, Waterman PD, Hartman C, Stoddard AM, Quinn MM, Sorensen G, Barbeau E. The inverse hazard law: blood pressure, sexual harassment, racial discrimination, workplace abuse and occupational exposures in the *United for Health* study of US low-income black, white, and Latino workers (Greater Boston Area, Massachusetts, United States, 2003–2004). *Soc Sci Med* 2008;67:1970–1981; Krieger N, Chen JT, Coull B, Waterman PD, Beckfield J. The unique impact of abolition of Jim Crow laws on reducing inequities in infant death rates and implications for choice of comparison groups in analyzing societal determinants of health. *Am J Public Health* 2013;103(12):2234–2244; Krieger N. Measures of racism, sexism, heterosexism, and gender binarism for health equity research: from

structural injustice to embodied harm—an ecosocial analysis. *Annu Rev Public Health* 2020 April 2;41:37–62. doi:10.1146/annurev-publhealth-040119-094017.

128. Krieger N. Embodying inequality: a review of concepts, measures, and methods for studying health consequences of discrimination. *Int J Health Serv* 1999;29:295–352.
129. Krieger N. Embodying inequality: a review of concepts, measures, and methods for studying health consequences of discrimination. *Int J Health Serv* 1999;29:295–352; Williams DR, Collins C. Racial residential segregation: a fundamental cause of racial disparities in health. *Public Health Rep* 2001;116(5):404–416.
130. Krieger N. Discrimination and health inequities. In: Berkman LF, Kawachi I, Glymour M (eds). *Social Epidemiology*. 2nd ed. New York: Oxford University Press, 2014; 63–125.
131. Williams DR, Lawrence JA, Davis BA. Racism and health: evidence and needed research. *Annu Rev Public Health* 2019;40:105–125. doi:10.1146/annurev-publhealth-040218-043750; Bailey ZD, Krieger N, Agénor M, Graves J, Linos N, Bassett M. Structural racism and health inequities: evidence and interventions. *Lancet* 2017;389:1453–1463; Paradies Y, Ben J, Denson N, Elias A, Priest N, Pieterse A, Gupta A, Kelaher M, Gee G. Racism as a determinant of health: a systematic review and meta-analysis. *PLoS One* 2015;10(9):e0138511. doi:10.1371/journal.pone.0138511.
132. Krieger N. Measures of racism, sexism, heterosexism, and gender binarism for health equity research: from structural injustice to embodied harm-an ecosocial analysis. *Annu Rev Public Health* 2020;41:37–62.
133. Krieger N. Discrimination and health inequities. In: Berkman LF, Kawachi I, Glymour M (eds). *Social Epidemiology*. 2nd ed. New York: Oxford University Press, 2014; 63–125; Bailey ZD, Krieger N, Agénor M, Graves J, Linos N, Bassett M. Structural racism and health inequities: evidence and interventions. *Lancet* 2017;389:1453–1463; Krieger N. Why epidemiologists must reckon with racism. In: Ford CL, Griffith D, Bruce M, Gilbert K (eds). *Racism: Science and Tools for the Public Health Professional*. Washington, DC: American Public Health Association Press, 2019; 249–266; Paradies Y, Ben J, Denson N, Elias A, Priest N, Pieterse A, Gupta A, Kelaher M, Gee G. Racism as a determinant of health: a systematic review and meta-analysis. *PLoS One* 2015;10(9):e0138511. doi:10.1371/journal.pone.0138511; Williams DR, Lawrence JA, Davis BA. Racism and health: evidence and needed research. *Annu Rev Public Health* 2019;40:105–125. doi:10.1146/annurev-publhealth-040218-043750.
134. Beckfield J. *Political Sociology and the People's Health*. New York: Oxford University Press, 2018.

135. Krieger N. Measures of racism, sexism, heterosexism, and gender binarism for health equity research: from structural injustice to embodied harm-An ecosocial analysis. *Annu Rev Public Health* 2020;41:37–62.
136. Anderson C. *One Person, No Vote: How Voter Suppression Is Destroying our Democracy*. New York: Bloomsbury Publishing, 2018; American Civil Liberties Union. Block the Vote: Voter Suppression in 2020. https://www.aclu.org/news/civil-liberties/block-the-vote-voter-suppression-in-2020/; accessed January 2, 2021.
137. Krieger N. Measures of racism, sexism, heterosexism, and gender binarism for health equity research: from structural injustice to embodied harm-an ecosocial analysis. *Annu Rev Public Health* 2020;41:37–62.
138. Stiglitz JE. *The Great Divide: Unequal Societies and What We Can Do About Them*. New York: W. W. Norton, 2015; Piketty T. *Capital in the Twenty-First Century*. Goldhammer A, Translator. Cambridge, MA: Harvard University Press, 2014; Boushey H, DeLong J, Steinbaum M (eds). *After Piketty: The Agenda for Economics and Inequality*. Cambridge, MA: Harvard University Press, 2018; Beckfield J. *Political Sociology and the People's Health*. New York: Oxford University Press, 2018.
139. Krieger N. Discrimination and health inequities. In: Berkman LF, Kawachi I, Glymour M (eds). *Social Epidemiology*. 2nd ed. New York: Oxford University Press, 2014; 63–125; Krieger N, Waterman PD, Gryparis A, Coull BA. Black carbon exposure, socioeconomic and racial/ethnic spatial polarization, and the Index of Concentration at the Extremes (ICE) [published online ahead of print June 18, 2015]. *Health Place* 2015;34:215–228. doi:10.1016/j.healthplace.2015.05.008; Krieger N, Waterman PD, Spasojevic J, Li W, Maduro G, Van Wye G. Public health monitoring of privilege and deprivation with the Index of Concentration at the Extremes [published online ahead of print December 21, 2015]. *Am J Public Health* 2016;106(2):256–263. doi:10.2105/AJPH.2015.302955.
140. Krieger N. Discrimination and health inequities. In: Berkman LF, Kawachi I, Glymour M (eds). *Social Epidemiology*. 2nd ed. New York: Oxford University Press, 2014; 63–125; Krieger N, Waterman PD, Gryparis A, Coull BA. Black carbon exposure, socioeconomic and racial/ethnic spatial polarization, and the Index of Concentration at the Extremes (ICE) [published online ahead of print June 18, 2015]. *Health Place* 2015;34:215–228. doi:10.1016/j.healthplace.2015.05.008; Krieger N, Waterman PD, Spasojevic J, Li W, Maduro G, Van Wye G. Public health monitoring of privilege and deprivation with the Index of Concentration at the Extremes [published online ahead of print December 21, 2015]. *Am J Public Health* 2016;106(2):256–263. doi:10.2105/AJPH.2015.302955.

141. Massey D, Denton N. *American Apartheid: Segregation and the Making of the Underclass.* Cambridge, MA: Harvard University Press, 1993; Massey D, Brodmann S. *Spheres of Influence: The Social Ecology of Racial and Class Inequality.* New York: Russell Sage Foundation, 2014; Massey DS, Denton NA. The dimensions of residential segregation. *Social Forces* 1988;67(2):281–315; Massey DS. Reflections on the dimensions of segregation. *Social Forces* 2012;91(1):39–43; Massey DS. The age of extremes: concentrated affluence and poverty in the twenty-first century. *Demography* 1996;33(4):395–412.

142. Massey DS. The prodigal paradigm returns: ecology comes back to sociology. In: Booth A, Crouter A (eds). *Does It Take a Village? Community Effects on Children, Adolescents, and Families.* Mahwah, NJ: Lawrence Erlbaum Associates, 2001; 41–48.

143. US Immigrations and Customs Enforcement (ICE). https://www.ice.gov/; accessed January 2, 2021; Davis JH, Shear MD. *Border Wars: Inside Trump's Assault on Immigration.* New York: Simon & Schuster, 2019; Hesson T, Kahn C. Trump pushes anti-immigrant message even as coronavirus dominates campaign. *US News,* August 14, 2020. https://www.usnews.com/news/top-news/articles/2020-08-14/trump-pushes-anti-immigrant-message-even-as-coronavirus-dominates-campaign; accessed February 15, 2021; Shear MD, Jordan M. Undoing Trump's anti-immigrant policies will mean looking at the fine print. *New York Times,* February 10, 2021. https://www.nytimes.com/2021/02/10/us/politics/trump-biden-us-immigration-system.html; accessed February 15, 2021; Krieger N, Huynh M, Li W, Waterman PD, Van Wye G. Severe sociopolitical stressors and preterm births in New York City: 1 September 2015 to 31 August 2017 [published online ahead of print October 16, 2018]. *J Epidemiol Community Health* 2018;72(12):1147–1152. doi:10.1136/jech-2018-211077; Gemmill A, Catalano R, Alcalá H, Karasek D, Casey JA, Bruckner TA. The 2016 presidential election and periviable births among Latina women [published online ahead of print October 3, 2020]. *Early Hum Dev* 2020;151:105203. doi:10.1016/j.earlhumdev.2020.105203.

144. Massey DS. The prodigal paradigm returns: ecology comes back to sociology. In: Booth A, Crouter A (eds). *Does It Take a Village? Community Effects on Children, Adolescents, and Families.* Mahwah, NJ: Lawrence Erlbaum Associates, 2001; 41–48.

145. Krieger N, Waterman PD, Gryparis A, Coull BA. Black carbon exposure, socioeconomic and racial/ethnic spatial polarization, and the Index of Concentration at the Extremes (ICE) [published online ahead of print June 18, 2015]. *Health Place* 2015 Jul;34:215–228. doi:10.1016/j.healthplace.2015.05.008. Do DP, Dubowitz T, Bird CE, Lurie N, Escarce JJ, Finch BK. Neighborhood context and ethnicity differences in body mass index: a multilevel analysis using the NHANES

III survey (1988–1994). *Econ Hum Biol* 2007;5(2):179–203; Carpiano RM, Lloyd JE, Hertzman C. Concentrated affluence, concentrated disadvantage, and children's readiness for school: a population-based, multi-level investigation. *Soc Sci Med* 2009;69(3):420–432; Finch BK, Phuong Do D, Heron M, Bird C, Seeman T, Lurie N. Neighborhood effects on health: concentrated advantage and disadvantage. *Health Place* 2010;16(5):1058–1060; Casciano R, Massey DS. Neighborhood disorder and anxiety symptoms: new evidence from a quasi-experimental study. *Health Place* 2012;18(2):180–190; Eastwood JG, Jalaludin BB, Kemp LA, Phung HN, Barnett BE. Immigrant maternal depression and social networks. A multilevel Bayesian spatial logistic regression in South Western Sydney, Australia. *Spat Spatiotemporal Epidemiol* 2013;6:49–58; Rudolph AE, Crawford ND, Latkin C, Fowler JH, Fuller CM. Individual and neighborhood correlates of membership in drug using networks with a higher prevalence of HIV in New York City (2006–2009). *Ann Epidemiol* 2013;23(5):267–274; Kramer M. Some thoughts on measuring place and health. Presentation at Health Places Research Group, Georgia Institute of Technology, March 26, 2013; in 2014, this was available at a site that no longer exists: https://www.cqgrd.gatech.edu/sites/files/cqgrd/measuringneightboods_march2013.pdf.

146. *Study that introduced the ICE for racialized economic segregation and for racial segregation:* Krieger N, Waterman PD, Gryparis A, Coull BA. Black carbon exposure, socioeconomic and racial/ethnic spatial polarization, and the Index of Concentration at the Extremes (ICE) [published online ahead of print June 18, 2015]. *Health Place* 2015;34:215–228. doi:10.1016/j.healthplace.2015.05.008; *Subsequent studies by Krieger et al. using the ICE, including in multilevel analyses (in chronological order):* Feldman JM, Waterman PD, Coull BA, Krieger N. Spatial social polarisation: using the Index of Concentration at the Extremes jointly for income and race/ethnicity to analyse risk of hypertension [published online ahead of print July 1, 2015]. *J Epidemiol Community Health* 2015;69(12):1199–1207. doi:10.1136/jech-2015-205728; Krieger N, Waterman PD, Spasojevic J, Li W, Maduro G, Van Wye G. Public health monitoring of privilege and deprivation with the Index of Concentration at the Extremes [published online ahead of print December 21, 2015]. *Am J Public Health* 2016;106(2):256–263. doi:10.2105/AJPH.2015.302955; Krieger N, Singh N, Waterman PD. Metrics for monitoring cancer inequities: residential segregation, the Index of Concentration at the Extremes (ICE), and breast cancer estrogen receptor status (USA, 1992–2012) [published online ahead of print August 8, 2016]. *Cancer Causes Control* 2016;27(9):1139–1151. doi:10.1007/s10552-016-0793-7; Krieger N, Feldman JM, Waterman PD, Chen JT, Coull BA, Hemenway D. Local residential segregation matters: stronger association of census tract compared to conventional city-level measures with fatal and non-fatal assaults

(total and firearm related), using the index of concentration at the extremes (ICE) for racial, economic, and racialized economic segregation, Massachusetts (US), 1995–2010. *J Urban Health* 2017;94(2):244–258. doi:10.1007/s11524-016-0116-z; Krieger N, Waterman PD, Batra N, Murphy JS, Dooley DP, Shah SN. Measures of local segregation for monitoring health inequities by local health departments [published online ahead of print April 20, 2017]. *Am J Public Health* 2017;107(6):903–906. doi:10.2105/AJPH.2017.303713; Huynh M, Spasojevic J, Li W, Maduro G, Van Wye G, Waterman PD, Krieger N. Spatial social polarization and birth outcomes: preterm birth and infant mortality—New York City, 2010–14 [published online ahead of print April 6, 2017]. *Scand J Public Health* 2018;46(1):157–166. doi:10.1177/1403494817701566; Krieger N, Kim R, Feldman J, Waterman PD. Using the Index of Concentration at the Extremes at multiple geographical levels to monitor health inequities in an era of growing spatial social polarization: Massachusetts, USA (2010–14). *Int J Epidemiol* 2018;47(3):788–819. doi:10.1093/ije/dyy004; Krieger N, Feldman JM, Kim R, Waterman PD. Cancer incidence and multilevel measures of residential economic and racial segregation for cancer registries. *JNCI Cancer Spectr* 2018;2(1):pky009. doi:10.1093/jncics/pky009; Scally BJ, Krieger N, Chen JT. Racialized economic segregation and stage at diagnosis of colorectal cancer in the United States [published online ahead of print April 27, 2018]. *Cancer Causes Control* 2018;29(6):527–537. doi:10.1007/s10552-018-1027-y; Feldman JM, Gruskin S, Coull BA, Krieger N. Police-related deaths and neighborhood economic and racial/ethnic polarization, United States, 2015–2016 [published online ahead of print January 24, 2019]. *Am J Public Health* 2019;109(3):458–464. doi:10.2105/AJPH.2018.304851; Krieger N, Waterman PD, Chen JT. COVID-19 and overall mortality inequities in the surge in death rates by zip code characteristics: Massachusetts, January 1 to May 19, 2020 [published online ahead of print October 15, 2020]. *Am J Public Health* 2020;110(12):1850–1852. doi:10.2105/AJPH.2020.305913; Chen JT, Krieger N. Revealing the unequal burden of COVID-19 by income, race/ethnicity, and household crowding: us county versus zip code analyses. *J Public Health Manag Pract* 2021 Jan/Feb;27(Suppl 1):S43–S56, COVID-19 and Public Health: Looking Back, Moving Forward. doi:10.1097/PHH.0000000000001263; Chin T, Kahn R, Li R, Chen JT, Krieger N, Buckee CO, Balsari S, Kiang MV. US-county level variation in intersecting individual, household and community characteristics relevant to COVID-19 and planning an equitable response: a cross-sectional analysis. *BMJ Open* 2020;10:e039886. doi:10.1136/bmjopen-2020-039886. https://pubmed.ncbi.nlm.nih.gov/31765272/; Krieger N, Wright E, Chen JT, Waterman PD, Huntley ER, Arcaya M. Cancer stage at diagnosis, historical redlining, and current neighborhood characteristics: breast, cervical, lung, and colorectal cancers, Massachusetts, 2001–2015. *Am J Epidemiol* 2020;189(10):1065–1075. doi:10.1093/aje/kwaa045; Krieger N, Nethery RC, Chen JT, Waterman PD, Wright E, Rushovich T, Coull BA. Impact of differential privacy and census tract data source (Decennial Census versus American Community Survey) for monitoring health inequities [published online ahead of print December 22, 2020]. *Am J Public Health* 2020;e1–e4. doi:10.2105/AJPH.2020.305989.

147. Krieger N, Chen JT, Testa C, Hanage WP. The changing political geographies of COVID-19 in the US. *Harvard Center for Population and Development Studies Working Paper Series* 2020 Oct 14;20(3). https://www.hsph.harvard.edu/population-development/research/working-papers/harvard-pop-center-working-paper-series/; accessed January 2, 2021.

148. Massey D, Denton N. *American Apartheid: Segregation and the Making of the Underclass.* Cambridge, MA: Harvard University Press, 1993; Massey D, Brodmann S. *Spheres of Influence: The Social Ecology of Racial and Class Inequality.* New York: Russell Sage Foundation, 2014; Massey DS. Reflections on the dimensions of segregation. *Social Forces* 2012;91(1):39–43; Krieger N, Waterman PD, Spasojevic J, Li W, Maduro G, Van Wye G. Public health monitoring of privilege and deprivation with the Index of Concentration at the Extremes [published online ahead of print December 21, 2015]. *Am J Public Health* 2016;106(2):256–263. doi:10.2105/AJPH.2015.302955.

149. Krieger N, Waterman PD, Gryparis A, Coull BA. Black carbon exposure, socioeconomic and racial/ethnic spatial polarization, and the Index of Concentration at the Extremes (ICE) [published online ahead of print June 18, 2015]. *Health Place* 2015;34:215–228. doi:10.1016/j.healthplace.2015.05.008; Krieger N, Waterman PD, Spasojevic J, Li W, Maduro G, Van Wye G. Public health monitoring of privilege and deprivation with the Index of Concentration at the Extremes [published online ahead of print December 21, 2015]. *Am J Public Health* 2016;106(2):256–263. doi:10.2105/AJPH.2015.302955; Krieger N, Chen JT, Waterman PD. Using the methods of the Public Health Disparities Geocoding Project to monitor COVID-19 inequities and guide action for health justice. May 15, 2020. https://www.hsph.harvard.edu/thegeocodingproject/covid-19-resources/; accessed January 2, 2021.

150. Krieger N, Waterman PD, Gryparis A, Coull BA. Black carbon exposure, socioeconomic and racial/ethnic spatial polarization, and the Index of Concentration at the Extremes (ICE) [published online ahead of print June 18, 2015]. *Health Place* 2015;34:215–228. doi:10.1016/j.healthplace.2015.05.008.

151. *Study that introduced the ICE for racialized economic segregation and for racial segregation:* Krieger N, Waterman PD, Gryparis A, Coull BA. Black carbon exposure, socioeconomic and racial/ethnic spatial polarization, and the Index of Concentration at the Extremes (ICE) [published online ahead

of print June 18, 2015]. *Health Place* 2015;34:215–228. doi:10.1016/j.healthplace.2015.05.008; *Subsequent studies by Krieger et al. using the ICE, including in multilevel analyses (in chronological order):* Feldman JM, Waterman PD, Coull BA, Krieger N. Spatial social polarisation: using the Index of Concentration at the Extremes jointly for income and race/ethnicity to analyse risk of hypertension [published online ahead of print July 1, 2015]. *J Epidemiol Community Health* 2015;69(12):1199–1207. doi:10.1136/jech-2015-205728; Krieger N, Waterman PD, Spasojevic J, Li W, Maduro G, Van Wye G. Public health monitoring of privilege and deprivation with the Index of Concentration at the Extremes [published online ahead of print December 21, 2015]. *Am J Public Health* 2016;106(2):256–263. doi:10.2105/AJPH.2015.302955; Krieger N, Singh N, Waterman PD. Metrics for monitoring cancer inequities: residential segregation, the Index of Concentration at the Extremes (ICE), and breast cancer estrogen receptor status (USA, 1992–2012) [published online ahead of print August 8, 2016]. *Cancer Causes Control* 2016;27(9):1139–1151. doi:10.1007/s10552-016-0793-7; Krieger N, Feldman JM, Waterman PD, Chen JT, Coull BA, Hemenway D. Local residential segregation matters: stronger association of census tract compared to conventional city-level measures with fatal and non-fatal assaults (total and firearm related), using the index of concentration at the extremes (ICE) for racial, economic, and racialized economic segregation, Massachusetts (US), 1995–2010. *J Urban Health* 2017;94(2):244–258. doi:10.1007/s11524-016-0116-z; Krieger N, Waterman PD, Batra N, Murphy JS, Dooley DP, Shah SN. Measures of local segregation for monitoring health inequities by local health departments [published online ahead of print April 20, 2017]. *Am J Public Health* 2017;107(6):903–906. doi:10.2105/AJPH.2017.303713; Huynh M, Spasojevic J, Li W, Maduro G, Van Wye G, Waterman PD, Krieger N. Spatial social polarization and birth outcomes: preterm birth and infant mortality—New York City, 2010–14 [published online ahead of print April 6, 2017]. *Scand J Public Health* 2018;46(1):157–166. doi:10.1177/1403494817701566; Krieger N, Kim R, Feldman J, Waterman PD. Using the Index of Concentration at the Extremes at multiple geographical levels to monitor health inequities in an era of growing spatial social polarization: Massachusetts, USA (2010–14). *Int J Epidemiol* 2018;47(3):788–819. doi:10.1093/ije/dyy004; Krieger N, Feldman JM, Kim R, Waterman PD. Cancer incidence and multilevel measures of residential economic and racial segregation for cancer registries. *JNCI Cancer Spectr* 2018;2(1):pky009. doi:10.1093/jncics/pky009; Scally BJ, Krieger N, Chen JT. Racialized economic segregation and stage at diagnosis of colorectal cancer in the United States [published online ahead of print April 27, 2018]. *Cancer Causes Control* 2018;29(6):527–537. doi:10.1007/s10552-018-1027-y; Feldman JM, Gruskin S, Coull BA, Krieger N. Police-related deaths and neighborhood economic and racial/ethnic polarization, United States, 2015–2016 [published online ahead of print January 24, 2019]. *Am J Public Health* 2019;109(3):458–464. doi:10.2105/AJPH.2018.304851; Krieger N, Waterman PD, Chen JT. COVID-19 and overall mortality inequities in the surge in death rates by zip code characteristics: Massachusetts, January 1 to May 19, 2020 [published online ahead of print October 15, 2020]. *Am J Public Health* 2020;110(12):1850–1852. doi:10.2105/AJPH.2020.305913; Chen JT, Krieger N. Revealing the unequal burden of COVID-19 by income, race/ethnicity, and household crowding: us county versus zip code analyses. *J Public Health Manag Pract* 2021 Jan/Feb;27(Suppl 1):S43–S56, COVID-19 and Public Health: Looking Back, Moving Forward. doi:10.1097/PHH.0000000000001263; Chin T, Kahn R, Li R, Chen JT, Krieger N, Buckee CO, Balsari S, Kiang MV. US-county level variation in intersecting individual, household and community characteristics relevant to COVID-19 and planning an equitable response: a cross-sectional analysis. *BMJ Open* 2020;10:e039886. doi:10.1136/bmjopen-2020-039886. https://pubmed.ncbi.nlm.nih.gov/31765272/; Krieger N, Wright E, Chen JT, Waterman PD, Huntley ER, Arcaya M. Cancer stage at diagnosis, historical redlining, and current neighborhood characteristics: breast, cervical, lung, and colorectal cancers, Massachusetts, 2001–2015. *Am J Epidemiol* 2020;189(10):1065–1075. doi:10.1093/aje/kwaa045; Krieger N, Nethery RC, Chen JT, Waterman PD, Wright E, Rushovich T, Coull BA. Impact of differential privacy and census tract data source (Decennial Census versus American Community Survey) for monitoring health inequities [published online ahead of print December 22, 2020]. *Am J Public Health* 2020;e1–e4. doi:10.2105/AJPH.2020.305989.

152. Krieger N, Feldman JM, Waterman PD, Chen JT, Coull BA, Hemenway D. Local residential segregation matters: stronger association of census tract compared to conventional city-level measures with fatal and non-fatal assaults (total and firearm related), using the index of concentration at the extremes (ICE) for racial, economic, and racialized economic segregation, Massachusetts (US), 1995–2010. *J Urban Health* 2017;94(2):244–258. doi:10.1007/s11524-016-0116-z; Krieger N, Kim R, Feldman J, Waterman PD. Using the Index of Concentration at the Extremes at multiple geographical levels to monitor health inequities in an era of growing spatial social polarization: Massachusetts, USA (2010–14). *Int J Epidemiol* 2018;47(3):788–819. doi:10.1093/ije/dyy004.
153. Krieger N, Feldman JM, Waterman PD, Chen JT, Coull BA, Hemenway D. Local residential segregation matters: stronger association of census tract compared to conventional city-level measures with fatal and non-fatal assaults (total and firearm related), using the index of concentration at the extremes (ICE) for racial, economic, and racialized economic segregation, Massachusetts (US), 1995–2010. *J Urban Health* 2017;94(2):244–258. doi:10.1007/s11524-016-0116-z; Krieger

N, Kim R, Feldman J, Waterman PD. Using the Index of Concentration at the Extremes at multiple geographical levels to monitor health inequities in an era of growing spatial social polarization: Massachusetts, USA (2010–14). *Int J Epidemiol* 2018;47(3):788–819. doi:10.1093/ije/dyy004.

154. *Examples of other studies that have used our ICE measures (as of September 21, 2020, in alphabetical order):* Bruzzese JM, Kingston S, Falletta KA, Bruzelius E, Poghosyan L. Individual and neighborhood factors associated with undiagnosed asthma in a large cohort of urban adolescents. *J Urban Health* 2019;96(2):252–261. doi:10.1007/s11524-018-00340-2; Chambers BD, Arabia SE, Arega HA, Altman MR, Berkowitz R, Feuer SK, Franck LS, Gomez AM, Kober K, Pacheco-Werner T, Paynter RA, Prather AA, Spellen SA, Stanley D, Jelliffe-Pawlowski LL, McLemore MR. Exposures to structural racism and racial discrimination among pregnant and early post-partum Black women living in Oakland, California [published online ahead of print January 23, 2020]. *Stress Health* 2020;36(2):213–219. doi:10.1002/smi.2922; Chambers BD, Baer RJ, McLemore MR, Jelliffe-Pawlowski LL. Using Index of Concentration at the Extremes as indicators of structural racism to evaluate the association with preterm birth and infant mortality-California, 2011–2012. *J Urban Health* 2019;96(2):159–170. doi:10.1007/s11524-018-0272-4; De Maio F, Ansell D. "As natural as the air around us": on the origin and development of the concept of structural violence in health research [published online ahead of print August 9, 2018]. *Int J Health Serv* 2018;48(4):749–759. doi:10.1177/0020731418792825; Feldman JM, Conderino S, Islam NS, Thorpe LE. Subgroup variation and neighborhood social gradients-an analysis of hypertension and diabetes among Asian Patients (New York City, 2014–2017) [published online ahead of print June 2, 2020]. *J Racial Ethn Health Disparities*. doi:10.1007/s40615-020-00779-7; Fong KC, Yitshak-Sade M, Lane KJ, Fabian MP, Kloog I, Schwartz JD, Coull BA, Koutrakis P, Hart JE, Laden F, Zanobetti A. Racial disparities in associations between neighborhood demographic polarization and birth weight. *Int J Environ Res Public Health* 2020;17(9):3076. doi:10.3390/ijerph17093076; Gourevitch MN, Athens JK, Levine SE, Kleiman N, Thorpe LE. City-level measures of health, health determinants, and equity to foster population health improvement: the City Health Dashboard [published online ahead of print February 21, 2019]. *Am J Public Health* 2019;109(4):585–592. doi:10.2105/AJPH.2018.304903; Ish J, Symanski E, Whitworth KW. Exploring disparities in maternal residential proximity to unconventional gas development in the Barnett Shale in North Texas. *Int J Environ Res Public Health* 2019 Jan 23;16(3):298. doi:10.3390/ijerph16030298; Janevic T, Zeitlin J, Egorova N, Hebert PL, Balbierz A, Howell EA. Neighborhood racial and economic polarization, hospital of delivery, and severe maternal morbidity. *Health Aff (Millwood)* 2020;39(5):768–776. doi:10.1377/hlthaff.2019.00735; Kandasamy V, Hirai AH, Kaufman

JS, James AR, Kotelchuck M. Regional variation in Black infant mortality: the contribution of contextual factors. *PLoS One* 2020;15(8):e0237314. doi:10.1371/journal.pone.0237314. PMID:32780762; Lange-Maia BS, De Maio F, Avery EF, Lynch EB, Laflamme EM, Ansell DA, Shah RC. Association of community-level inequities and premature mortality: Chicago, 2011–2015 [published online ahead of print August 31, 2018]. *J Epidemiol Community Health* 2018;72(12):1099–1103. doi:10.1136/jech-2018-210916; Barrozo LG, Desigualdades na mortalidade infantil no Município de São Paulo: em busca do melhor indicador. *Confins* 2018 3 Oct. https://doi-org.ezp-prod1.hul.harvard.edu/10.4000/confins.15010; Linton SL, Cooper HLF, Chen YT, Khan MA, Wolfe ME, Ross Z, Des Jarlais DC, Friedman SR, Tempalski B, Broz D, Semaan S, Wejnert C, Paz-Bailey G. Mortgage discrimination and racial/ethnic concentration are associated with same-race/ethnicity partnering among people who inject drugs in 19 US Cities. *J Urban Health* 2020;97(1):88–104. doi:10.1007/s11524-019-00405-w; Nesoff ED, Branas CC, Martins SS. The geographic distribution of fentanyl-involved overdose deaths in Cook County, Illinois [published online ahead of print November 14, 2019]. *Am J Public Health* 2020;110(1):98–105. doi:10.2105/AJPH.2019.305368; Palakshappa D, Lenoir K, Brown CL, Skelton JA, Block JP, Taveras EM, Lewis KH. Identifying geographic differences in children's sugar-sweetened beverage and 100% fruit juice intake using health system data [published online ahead of print June 17, 2020]. *Pediatr Obes* 2020;15(11):e12663. doi:10.1111/ijpo.12663; Shrimali BP, Pearl M, Karasek D, Reid C, Abrams B, Mujahid M. Neighborhood privilege, preterm delivery, and related racial/ethnic disparities: an intergenerational application of the Index of Concentration at the Extremes. *Am J Epidemiol* 2020;189(5):412–421. doi:10.1093/aje/kwz279; Shumate C, Hoyt A, Liu C, Kleinert A, Canfield M. Understanding how the concentration of neighborhood advantage and disadvantage affects spina bifida risk among births to non-Hispanic white and Hispanic women, Texas, 1999–2014 [published online ahead of print September 10, 2018]. *Birth Defects Res* 2019;111(14):982–990. doi:10.1002/bdr2.1374; Tumin D, Horan J, Shrider EA, Smith SA, Tobias JD, Hayes D Jr, Foraker RE. County socioeconomic characteristics and heart transplant outcomes in the United States [published online ahead of print June 3, 2017]. *Am Heart J* 2017;190:104–112. doi:10.1016/j.ahj.2017.05.013; Wallace ME, Crear-Perry J, Green C, Felker-Kantor E, Theall K. Privilege and deprivation in Detroit: infant mortality and the Index of Concentration at the Extremes. *Int J Epidemiol* 2019;48(1):207–216. doi:10.1093/ije/dyy149; Ward JB, Albrecht SS, Robinson WR, Pence BW, Maselko J, Haan MN, Aiello AE. Neighborhood language isolation and depressive symptoms among elderly U.S. Latinos [published online ahead of print August 24, 2018]. *Ann Epidemiol* 2018;28(11):774–782. doi:10.1016/j.annepidem.2018.08.009; Westrick AC, Bailey ZD, Schlumbrecht

M, Hlaing WM, Kobetz EE, Feaster DJ, Balise RR. Residential segregation and overall survival of women with epithelial ovarian cancer [published online ahead of print June 2, 2020]. *Cancer* 2020;126(16):3698–3707. doi:10.1002/cncr.32989; Wiese D, Stroup AM, Crosbie A, Lynch SM, Henry KA. The impact of neighborhood economic and racial inequalities on the spatial variation of breast cancer survival in New Jersey [published online ahead of print October 24, 2019]. *Cancer Epidemiol Biomarkers Prev* 2019;28(12):1958–1967. doi:10.1158/1055-9965.EPI-19-0416.

155. Rothstein R. *The Color of Law: A Forgotten History of How Our Government Segregated America*. New York: Liveright, 2017; Metzger MW, Webber HS (eds). *Facing Segregation: Housing Policy Solutions for a Stronger Society*. New York: Oxford University Press, 2018; Ellen IG, Steil JP (eds). *The Dream Revisited: Contemporary Debates About Housing, Segregation, and Opportunity in the Twenty-First Century*. New York: Columbia University Press, 2019; Nelson RK, Winling L, Marciano R . Mapping inequality. In: Nelson RK, Ayers EL (eds). *American Panorama*. https://dsl.richmond.edu/panorama/redlining/#loc=4/36.71/-96.93; accessed January 2, 2021; Massey D, Denton NA. *American Apartheid: Segregation and the Making of the Underclass*. Cambridge, MA: Harvard University Press, 1993.
156. Rothstein R. *The Color of Law: A Forgotten History of How Our Government Segregated America*. New York: Liveright, 2017; Metzger MW, Webber HS (eds). *Facing Segregation: Housing Policy Solutions for a Stronger Society*. New York: Oxford University Press, 2018; Ellen IG, Steil JP (eds). *The Dream Revisited: Contemporary Debates About Housing, Segregation, and Opportunity in the Twenty-First Century*. New York: Columbia University Press, 2019; Nelson RK, Winling L, Marciano R . Mapping inequality. In: Nelson RK, Ayers EL (eds). *American Panorama*. https://dsl.richmond.edu/panorama/redlining/#loc=4/36.71/-96.93; accessed January 2, 2021; Massey D, Denton NA. *American Apartheid: Segregation and the Making of the Underclass*. Cambridge, MA: Harvard University Press, 1993.
157. Rothstein R. *The Color of Law: A Forgotten History of How Our Government Segregated America*. New York: Liveright, 2017; Metzger MW, Webber HS (eds). *Facing Segregation: Housing Policy Solutions for a Stronger Society*. New York: Oxford University Press, 2018; Ellen IG, Steil JP (eds). *The Dream Revisited: Contemporary Debates About Housing, Segregation, and Opportunity in the Twenty-First Century*. New York: Columbia University Press, 2019; Nelson RK, Winling L, Marciano R . Mapping inequality: Introduction. In: Nelson RK, Ayers EL (eds). *American Panorama*. https://dsl.richmond.edu/panorama/redlining/#loc=5/39.1/-94.58&text=intro; accessed January 2, 2021; Massey D, Denton NA. *American Apartheid: Segregation and the Making of the Underclass*. Cambridge, MA: Harvard University Press, 1993.

158. Nelson RK, Winling L, Marciano R . Mapping inequality: Introduction. In: Nelson RK, Ayers EL (eds). *American Panorama*. https://dsl.richmond.edu/panorama/redlining/#loc=5/39.1/-94.58&text=intro; accessed January 2, 2021.
159. Krieger N, Van Wye G, Huynh M, Waterman PD, Maduro G, Li W, Gwynn C, Barbot O, Bassett MT. Historical redlining, structural racism, and preterm birth risk in New York City (2013–2017). *Am J Public Health* 2020;110(7):1046–1053.
160. Krieger N, Wright E, Chen JT, Waterman PD, Huntley ER, Arcaya M. Cancer stage at diagnosis, historical redlining, and current neighborhood characteristics: breast, cervical, lung, and colorectal cancer, Massachusetts, 2001–2015 [published online ahead of print March 27, 2020]. *Am J Epidemiol* 2020 Oct 1;189(10):1065-1057. doi:10.1093/aje/kwaa045.
161. Huggins JC. A cartographic perspective on the correlation between redlining and public health in Austin, Texas—1951. *Cityscape (Wash, DC)* 2017;19(2):267–280; Jacoby SF, Dong B, Beard JH, Wiebe DJ, Morrison CN. The enduring impact of historical and structural racism on urban violence in Philadelphia. *Soc Sci Med* 2018;199:87–95; McClure E, Feinstein L, Cordoba E, Douglas C, Emch M, Robinson W, Galea S, Aiello AE. The legacy of redlining in the effect of foreclosures on Detroit residents' self-rated health. *Health Place* 2019;55:9–19; Trangenstein PJ, Gray C, Rossheim ME, Sadler R, Jernigan DL. Alcohol outlet clusters and population disparities. *J Urban Health* 2020;97(1):123–136; Namin S, Xu W, Zhou Y, Beyer K. The legacy of the Home Owners' Loan Corporation and the political ecology of urban trees and air pollution in the United States [published online ahead of print December 20, 2019]. *Soc Sci Med* 2020;246:112758. doi:10.1016/j.socscimed.2019.112758; Benns M, Ruther M, Nash N, Bozeman M, Harbrecht B, Miller K. The impact of historical racism on modern gun violence: redlining in the city of Louisville, KY [published online ahead of print June 25, 2020]. *Injury* 2020;51(10):2192–2198. doi:10.1016/j.injury.2020.06.042; Nardone AL, Casey JA, Rudolph KE, Karasek D, Mujahid M, Morello-Frosch R. Associations between historical redlining and birth outcomes from 2006 through 2015 in California. *PLoS One* 2020;15(8):e0237241. doi:10.1371/journal.pone.0237241; Nardone A, Casey JA, Morello-Frosch R, Mujahid M, Balmes JR, Thakur N. Associations between historical residential redlining and current age-adjusted rates of emergency department visits due to asthma across eight cities in California: an ecological study. *Lancet Planet Health* 2020;4(1):e24–e31. doi:10.1016/S2542-5196(19)30241-4; Lee JP, Ponicki W, Mair C, Gruenewald P, Ghanem L. What explains the concentration of off-premise alcohol outlets in Black neighborhoods? *SSM Popul Health* 2020;12:100669. doi:10.1016/j.ssmph.2020.100669; Bertocchi G, Dimico A. COVID-19, race, and redlining. *medRxiv*, July 20, 2020. https://doi.org/10.1101/2020.07/11.20148486.

162. University of Richmond Digital Scholarship Lab and the National Community Reinvestment Coalition. Not Even Past: Social Vulnerability and the Legacy of Redlining. Latest Maps, 2020. https://dsl.richmond.edu/socialvulnerability/; accessed January 2, 2021.
163. National Community Reinvestment Coalition. https://www.ncrc.org/; accessed January 2, 2021.
164. National Community Reinvestment Coalition. About. https://www.ncrc.org/about/; accessed January 2, 2021.
165. McClure K. What should be the future of the low-income housing tax credit program? *Hous Policy Debate* 2019;29(1):65–81. doi:10.1080/10511482.2018.1469526; *Texas Dept. of Housing and Community Affairs v. Inclusive Communities Project, Inc.* (SUPREME COURT OF THE UNITED STATES 2015).
166. Tegeler P. Affirmatively furthering fair housing and the inclusive communities project case: bringing the fair housing act into the twenty-first century. In: Metzger MW, Webber HS (eds). *Facing Segregation: Housing Policy Solutions for a Stronger Society*. New York: Oxford University Press, 2018; 77–91; Office of the Comptroller of the Currency. Reforming the Community Reinvestment Act Regulatory Framework. *Fed Regist* 2018;83:FR 45053(Docket ID OCC-2018-0008):45053-45059.
167. Sard B, Tegeler P. Children and housing vouchers. In: Ellen IG, Steil JP (eds). *The Dream Revisited: Contemporary Debates About Housing, Segregation, and Opportunity*. New York: Columbia University Press, 2019; 298–303; US Department of Housing and Urban Development. Establishing a more effective fair market rent system using small area fair market rents in the housing choice voucher program instead of the current 50th percentile FMRs. *Fed Regist* 2016;81(221):80567–80587; *Open Communities Alliance v. Carson* (D.D.C. 2017).
168. Steil J, Kelly N. The fairest of them all: analyzing affirmatively furthering fair housing compliance. *Hous Policy Debate* 2019;29(1):85–105. doi:10.1080/10511482.2018.1469527; Steil J. Antisubordination planning. *J Plan Educ Res* 2018 Dec;1–10; Reconsideration of HUD's implementation of the Fair Housing Act's Disparate Impact Standard. *Fed Regist* 2018;83:28560–28562; Affirmatively Furthering Fair Housing Rule. *Fed Regist* 2015;80:42271–42371.
169. City of Boston Planning and Development Agency. *PLAN: JP/Rox*, Boston, MA, 2018; 51–52. http://www.bostonplans.org/planning/planning-initiatives/plan-jp-rox; accessed January 2, 2021; City of Somerville Office of Housing Stability. *Tenant Selection Plan for the 100 Homes Program*. Somerville, MA; 2018:2–3; City of Portland Housing Bureau. *North/Northeast Housing Strategy Preference Policy*. 2015. https://www.portlandoregon.gov/phb/article/

671059#what-is-the-preference-policy; accessed January 2, 2021; City of Seattle. *Executive Order 02/2019*. 2019. https://www.capitolhillseattle.com/2019/08/community-preference-a-new-anti-displacement-policy-could-have-big-impacts-for-the-central-district-and-capitol-hill/; accessed January 2, 2021.

170. Department of Housing and Urban Development [Docket No. FR–5173–N–07]. Affirmatively furthering fair housing assessment tool: announcement of final approved document. *Fed Regist* 2015;80(251):81840–81856. https://www.govinfo.gov/content/pkg/FR-2015-12-31/pdf/2015-32680.pdf; accessed January 2, 2021.
171. National Low Income Housing Coalition. HUD indefinitely suspends AFFH rule, withdraws assessment tool. May 21, 2018. https://nlihc.org/resource/hud-indefinitely-suspends-affh-rule-withdraws-assessment-tool; accessed January 2, 2021; US Department of Housing and Urban Development (HUD). Secretary Carson terminates 2015 AFFH rule. HUD No. 20-19. July 23, 2020. https://www.hud.gov/press/press_releases_media_advisories/HUD_No_20_109; accessed January 2, 2021; Logan T. Trump vows to save suburbs. Some ask, from what? *Boston Globe*, August 18, 2020, A1, A7. https://www.bostonglobe.com/2020/08/17/nation/trump-says-biden-would-destroy-suburbs-what-is-he-talking-about/; accessed January 2, 2021.
172. Department of Housing and Urban Development [Docket No. FR–5173–N–07]. Affirmatively furthering fair housing assessment tool: announcement of final approved document. *Fed Regist* 2015;80(251):81840–81856. https://www.govinfo.gov/content/pkg/FR-2015-12-31/pdf/2015-32680.pdf; accessed January 2, 2021.
173. Affirmatively Furthering Fair Housing Rule. *Fed Regist* 2015;80:42271–42371.
174. *24 CFR § 5.162(d).*
175. Affirmatively furthering fair housing: withdrawal of the assessment tool for local governments. *Fed Regist* 2018;83:23922–23927.
176. Steil J, Kelly N. The fairest of them all: analyzing affirmatively furthering fair housing compliance. *Hous Policy Debate* 2019;29(1):85–105. doi:10.1080/10511482.2018.1469527; Affirmatively furthering fair housing: streamlining and enhancements. *Fed Regist* 2018;83:40713–40715.
177. The Biden Plan for Investing in Our Communities Through Housing. https://joebiden.com/housing/; accessed January 2, 2021; Birenbaum G. Democrats say affordable housing would be a top priority in the Biden administration. *The Hill*, August 18, 2020. https://thehill.com/policy/finance/512599-democrats-say-affordable-housing-would-be-a-top-priority-in-a-biden; accessed January 2, 2021.
178. National Fair Housing Alliance Issues Policy Roadmap for Biden Administration and 117th Congress. https://nationalfairhousing.org/2020/12/18/national-fair-housing-alliance-issues-policy-roadmap-for-biden-administration-and-117th-congress/; accessed January 2, 2021.

179. Fleetwood S. How Biden's housing plan would destroy America's suburbs. *The Federalist*, August 13, 2020. https://thefederalist.com/2020/08/13/how-bidens-housing-plans-would-destroy-americas-suburbs/; accessed January 2, 2020; Creitz C. Stanley Kurtz: Obama-era "Affirmatively Furthering Fair Housing Rule" an attack on suburbs. *Fox News*, July 26, 2020. https://www.foxnews.com/media/stanley-kurtz-obama-era-affirmatively-furthering-fair-housing-rule-attack-on-suburbs; accessed January 2, 2020.
180. Shiman LJ, Freeman K, Bedell J, Bassett MT. Making injustice visible: how a health department can demonstrate the connection between structural racism and the health of whole neighborhoods [published online ahead of print September 9, 2020]. *J Public Health Manag Pract*. doi:10.1097/PHH.0000000000001259.
181. For a continually updated listing of my PubMed publications and also other publications (including books), see https://www.hsph.harvard.edu/nancy-krieger/.
182. Davey Smith G. The end of the beginning for chronic disease epidemiology. *Int J Epidemiol* 2010;39(1):1–3; Davey Smith G. The uses of "Uses of Epidemiology." *Int J Epidemiol* 2001;30:1146–1155; Krieger N. Commentary: Ways of asking and ways of living: reflections on the 50th anniversary of Morris' ever-useful *Uses of Epidemiology*. *Int J Epidemiol* 2010;36:1173–1180; Watts G. Jeremy Noah Morris. *Lancet* 2010;375(9712):370. DOI:https://doi.org/10.1016/S0140-6736(10)60157-2; Oakley A. Appreciation: Jerry [Jeremiah Noah] Morris, 1910–2009. *Int J Epidemiol* 2010;39(1):274–276.
183. Morris JN. *Uses of Epidemiology*. Edinburgh: E. & S. Livingston Ltd., 1957.
184. Morris JN. *Uses of Epidemiology*. Edinburgh: E. & S. Livingston Ltd., 1957; 3.
185. Morris JN. *Uses of Epidemiology*. Edinburgh: E. & S. Livingston Ltd., 1957; 96.

CHAPTER 3

1. Krieger N. ENOUGH: COVID-19, structural racism, police brutality, plutocracy, climate change-and time for health justice, democratic governance, and an equitable, sustainable future [published online ahead of print August 20, 2020]. *Am J Public Health* 2020; 110(11):1620–1623. doi:10.2105/AJPH.2020.305886; Roy A. The pandemic is a portal. *Financial Times*, April 3, 2020. https://www.ft.com/content/10d8f5e8-74eb-11ea-95fe-fcd274e920ca; accessed January 3, 2021; Pilkington E. As 100,000 die, the virus lays bare America's brutal fault lines—race, gender, poverty, and broken politics. *The Guardian*, May 28, 2020. https://www.theguardian.com/us-news/2020/may/28/us-coronavirus-death-toll-racial-disparity-inequality; accessed January 3, 2021; Freedland J. The magnifying glass: how Covid revealed the truth about our world. *The Guardian*,

December 11, 2020. https://www.theguardian.com/world/2020/dec/11/covid-upturned-planet-freedland; accessed February 14, 2021; Wallace R. *Dead Epidemiologists: On the Origins of COVID-19*. New York: Monthly Review Press, 2020; Diez Roux AV. Population health in the time of COVID-19: confirmations and revelations [published online ahead of print August 18, 2020]. *Milbank Q* 2020; 98(3):629–640. doi:10.1111/1468-0009.12474; Rohland E. COVID-19, climate, and white supremacy: multiple crises or one? *J Hist Env Society* 2020; 5:23–32; Bailey ZD, Moon JR. Racism and the political economy of COVID-19: will we continue to resurrect the past? *J Health Polit Policy Law* 2020;45(6):937–950. doi:10.1215/03616878-8641481; Gravlee CC. Systemic racism, chronic health inequities, and COVID-19: a syndemic in the making? [published online ahead of print August 4, 2020]. *Am J Hum Biol* 2020;32(5):e23482. doi:10.1002/ajhb.23482; Abbasi J. Taking a closer look at COVID-19, health inequities, and racism. *JAMA* 2020;324(5):427–429. doi:10.1001/jama.2020.11672; Oliveira RG, Cunha APD, Gadelha AGDS, Carpio CG, Oliveira RB, Corrêa RM. Racial inequalities and death on the horizon: COVID-19 and structural racism. *Cad Saude Publica* 2020;36(9):e00150120. doi:10.1590/0102-311X00150120; Hardeman RR, Medina EM, Boyd RW. Stolen breaths [published online ahead of print June 10, 2020]. *N Engl J Med* 2020;383(3):197–199. doi:10.1056/NEJMp2021072; Egede LE, Walker RJ. Structural racism, social risk factors, and Covid-19—a dangerous convergence for black Americans [published online ahead of print July 22, 2020]. *N Engl J Med* 2020;383(12):e77. doi:10.1056/NEJMp2023616; Bambra C, Riordan R, Ford J, Matthews F. The COVID-19 pandemic and health inequalities [published online ahead of print June 13, 2020]. *J Epidemiol Community Health* 2020;74(11):964–968. doi:10.1136/jech-2020-214401; Minkler M, Griffin J, Wakimoto P. Seizing the moment: policy advocacy to end mass incarceration in the time of COVID-19 [published online ahead of print June 9, 2020]. *Health Educ Behav* 2020;47(4):514–518. doi:10.1177/1090198120933281.

2. Baker P. More than ever, Trump casts himself as the defender of white America. *New York Times*, September 6, 2020. https://www.nytimes.com/2020/09/06/us/politics/trump-race-2020-election.html; accessed September 7, 2020; Schwartz MS. Trump tells agencies to end trainings on "white privilege" and "critical race theory." *NPR*, September 5, 2020. https://www.npr.org/2020/09/05/910053496/trump-tells-agencies-to-end-trainings-on-white-privilege-and-critical-race-theor; accessed September 7, 2020; Beggin R. Trump orders federal anti-racism training to be ended, calling it a "sickness." *Vox*, September 5, 2020. https://www.vox.com/2020/9/5/21423969/trump-federal-antiracism-training-critical-race-theory-white-privilege; accessed September 7, 2020; Pietsch B. Here's what we know about what happened in Portland: a man was shot and killed

after supporters of President Trump clashed with counterprotesters. *New York Times*, August 30, 2020; updated September 4, 2020. https://www.nytimes.com/2020/08/30/us/portland-shooting-explained.html; accessed September 7, 2020; Seymour R. How did the US's mainstream right end up openly supporting vigilante terror? *The Guardian*, September 1, 2020. https://www.theguardian.com/commentisfree/2020/sep/01/us-mainstream-right-vigilante-terror; accessed September 7, 2020; Levin S. White supremacists and militias have infiltrated police across the US, report says. *The Guardian*, August 27, 2020. https://www.theguardian.com/us-news/2020/aug/27/white-supremacists-militias-infiltrate-us-police-report; accessed September 7, 2020; Holpuch A. "White supremacy won today": critics condemn Trump acquittal as racist vote. *The Guardian*, February 14, 2021. https://www.theguardian.com/us-news/2021/feb/14/trump-acquittal-white-supremacy-racist-vote; accessed February 14, 2021.

3. Krieger N. ENOUGH: COVID-19, structural racism, police brutality, plutocracy, climate change-and time for health justice, democratic governance, and an equitable, sustainable future [published online ahead of print August 20, 2020]. *Am J Public Health* 2020;110(11):1620–1623. doi:10.2105/AJPH.2020.305886. *Searches for articles on racism, health, and COVID-19 for 2020 in major media:* Dow Jones Factiva (media search), for "racism AND health AND covid": *N* = 31,480 (for publications only, nonduplicates); *N* = 10,647 (for web news only, nonduplicates); https://global-factiva-com.ezp-prod1.hul.harvard.edu/ha/default.aspx#./!?&_suid=16096949816740848669570020864 6; accessed January 3, 2021; Nexis Uni (news search), for "racism AND health AND covid": *N* = 5,797; https://advance-lexis-com.ezp-prod1.hul.harvard.edu/search/?pdmfid=1516831&crid=be7e9c87-992c-4d7d-95ee-0cb55a1dbe5b&pdpsf=&pdpost=&pdstartin=&pdsearchterms=racism+and+health+and+covid&pdsearchtype=SearchBox&pdtypeofsearch=searchboxclick&pdsf=&pdquerytemplateid=&pdtimeline=01%2F01%2F2020%7Cdateafter&pdfromadvancedsearchpage=true&ecomp=pyLg9kk&earg=pdpsf&prid=32732b33-6e9f-42df-b5f4-93651a26589a; accessed January 3, 2021; Ethnic News Watch, for "racism AND health AND covid": *N* = 977 results; https://search-proquest-com.ezp-prod1.hul.harvard.edu/ethnicnewswatch/results/D258EA10FF64AECPQ/1?accountid=11311; accessed January 3, 2021; *New York Times*, for "racism health covid": *N* = 516 results; https://www.nytimes.com/search?dropmab=false&endDate=20210103&query=racism%20health%20covid&sort=best&startDate=20200103; accessed January 3, 2021; *Washington Post*, for "racism health covid": *N* = 591 results; https://www.washingtonpost.com/newssearch/?query=racism%20health%20covid&btn-search=&sort=Relevance&datefilter=All%20Since%202005; accessed January 3, 2021.

4. Krieger N. ENOUGH: COVID-19, structural racism, police brutality, plutocracy, climate change-and time for health justice, democratic governance,

and an equitable, sustainable future [published online ahead of print August 20, 2020]. *Am J Public Health* 2020;110(11):1620–1623. doi:10.2105/AJPH.2020.305886; Chavez N. 2020: the year America confronted racism. *CNN*, December 2020. https://www.cnn.com/interactive/2020/12/us/america-racism-2020/; accessed January 3, 2021; Fannin M. The truth in Black and White: an apology from The Kansas City Star. *KC Star*, December 20, 2020. https://www.kansascity.com/news/local/article247928045.html; accessed January 3, 2021; Editorial: An examination of The Times' failures on race, our apology and a path forward. *LA Times*, September 27, 2020. https://www.latimes.com/opinion/story/2020-09-27/los-angeles-times-apology-racism; accessed January 3, 2021; Associated Press. More churches are committing to racism-linked reparations. *US News*, December 13, 2020. https://www.usnews.com/news/us/articles/2020-12-13/more-us-churches-are-committing-to-racism-linked-reparations; accessed January 3, 2021; Brunt C. Athletes act: stars rise up against racial injustice in 2020. *AP Press, Atlantic Broadband*, December 30, 2020. https://myworld.atlanticbb.com/news/read/category/news/article/the_associated_press-athletes_act_stars_rise_up_against_racial_injustic-ap; accessed January 3, 2021; Weiser S. What Big Business said in all those anti-racism statements: not much, says our analysis. *Colorlines*, October 8, 2020. https://www.colorlines.com/articles/what-big-business-said-all-those-anti-racism-statements-not-much-says-our-analysis; accessed January 3, 2021; Bartlett T. Mistakes were made: reviewing a curious year of contrition in higher education. *Chronicle Higher Education*, December 7, 2020. https://www.chronicle.com/article/mistakes-were-made; accessed January 3, 2021; Dow Jones Factiva (media search), for "racism AND apology" for 1/1/20–12/31/20: N = 7,719 (for publications only, nonduplicates); N = 1,618 (for web news only, nonduplicates); https://global-factiva-com.ezp-prod1.hul.harvard.edu/ha/default.aspx#./!?&_suid=16096974931970505329366466228; accessed January 3, 2021; Nexis Uni (news search), for "racism AND apology" for 1/1/20–12/31/20": N = 7,410; https://advance-lexis-com.ezp-prod1.hul.harvard.edu/search/?pdmfid=1516831&crid=f8f38b6d-971b-4091-9500-bfd17e6c739a&pdsearchterms=racism+and+apology&pdstartin=hlct%3A1%3A1&pdtypeofsearch=searchboxclick&pdsearchtype=SearchBox&pdqttype=and&undefined=&pdquerytemplateid=&ecomp=_z-2k&prid=be7e9c87-992c-4d7d-95ee-0cb55a1dbe5b; accessed January 3, 2021; Nexis Uni (news search), for "racism AND apology AND government," for 1/1/20–12/31/20: N = 2, 433; https://advance-lexis-com.ezp-prod1.hul.harvard.edu/search/?pdmfid=1516831&crid=166d8a07-0c4f-48b6-85f2-29799b70642c&pdsearchterms=racism+and+apology+and+government&pdstartin=hlct%3A1%3A1&pdtypeofsearch=searchboxclick&pdsearchtype=SearchBox&pdqttype=and&undefined=&pdquerytemplateid=&ecomp=_z-2k&prid=3623638b-8801-4266-916b-6cca22cc3a87; accessed January 3, 2021; Nexis Uni (news search), for "racism AND apology AND university," for 1/1/20–12/31/20; *N* = 1,596; https://advance-lexis-com.ezp-prod1.hul.harvard.edu/search/?pdmfid=1516831&crid=b7b4f289-5ef3-4315-a5b4-888962733e62&pdsearchterms=racism+and+apology+and+university&pdstartin=hlct%3A1%3A1&pdtypeofsearch=searchboxclick&pdsearchtype=SearchBox&pdqttype=and&undefined=&pdquerytemplateid=&ecomp=_z-2k&prid=8e815b0b-6907-4939-add8-d8a6e892b452; accessed January 3, 2021.

5. Krieger N. ENOUGH: COVID-19, structural racism, police brutality, plutocracy, climate change-and time for health justice, democratic governance, and an equitable, sustainable future [published online ahead of print August 20, 2020]. *Am J Public Health* 2020;110(11):1620–1623. doi:10.2105/AJPH.2020.305886; American Public Health Association. Declarations of Racism as a Public Health Issue. https://www.apha.org/topics-and-issues/health-equity/racism-and-health/racism-declarations; accessed January 3, 2021; Singh M. "Long overdue": lawmakers declare racism a public health crisis. *The Guardian*, June 12, 2020. https://www.theguardian.com/society/2020/jun/12/racism-public-health-black-brown-coronavirus; accessed January 3, 2021; Vestal C. Racism is a public health crisis, say cities and counties. *Pew Stateline*, June 15, 2020. https://www.pewtrusts.org/en/research-and-analysis/blogs/stateline/2020/06/15/racism-is-a-public-health-crisis-say-cities-and-counties; accessed January 3, 2021; Holmes TE. Toward a cure: cities declare racism a public health crisis. *Colorlines*, September 2, 2020. https://www.colorlines.com/articles/toward-cure-cities-declare-racism-public-health-crisis; accessed September 6, 2020.
6. American Public Health Association. Declarations of Racism as a Public Health Issue. https://www.apha.org/topics-and-issues/health-equity/racism-and-health/racism-declarations; accessed January 3, 2021.
7. Krieger N. ENOUGH: COVID-19, structural racism, police brutality, plutocracy, climate change-and time for health justice, democratic governance, and an equitable, sustainable future [published online ahead of print August 20, 2020]. *Am J Public Health* 2020;110(11):1620–1623. doi:10.2105/AJPH.2020.305886; American Public Health Association. Racism is an ongoing public health crisis that needs our attention now. May 29, 2020. https://apha.org/news-and-media/news-releases/apha-news-releases/2020/racism-is-a-public-health-crisis; accessed January 3, 2021; Society for Epidemiologic Research. A statement on racism from SER. June 3, 2020. https://epiresearch.org/about-us/ser-statement; accessed January 3, 2021; American Medical Association. AMA Board of Trustees pledges action against racism, police brutality. https://www.ama-assn.org/press-center/ama-statements/ama-board-trustees-pledges-action-against-racism-police-brutality; accessed January 3, 2021; Churchwell K, Elkind MSV, Benjamin RM, Carson AP, Chang EK, Lawrence W, Mills A, Odom TM, Rodriguez CJ, Rodriguez F,

Sanchez E, Sharrief AZ, Sims M, Williams O; American Heart Association. Call to action: structural racism as a fundamental driver of health disparities: a presidential advisory from the American Heart Association [published online ahead of print November 2020]. *Circulation* 2020 Dec 15;142(24):e454–e468. doi:10.1161/CIR.0000000000000936.

8. Weiser S. What Big Business said in all those anti-racism statements: not much, says our analysis. *Colorlines,* October 8, 2020. https://www.colorlines.com/articles/what-big-business-said-all-those-anti-racism-statements-not-much-says-our-analysis; accessed January 3, 2021; Bartlett T. Mistakes were made: reviewing a curious year of contrition in higher education. *Chronicle Higher Education,* December 7, 2020. https://www.chronicle.com/article/mistakes-were-made; accessed January 3, 2021.
9. Hammonds E, Herzig RM. *The Nature of Difference: The Science of Race in the United States from Jefferson to Genomics.* Cambridge, MA: MIT Press, 2009; Hammonds EM, Reverby SM. Toward a historically informed analysis of racial health disparities since 1619. *Am J Public Health* 2019;109(10):1348–1349. doi:10.2105/AJPH.2019.305262; Kendi IX. *Stamped from the Beginning: The Definitive History of Racist Ideas in America.* New York: Nation Books, 2016; Yudell M. *Race Unmasked: Biology and Race in the Twentieth Century.* New York: Columbia University Press, 2014; Ernst W, Harris B (eds). *Race, Science, and Medicine, 1700–1960.* London: Routledge, 1999; Birn AE, Pillay Y, Holtz T. *Textbook of Global Health.* New York: Oxford University Press, 2017; Krieger N. *Epidemiology and the People's Health: Theory and Context.* New York: Oxford University Press, 2011; Krieger N. Shades of difference: theoretical underpinnings of the medical controversy on black/white differences in the United States, 1830–1870. *Int J Health Serv* 1987;17:259–278; Eglash R. Anti-racist technoscience: a generative tradition. In: Benjamin R (ed). *Captivating Technology: Race, Carceral Technoscience, and the Liberatory Imagination in Everyday Life.* Durham, NC: Duke University Press, 2019; 227–251; Ferry G. Rebecca Lee Crumpler: first Black woman physician in the USA. *Lancet* 2021;397:572; Besak J. W.E.B. Du Bois embraced science to fight racism as editor of NAACP's magazine The Crisis. *The Conversation,* December 14, 2020. https://theconversation.com/w-e-b-du-bois-embraced-science-to-fight-racism-as-editor-of-naacps-magazine-the-crisis-150825; accessed February 14, 2021.
10. Krieger N. *Epidemiology and the People's Health: Theory and Context.* New York: Oxford University Press, 2011; Krieger N. Shades of difference: theoretical underpinnings of the medical controversy on black/white differences in the United States, 1830–1870. *Int J Health Serv* 1987;17:259–278; Hammonds E, Herzig RM. *The Nature of Difference: The Science of Race in the United States from Jefferson to Genomics.* Cambridge, MA: MIT Press, 2009; Hammonds EM, Reverby SM.

Toward a historically informed analysis of racial health disparities since 1619. *Am J Public Health* 2019;109(10):1348–1349. doi:10.2105/AJPH.2019.305262; Kendi IX. *Stamped from the Beginning: The Definitive History of Racist Ideas in America*. New York: Nation Books, 2016; Yudell M. *Race Unmasked: Biology and Race in the Twentieth Century*. New York: Columbia University Press, 2014; Birn AE, Brown TM (eds). *Comrades in Health: U.S. Internationalists, Abroad and at Home*. New Brunswick, NJ: Rutgers University Press, 2013; Eglash R. Anti-racist technoscience: a generative tradition. In: Benjamin R (ed). *Captivating Technology: Race, Carceral Technoscience, and the Liberatory Imagination in Everyday Life*. Durham, NC: Duke University Press, 2019; 227–251.

11. Krieger N. Researching critical questions on social justice and public health: an ecosocial perspective. In: Levy BS, Sidel VW (eds). *Social Injustice and Public Health*. New York: Oxford University Press, 2006; 460–479.
12. Krieger N. ENOUGH: COVID-19, structural racism, police brutality, plutocracy, climate change-and time for health justice, democratic governance, and an equitable, sustainable future [published online ahead of print August 20, 2020]. *Am J Public Health* 2020;110(11):1620–1623. doi:10.2105/AJPH.2020.305886; New York Times. What We Have Lost, Sunday review section. *New York Times*, November 1, 2020; New York Times. Notes from the Newsroom: overnight emails tell the story of 2020 (special section). *New York Times*, January 3, 2021; CNN Editorial Research. 2020 in review: fast facts. *CNN*, December 31, 2020. https://www.cnn.com/2020/12/10/us/2020-in-review-fast-facts/index.html; accessed January 3, 2021.
13. Leung S, Edelman L. Coronavirus made the wealth gap worse. How long can a divided economy stand? *Boston Globe*, September 6, 2020. https://www.bostonglobe.com/2020/09/05/business/coronavirus-made-wealth-gap-worse-how-long-can-divided-economy-stand/; accessed January 3, 2021; Collins C, Ocampo O, Palaski S. *Billionaire Bonanza 2020: Wealth Windfalls, Tumbling Taxes, and Pandemic Profiteers*. Washington, DC: Institute for Policy Studies, November 2020. https://ips-dc.org/billionaire-bonanza-2020/; accessed January 3, 2021; Gneiting U, Lusiani N, Tamir I. *Profits and the Pandemic: From Corporate Extraction for the Few to an Economy that Works for All*. Oxford: Oxfam BG, September 2020. https://oxfamilibrary.openrepository.com/handle/10546/621044; accessed January 3, 2021; Inequality.org. COVID-19 and inequality. Institute for Policy Studies. https://inequality.org/facts/inequality-and-covid-19/ ; accessed January 3, 2021.
14. Buchanan L, Bui Q, Patel JK. Black Lives Matter may be the largest movement in U.S. history. *New York Times*, July 3, 2020. https://www.nytimes.com/interactive/2020/07/03/us/george-floyd-protests-crowd-size.html; accessed January 3, 2021; Vera Institute. *Annual Report 2020: Reckoning with Justice*. Vera

Institute. https://www.vera.org/annual-report-2020-reckoning-with-justice; accessed January 3, 2021; American Civil Liberties Union. *ACLU Annual Report 2020*. ACLU. https://www.vera.org/annual-report-2020-reckoning-with-justice; accessed January 3, 2021.

15. Rutenberg J. The attack on voting. *New York Times*, September 30, 2020. https://www.nytimes.com/2020/09/30/magazine/trump-voter-fraud.html; accessed January 3, 2021; Anderson C. Millions of Americans have risen up and said: democracy won't die on my watch. *The Guardian*, November 7, 2020. https://www.theguardian.com/commentisfree/2020/nov/07/americans-democracy-trump; accessed January 3, 2020; Hyde SD, Saunders EN. Trump didn't break our democracy. But did he fatally weaken it? *New York Times*, December 15, 2020. https://www.nytimes.com/2020/12/15/opinion/trump-democracy-america.html; accessed January 3, 2021; Gessen M. The coup stage of Donald Trump's presidency. *New Yorker*, November 20, 2020. https://www.newyorker.com/news/our-columnists/the-coup-stage-of-donald-trumps-presidency; accessed January 3, 2021; Anderson C. *One Person, No Vote: How Voter Suppression Is Destroying Our Democracy*. New York: Bloomsbury Press, 2018; Krieger N. Climate crisis, health equity, and democratic governance: the need to act together. *J Public Health Policy* 2020;41(1):4–10. doi:10.1057/s41271-019-00209-x.
16. Krieger N. The US Census and the people's health: public health engagement from enslavement and "Indians not taxed" to census tracts and health equity (1790–2018) [published online ahead of print June 20, 2019]. *Am J Public Health* 2019;109(8):1092–1100. doi:10.2105/AJPH.2019.305017; Mervis J. Census experts fear rush to finish tally will yield flawed data. *Science* 2020;369(6509):1285–1286; Editorial Board. Trump is plotting against the Census. Here's why. *New York Times*, August 5, 2020. https://www.nytimes.com/2020/08/05/opinion/trump-census-2020.html; accessed January 3, 2021; Wines M. A federal judge blocked the Trump administration from ending the census early. *New York Times*, September 25, 2020. https://www.nytimes.com/2020/09/25/us/a-federal-judge-blocked-the-trump-administration-from-ending-the-census-early.html; accessed January 3, 2021; Wines M. Ruling against shortened count adds to questions about the Census. *New York Times*, September 25, 2020. https://www.nytimes.com/2020/09/25/us/trump-census-deadline.html; accessed January 3, 2021; Wezerek G, Whitby A. Trump's desperate plan to cut short the Census could backfire. *New York Times*, September 23, 2020. https://www.nytimes.com/interactive/2020/09/23/opinion/trump-census-2020.html; accessed January 3, 2021; Levine S. Officials outside bureau made decision to speed US census, report finds. *The Guardian*, September 22, 2020. https://www.theguardian.com/us-news/2020/sep/22/us-census-inspector-general-report; accessed January 3, 2021.

17. Pomerantsev P. The disinformation age: a revolution in propaganda. *The Guardian*, July 27, 2019. https://www.theguardian.com/books/2019/jul/27/the-disinformation-age-a-revolution-in-propaganda; accessed January 3, 2021; Alba D, Frenkel F. From voter fraud to vaccine lies: misinformation peddlers shift gears. *New York Times*, December 16, 2020, updated December 19, 2020. https://www.nytimes.com/2020/12/16/technology/from-voter-fraud-to-vaccine-lies-misinformation-peddlers-shift-gears.html?searchResultPosition=9; accessed January 3, 2021; Cadwalladr C. Fresh Cambridge Analytica leak "shows global manipulation is out of control." *The Guardian*, January 4, 2020. https://www.theguardian.com/uk-news/2020/jan/04/cambridge-analytica-data-leak-global-election-manipulation; accessed January 4, 2021; Cadwalladr C. If you're not terrified about Facebook, you haven't been paying attention. *The Guardian*, July 26, 2020. https://www.theguardian.com/commentisfree/2020/jul/26/with-facebook-we-are-already-through-the-looking-glass; accessed January 4, 2021; Jane E. *Misogyny Online: A Short (and Brutish) History*. London: SAGE Publications, 2017; Stern A. *Proud Boys and the White Ethnostate: How the Alt-Right Is Warping the American Imagination*. Boston: Beacon Press, 2019; Mayer J. *Dark Money: The Hidden History of the Billionaires Behind the Rise of the Radical Right*. New York: Doubleday, 2016.
18. Friel S. *Climate Change and the People's Health*. New York: Oxford University Press, 2019; Popovich N, Albekc-Ripka L, Pierre-Louis K. The Trump administration is reversing more than 100 environmental rules. Here's the full list. *New York Times*, November 10, 2020. https://www.nytimes.com/interactive/2020/climate/trump-environment-rollbacks-list.html; accessed January 3, 2021; Lipton E. A regulatory rush by federal agencies to secure Trump's legacy. *New York Times*, October 16, 2020; updated November 3, 2020. https://www.nytimes.com/2020/10/16/us/politics/regulatory-rush-federal-agencies-trump.html; accessed January 3, 2021; Davenport C. What will Trump's most profound legacy be? Possibly climate damage. *New York Times*, November 9, 2020; updated November 3, 2020. https://www.nytimes.com/2020/11/09/climate/trump-legacy-climate-change.html; accessed January 3, 2021; Chang A, Holden E, Milman O, Yachot N. 75 ways Trump made America dirtier and the planet warmer. *The Guardian*, October 20, 2020. https://www.theguardian.com/us-news/ng-interactive/2020/oct/20/trump-us-dirtier-planet-warmer-75-ways; accessed January 3, 2021.
19. Garzia A. *The Purpose of Power: How We Come Together When We Fall Apart*. New York: One World, Random House, 2019; Miller C, Crane J (eds). *The Nature of Hope: Grassroots Organizing, Environmental Justice, and Political Change*. Louisville: University Press of Colorado, 2019; Pastor M, Terriquez V, Lin M. How community organizing promotes health equity, and how health equity affects organizing. *Health Aff (Project Hope)* 2018;37(3):358–363; Birn AE, Pillay Y,

Holtz TH. *Textbook of Global Health*. New York: Oxford University Press, 2017; People's Health Movement. https://phmovement.org/; accessed January 3, 2021; Movement for Black Lives. Vision for Black Lives: 2020 policy platform. https://m4bl.org/policy-platforms/; accessed January 3, 2021.

20. Peltier E. In landmark ruling, air pollution recorded as a cause of death for British girl. *New York Times*, December 16, 2020. https://www.nytimes.com/2020/12/16/world/europe/britain-air-pollution-death.html; accessed January 3, 2021.
21. Peltier E. In landmark ruling, air pollution recorded as a cause of death for British girl. *New York Times*, December 16, 2020. https://www.nytimes.com/2020/12/16/world/europe/britain-air-pollution-death.html; accessed January 3, 2021.
22. Chakradhar S. Pediatricians see evidence of climate change in their patients. *Boston Globe*, December 25, 2020, C6–C7 (print edition). https://epaper.bostonglobe.com/infinity/article_popover_share.aspx?guid=0de1a008-25b3-4c79-9dee-52af2378633e; accessed January 3, 2021.
23. Hamlin C. Could you starve to death in England in 1839? The Chadwick-Farr controversy and the loss of the "social" in public health. *Am J Public Health* 1995;85(6):856–866; Hamlin C. *Public Health and Social Justice in the Age of Chadwick: Britain, 1800–1854* (Cambridge History of Medicine). Cambridge: Cambridge University Press, 1998.
24. Chadwick E, Great Britain Poor Law Commissioners. *Report on The Sanitary Condition of The Labouring Population of Great Britain: A Supplementary Report on the Results of a Special Inquiry Into The Practice of Interment In Towns.* Made at the request of Her Majesty's principal Secretary of State for the Home Department. London: Printed by W. Clowes and Sons, for H.M.S.O., 1843.
25. Hamlin C. Could you starve to death in England in 1839? The Chadwick-Farr controversy and the loss of the "social" in public health. *Am J Public Health* 1995;85(6):856–866; Hamlin C. *Public Health and Social Justice in the Age of Chadwick: Britain, 1800–1854* (Cambridge History of Medicine). Cambridge: Cambridge University Press, 1998.
26. "theory, n." *OED Online*, Oxford University Press. https://www.oed.com/view/Entry/200431; accessed December 29, 2020; Krieger N. *Epidemiology and the People's Health: Theory and Context*. New York: Oxford University Press, 2011; 17–28; Krieger N. Got theory? On the 21 st c. ce rise of explicit use of epidemiologic theories of disease distribution: a review and ecosocial analysis. *Curr Epidemiol Rep* 2014;1(1):45–56.
27. Vitale AS. *The End of Policing*. London: Verso, 2017; Nelson A. The long durée of Black Lives Matter. *Am J Public Health* 2016;106(10):1734–1737; Oreskes N, Conway EM. *Merchants of Doubt: How a Handful of Scientists Obscured the Truth on Issues from Tobacco Smoke to Global Warming.* New York: Bloomsbury, 2015; Intergovernmental Panel on Climate Change (IPCC), United Nations.

About: History of the IPCC. https://www.ipcc.ch/about/history/; accessed January 3, 2021.

28. Vitale AS. *The End of Policing*. London: Verso, 2017; Johnson MS. *Street Justice: A History of Police Violence in New York City*. Boston: Beacon Press, 2003; Skolnick JH. Policing. In: Smelser N, Baltes P (eds). *International Encyclopedia of the Social and Behavioral Sciences*. 1st ed. Amsterdam, New York: Elsevier, 2001; 11535–11541; Deflem M, Hauptman S. Policing. In: Wright JD (ed). *International Encyclopedia of the Social and Behavioral Sciences*. 2nd ed. Vol. 18. London: Elsevier, 2015; 261–265.
29. Vitale AS. *The End of Policing*. London: Verso, 2017; Johnson MS. *Street Justice: A History of Police Violence in New York City*. Boston: Beacon Press, 2003; Fogelson RM, Rubenstein RD (eds). *The Complete Report of Mayor LaGuardia's Commission on the Harlem Riot of March 19, 1935*. New York: Arno Press and The New York Times, 1969; Walker S. The urban police in American history: a review of the literature. *J Political Sci Admin* 1976;4:252–260; Monkkonen EH. From cop history to social history: the significance of the police in American history. *J Soc Hist* 1982;15:575–591; Barker K, Keller MH, Eder S. How cities lost control of police discipline. *New York Times*, December 22, 2020. https://www.nytimes.com/2020/12/22/us/police-misconduct-discipline.html; accessed December 23, 2020.
30. Oreskes N, Conway EM. *Merchants of Doubt: How a Handful of Scientists Obscured the Truth on Issues from Tobacco Smoke to Global Warming*. New York: Bloomsbury, 2015;Intergovernmental Panel on Climate Change (IPCC), United Nations. About: History of the IPCC. https://www.ipcc.ch/about/history/; accessed January 3, 2021; Marshall E. EPA's plan for cooling the global greenhouse. *Science* 1989;243:1544–1555.
31. Stone D. Causal stories and the formation of policy agendas. *Political Sci Q* 1989;104(2):281–300; Blyth M. The new ideas scholarship in the mirror of historical institutionalism: a case of old whines in new bottles? *J European Policy* 2016;23(3):464–471; see, for subsequent analyses building on the 1989 article:Gamble VN, Stone D. US policy on health inequities: the interplay of politics and research. *J Health Politics Policy Law* 2006;31(1):93–126; Stone D. *Policy Paradox: The Art of Political Decision Making*. 3rd. ed. New York: W. W. Norton, 2012 (note: 1st ed. was in 1997, 2nd in 2002).
32. Kingdon JW. *Agendas, Alternatives, and Public Policies*. Boston: Little, Brown, 1984 (note: 2nd ed. was in 1995; reissued in 2003; updated in 2011);Greer S. John W. Kingdon, agendas, alternatives, and public policies. In: Lodge M, Page EC, Balla SJ (eds). *The Oxford Handbook of Classics in Public Policy and Administration*. New York: Oxford University Press, 2015:417–432; online publication: July 2016. doi:10.1093/oxfordhb/9780119646135.013.18; Howlett M, McConnell A, Peel A. Streams and stages: reconciling Kingdon

and policy process theory. *Eur J Political Res* 2015;54(3):419–434; see, for subsequent analyses building on the 1984 book:Kingdon J, Thurber J. *Agendas, Alternatives, and Public Policies* (Updated 2nd ed., Longman Classics in Political Science). Boston: Longman, 2011.

33. Stone D. Causal stories and the formation of policy agendas. *Political Sci Q* 1989;104(2):281–300; quote: p. 283.
34. Stone D. Causal stories and the formation of policy agendas. *Political Sci Q* 1989;104(2):281–300; quote: p. 295.
35. Stone D. Causal stories and the formation of policy agendas. *Political Sci Q* 1989;104(2):281–300; quote: p. 299.
36. Stone D. Causal stories and the formation of policy agendas. *Political Sci Q* 1989;104(2):281–300; quote: p. 300.
37. PubMed. Advanced Search on 9/26/20 using terms "(police) AND (brutality)." https://pubmed.ncbi.nlm.nih.gov/?term=%28%28police%29+AND+%28brutality%29%29+&sort=pubdate; accessed September 26, 2020.
38. PubMed. Advanced Search on 9/26/20 using terms "(police) AND (brutality)." https://pubmed.ncbi.nlm.nih.gov/?term=%28%28police%29+AND+%28brutality%29%29+&sort=pubdate; accessed September 26, 2020.
39. PubMed. Advanced Search on 9/26/20 using terms "(police) AND (racism)." https://pubmed.ncbi.nlm.nih.gov/?term=%28police%29+AND+%28racism%29&sort=pubdate; accessed September 26, 2020.
40. PubMed. Advanced Search on 9/26/20 using terms "(police) AND (racism)." https://pubmed.ncbi.nlm.nih.gov/?term=%28police%29+AND+%28racism%29&sort=pubdate; accessed September 26, 2020.
41. Web of Science, all databases. TS = (police AND (brutality OR racism). *N* = 1,211. http://apps.webofknowledge.com.ezp-prod1.hul.harvard.edu/summary.do?product=UA&doc=1&qid=1&SID=5DYBUnmtfjJbDQ5OXWa&search_mode=AdvancedSearch&update_back2search_link_param=yes; accessed September 26, 2020.
42. Black Lives Matter. Herstory. https://blacklivesmatter.com/herstory/; accessed January 3, 2021; Taylor K-Y. *From #BlackLivesMatter to Black Liberation*. Chicago: Haymarket Books, 2016; Garza A. *The Purpose of Power: How We Come Together When We Fall Apart*. New York: One World, Penguin Random House, 2020.
43. Black Lives Matter. Herstory. https://blacklivesmatter.com/herstory/; accessed January 3, 2021; Taylor K-Y. *From #BlackLivesMatter to Black Liberation*. Chicago: Haymarket Books, 2016; Garza A. *The Purpose of Power: How We Come Together When We Fall Apart*. New York: One World, Penguin Random House, 2020.

44. Black Lives Matter. Herstory. https://blacklivesmatter.com/herstory/; accessed January 3, 2021; Taylor K-Y. *From #BlackLivesMatter to Black Liberation*. Chicago: Haymarket Books, 2016; Garza A. *The Purpose of Power: How We Come Together When We Fall Apart*. New York: One World, Penguin Random House, 2020.
45. Fatal Encounters. https://fatalencounters.org/; accessed January 3, 2021;Lozano AV. Fatal Encounters. One man is tracking every officer-involved killing in the U.S. *NBC News*, July 11, 2020. https://www.nbcnews.com/news/us-news/fatal-encounters-one-man-tracking-every-officer-involved-killing-u-n1233188; accessed January 3, 2021.
46. Washington Post. Fatal Force, police shootings database 2015–2020. https://www.washingtonpost.com/graphics/investigations/police-shootings-database/; accessed January 3, 2021.
47. The Guardian. The counted: people killed by the police in the US, recorded by the Guardian—with your help. https://www.theguardian.com/us-news/series/counted-us-police-killings; accessed January 3, 2021.
48. Krieger N, Chen JT, Waterman PD, Kiang MV, Feldman J. Police killings and police deaths are public health data and can be counted. *PLoS Med* 2015 Dec 8;12(12):e1001915. doi:10.1371/journal.pmed.1001915; Vitale AS. *The End of Policing*. London: Verso, 2017; Zimring FE. Police killings as a problem of governance. *Ann Am Acad Pol Soc Sci* 2020;687(1):114–123. doi:10.1177/0002716219888627;Loftin C, McDowall D, Xie M. Underreporting of homicides by police in the United States, 1976–2013. *Homicide Stud* 2017;21:159–174. doi:10.1177/1088767917693358.
49. Tran M. FBI chief: "unacceptable" that the Guardian has better data on police violence. *The Guardian*, October 8, 2015. https://www.theguardian.com/us-news/2015/oct/08/fbi-chief-says-ridiculous-guardian-washington-post-better-information-police-shootings; accessed January 3, 2021.
50. Fatal Encounters. https://fatalencounters.org/; accessed January 3, 2021;Washington Post. Fatal Force, police shootings database 2015–2020. https://www.washingtonpost.com/graphics/investigations/police-shootings-database/; accessed January 3, 2021; The Guardian. The Counted: People killed by the police in the US, recorded by the Guardian—with your help. https://www.theguardian.com/us-news/series/counted-us-police-killings; accessed January 3, 2021.
51. Krieger N, Chen JT, Waterman PD, Kiang MV, Feldman J. Police killings and police deaths are public health data and can be counted. *PLoS Med* 2015 Dec 8;12(12):e1001915. doi:10.1371/journal.pmed.1001915; Laurencin CT, Walker JM. Racial profiling is a public health and health disparities issue [published online ahead of print April 6, 2020]. *J Racial Ethn Health Disparities* 2020;7(3):393–397. doi:10.1007/s40615-020-00738-2; Ford CL. *Graham*, police violence, and health

through a public health lens. *Boston Univ Law Rev* 2020;100(3):1093–1110; Duarte CD, Alson JG, Garakani OB, Mitchell CM. Applications of the American Public Health Association's statement on addressing law enforcement violence as a public health issue. *Am J Public Health* 2020;110(S1):S30–S32. doi:10.2105/AJPH.2019.305447.

52. Feldman JM, Gruskin S, Coull BA, Krieger N. Killed by police: validity of media-based data and misclassification of death certificates in Massachusetts, 2004–2016 [published online ahead of print August 17, 2017]. *Am J Public Health* 2017;107(10):1624–1626. doi:10.2105/AJPH.2017.303940; Feldman JM, Gruskin S, Coull BA, Krieger N. Quantifying underreporting of law-enforcement-related deaths in United States vital statistics and news-media-based data sources: a capture-recapture analysis. *PLoS Med* 2017;14(10):e1002399. doi:10.1371/journal.pmed.1002399. Erratum in: *PLoS Med* 2017 Oct 26;14 (10):e1002449; Ozkan T, Worrall JL, Zettler H. Validating media-driven and crowdsourced police shooting data: a research note. *J Crime Justice* 2018;41(3):334–345. doi:10.1080/0735648X.2017.1326831;Finch BK, Beck A, Burghart DB, Johnson R, Klinger D, Thomas K. Using crowd-sourced data to explore police-related-deaths in the United States (2000–2017): the case of Fatal Encounters. *Open Health Data* 2019 May 7;6(1). http://doi.org/10.5334.ohd.30;Baćak V, Mausolf JG, Schwarz C. How comprehensive are media-based data on police officer-involved shootings? [published online ahead of print July 11, 2019] *J Interpers Violence* 2019;886260519860897. doi:10.1177/0886260519860897.

53. Barber C, Azrael D, Cohen A, Miller M, Thymes D, Wang DE, Hemenway D. Homicides by police: comparing counts from the national violent death reporting system, vital statistics, and supplementary homicide reports [published online ahead of print March 17, 2016]. *Am J Public Health* 2016;106(5):922–927. doi:10.2105/AJPH.2016.303074; Conner A, Azrael D, Lyons VH, Barber C, Miller M. Validating the National Violent Death Reporting System as a source of data on fatal shootings of civilians by law enforcement officers [published online ahead of print February 21, 2019]. *Am J Public Health* 2019;109(4):578–584. doi:10.2105/AJPH.2018.304904.

54. Feldman JM, Chen JT, Waterman PD, Krieger N. Temporal trends and racial/ethnic inequalities for legal intervention injuries treated in emergency departments: US men and women age 15–34, 2001–2014. *J Urban Health* 2016;93(5):797–807; Miller TR, Lawrence BA, Carlson NN, Hendrie D, Randall S, Rockett IR, Spicer RS. Perils of police action: a cautionary tale from US data sets [published online ahead of print July 25, 2016]. *Inj Prev* 2017;23(1):27–32. doi:10.1136/injuryprev-2016-042023; Mooney AC, McConville S, Rappaport AJ, Hsia RY. Association of legal intervention injuries with race and ethnicity among patients treated in emergency departments

in California. *JAMA Netw Open* 2018 Sep 7;1(5):e182150. doi:10.1001/jamanetworkopen.2018.2150; Holloway-Beth A, Rubin R, Joshi K, Murray LR, Friedman L. A 5-year retrospective analysis of legal intervention injuries and mortality in Illinois [published online ahead of print March 21, 2019]. *Int J Health Serv* 2019;49(3):606–622. doi:10.1177/0020731419836080.

55. Feldman JM, Gruskin S, Coull BA, Krieger N. Quantifying underreporting of law-enforcement-related deaths in United States vital statistics and news-media-based data sources: a capture-recapture analysis. *PLoS Med* 2017;14(10):e1002399. doi:10.1371/journal.pmed.1002399. Erratum in: *PLoS Med* 2017 Oct 26;14(10):e1002449; Loftin C, McDowall D, Xie M. Underreporting of homicides by police in the United States, 1976–2013. *Homicide Stud* 2017;21(2):159–174; Zimring FE. Police killings as a problem of governance. *Ann Am Acad Pol Soc Sci* 2020;687(1):114–123. doi:10.1177/0002716219888627.
56. Alang S, McAlpine D, McCreedy E, Hardeman R. Police brutality and black health: setting the agenda for public health scholars [published online ahead of print March 21, 2019]. *Am J Public Health* 2017;107(5):662–665. doi:10.2105/AJPH.2017.303691; Boyd RW. Police violence and the built harm of structural racism [published online ahead of print June 21, 2018]. *Lancet* 2018;392(10144):258–259. doi:10.1016/S0140-6736(18)31374-6.
57. Sewall AA, Feldman JM, Ray R, Gilbert KL, Jefferson KA, Lee H. Illness spillovers of lethal police violence: the significance of gendered marginalization [published online ahead of print July 22, 2020]. *Ethnic Racial Stud.* https://doi.org/10.1080/01419870.2020.1781913.
58. Bor J, Venkataramani AS, Williams DR, Tsai AC. Police killings and their spillover effects on the mental health of black Americans: a population-based, quasi-experimental study [published online ahead of print June 21, 2018]. *Lancet* 2018;392(10144):302–310. doi:10.1016/S0140-6736(18)31130-9; Lee JRS, Robinson MA. "That's my number one fear in life. It's the police": examining young black men's exposures to trauma and loss resulting from police violence and police killings. *J Black Psychol* 2019;45(3):143–184; Outland RL, Noel T Jr, Rounsville K, Boatwright T, Waleed C, Abraham A. Living with trauma: impact of police killings on the lives of the family and community of child and teen victims. *Curr Psychol* 2020. https://doi.org/10.1007/s12144-020-01129-w.
59. Ibragimov U, Beane S, Friedman SR, Smith JC, Tempalski B, Williams L, Adimora AA, Wingood GM, McKetta S, Stall RD, Cooper HL. Police killings of Black people and rates of sexually transmitted infections: a cross-sectional analysis of 75 large US metropolitan areas, 2016 [published online ahead of print August 23, 2019]. *Sex Transm Infect* 2020;96(6):429–431. doi:10.1136/sextrans-2019-054026.
60. Liu SY, Lim S, Gould LH. Impact of law enforcement-related deaths of unarmed black New Yorkers on emergency department rates, New York 2013–2016

[published online ahead of print October 7, 2020]. *J Epidemiol Community Health* 2020;jech–2020–214089. doi:10.1136/jech-2020-214089.

61. Livingston JD. Contact between police and people with mental disorders: a review of rates [published online ahead of print April 15, 2016]. *Psychiatr Serv* 2016;67(8):850–857. doi:10.1176/appi.ps.201500312; Saleh AZ, Appelbaum PS, Liu X, Scott Stroup T, Wall M. Deaths of people with mental illness during interactions with law enforcement [published online ahead of print April 21, 2018]. *Int J Law Psychiatry* 2018;58:110–116. doi:10.1016/j.ijlp.2018.03.003; Appelbaum PS. Can the Americans With Disabilities Act reduce the death toll from police encounters with persons with mental illness? *Psychiatr Serv* 2015;66(10):1012–1014. doi:10.1176/appi.ps.661005; Lane-McKinley K, Tsungmey T, Roberts LW. The Deborah Danner story: officer-involved deaths of people living with mental illness [published online ahead of print June 27, 2018]. *Acad Psychiatry* 2018;42(4):443–450. doi:10.1007/s40596-018-0945-z; Farkas K, Matthay EC, Rudolph KE, Goin DE, Ahern J. Mental and substance use disorders among legal intervention injury cases in California, 2005–2014 [published online ahead of print February 10, 2019]. *Prev Med* 2019;121:136–140. doi:10.1016/j.ypmed.2019.01.003; Kesic D, Thomas SD, Ogloff JR. Estimated rates of mental disorders in, and situational characteristics of, incidents of nonfatal use of force by police [published online ahead of print June 29, 2012]. *Soc Psychiatry Psychiatr Epidemiol* 2013;48(2):225–232. doi:10.1007/s00127-012-0543-4; Holloway-Beth A, Forst L, Lippert J, Brandt-Rauf S, Freels S, Friedman L. Risk factors associated with legal interventions [published online ahead of print January 15, 2016]. *Inj Epidemiol* 2016;3(1):2. doi:10.1186/s40621-016-0067-6.
62. American Public Health Association. Addressing Law Enforcement Violence as a Public Health Issue. Policy Number 201811. November 13, 2018. https://apha.org/policies-and-advocacy/public-health-policy-statements/policy-database/2019/01/29/law-enforcement-violence; accessed January 3, 2021; Duarte CD, Alson JG, Garakani OB, Mitchell CM. Applications of the American Public Health Association's statement on addressing law enforcement violence as a public health issue. *Am J Public Health* 2020;110(S1):S30–S32. doi:10.2105/AJPH.2019.305447; Human Impact Partners. How health departments can address police violence as a public health issue. September 2020. https://humanimpact.org/hipprojects/how-health-departments-can-address-police-violence-as-a-public-health-issue/; accessed January 3, 2021; Black Lives Matter. #Defund the police. May 30, 2020. https://blacklivesmatter.com/defundthepolice/; accessed January 3, 2021; People's Budget LA. Defund the Police. Reimagine Public Safety. June 2020. https://peoplesbudgetla.com/; accessed January 3, 2021; Lowrey A. Defund the police: America needs to rethink its priorities for the whole criminal-justice system. *The Atlantic,* June 5, 2020. https://www.theatlantic.com/ideas/archive/2020/

06/defund-police/612682/; accessed January 3, 2021; Levin S. The movement to defund the police has won historic victories across the U.S. What's next? https://www.theguardian.com/us-news/2020/aug/15/defund-police-movement-us-victories-what-next; accessed January 3, 2021.

63. People's Budget LA. Defund the Police. Reimagine Public Safety. June 2020. https://peoplesbudgetla.com/; accessed January 3, 2021; Lowrey A. Defund the police: America needs to rethink its priorities for the whole criminal-justice system. *The Atlantic*, June 5, 2020. https://www.theatlantic.com/ideas/archive/2020/06/defund-police/612682/; accessed January 3, 2021; Levin S. The movement to defund the police has won historic victories across the U.S. What's next? *The Guardian*, August 15, 2020. https://www.theguardian.com/us-news/2020/aug/15/defund-police-movement-us-victories-what-next; accessed January 3, 2021; Hendrickson C. Fact-checking Rep. Rashida Tlaib's claim on Detroit's police spending vs. health care. *Detroit Free Press*, June 16, 2020. https://www.freep.com/story/news/local/michigan/detroit/2020/06/16/fact-check-detroits-police-vs-health-spending-gap/3194086001/; accessed January 3, 2021; Krause K. Dallas city council passes budget that leaves police funding in place, cuts overtime. *Dallas Morning News*, September 24, 2020. https://www.dallasnews.com/news/politics/2020/09/24/dallas-city-council-passes-budget-that-leaves-police-funding-in-place/; accessed January 3, 2021; Billings R. Portland councilors to vote on city budgets that cut 65 positions. Budgets for police and human services have undergone the most scrutiny this year, because of the Black Lives Matter movement. *Portland Press Herald*, September 21, 2020. https://www.pressherald.com/2020/09/21/portland-councilors-to-vote-on-city-budget-that-cuts-65-positions/; accessed January 3, 2021; Esposito S. Activists call for alternative Cook County "Budget for Black Lives." *Chicago Sun Times*, September 9, 2020. https://chicago.suntimes.com/politics/2020/9/9/21429296/activists-alternative-cook-county-budget-black-lives; accessed January 3, 2021; Montgomery D. Texas governor proposes freezing taxes in cities that "defund" the police. *New York Times*, August 18, 2020. https://www.nytimes.com/2020/08/18/us/texas-abbott-police-defund-austin.html; accessed January 3, 2021; Evelyn K. Barack Obama criticizes "Defund the Police" slogan but faces backlash. *The Guardian*, December 2, 2020. https://www.theguardian.com/us-news/2020/dec/02/barack-obama-criticizes-defund-the-police-slogan-backlash; accessed January 3, 2021; Associated Press. Minneapolis switches $8 million from police budget to violence prevention. *The Guardian*, December 10, 2020. https://www.theguardian.com/us-news/2020/dec/10/minneapolis-defund-switches-police-budget-to-violence-prevention; accessed January 3, 2021.

64. Duarte CD, Alson JG, Garakani OB, Mitchell CM. Applications of the American Public Health Association's statement on addressing law enforcement violence as a public health issue. *Am J Public Health* 2020;110(S1):S30–S32. doi:10.2105/

AJPH.2019.305447; Human Impact Partners. How health departments can address police violence as a public health issue. September 2020. https://humanimpact.org/hipprojects/how-health-departments-can-address-police-violence-as-a-public-health-issue/; accessed January 3, 2021.

65. Krieger N, Chen JT, Waterman PD, Kiang MV, Feldman J. Police killings and police deaths are public health data and can be counted. *PLoS Med* 2015;12(12):e1001915. doi 10.1371/journal.pmed.1001915; Charles D, Himmelstein K, Keenan W, Barcelo N; White Coats for Black Lives National Working Group. White Coats for Black Lives: medical students responding to racism and police brutality. *J Urban Health* 2015;92(6):1007–1010. doi:10.1007/s11524-015-9993-9;Obasogie OK, Newman Z. Police violence, use of force policies, and public health. *Am J Law Med* 2017;43(2-3):279–295. doi:10.1177/0098858817723665; Cooper HL, Fullilove M. Editorial: Excessive police violence as a public health issue. *J Urban Health* 2016;93(Suppl 1):1–7. doi:10.1007/s11524-016-0040-2; Alang S, McAlpine D, McCreedy E, Hardeman R. Police brutality and black health: setting the agenda for public health scholars [published online ahead of print March 21, 2017]. *Am J Public Health* 2017;107(5):662–665. doi:10.2105/AJPH.2017.303691;American Public Health Association. Addressing Law Enforcement Violence as a Public Health Issue. Policy Number 201811. November 13, 2018. https://apha.org/policies-and-advocacy/public-health-policy-statements/policy-database/2019/01/29/law-enforcement-violence; accessed January 3, 2021; Boyd RW. Police violence and the built harm of structural racism [published online ahead of print June 21, 2018]. *Lancet* 2018;392(10144):258–259. doi:10.1016/S0140-6736(18)31374-6; Laurencin CT, Walker JM. Racial profiling is a public health and health disparities issue [published online ahead of print April 6, 2020]. *J Racial Ethn Health Disparities* 2020;7(3):393–397. doi:10.1007/s40615-020-00738-2; Ford CL. Graham, police violence, and health through a public health lens. *Boston Univ Law Rev* 2020;100(3):1093–1110; Krieger N. ENOUGH: COVID-19, structural racism, police brutality, plutocracy, climate change-and time for health justice, democratic governance, and an equitable, sustainable future [published online ahead of print August 20, 202]. *Am J Public Health* 2020;110(11):1620–1623. doi:10.2105/AJPH.2020.305886.

66. van Dijk AJ, Herrington V, Crofts N, Breunig R, Burris S, Sullivan H, Middleton J, Sherman S, Thomson N. Law enforcement and public health: recognition and enhancement of joined-up solutions. *Lancet* 2019;393(10168):287–294; Scheibe A, Howell S, Müller A, Katumba M, Langen B, Artz L, Marks M. Finding solid ground: law enforcement, key populations and their health and rights in South Africa. *J Int AIDS Soc* 2016;19(4 Suppl 3):20872. doi:10.7448/IAS.19.4.20872; Hayashi K, Small W, Csete J, Hattirat S, Kerr T. Experiences with policing among

people who inject drugs in Bangkok, Thailand: a qualitative study. *PLoS Med* 2013;10(12):e1001570.

67. Stone D. Causal stories and the formation of policy agendas. *Political Sci Q* 1989;104(2):281–300; quote: p. 300.
68. Web of Science search: 11/27/20. http://wcs.webofknowledge.com.ezp-prod1.hul.harvard.edu/RA/analyze.do?product=UA&SID=5F31tCvwnlgCPx6WZHU&field=SJ_ResearchArea_ResearchArea_en&yearSort=false.
69. Web of Science search: 11/27/20. http://apps.webofknowledge.com.ezp-prod1.hul.harvard.edu/summary.do?product=UA&search_mode=AdvancedSearch&doc=1&qid=3&SID=5F31tCvwnlgCPx6WZHU.
70. Web of Science search: 11/27/20. http://wcs.webofknowledge.com.ezp-prod1.hul.harvard.edu/RA/analyze.do?product=UA&SID=5F31tCvwnlgCPx6WZHU&field=SJ_ResearchArea_ResearchArea_en&yearSort=false.
71. Web of Science search: 11/27/20. http://apps.webofknowledge.com.ezp-prod1.hul.harvard.edu/summary.do?product=UA&search_mode=AdvancedSearch&doc=1&qid=3&SID=5F31tCvwnlgCPx6WZHU.
72. Web of Science search: 11/27/20. http://wcs.webofknowledge.com.ezp-prod1.hul.harvard.edu/RA/analyze.do?product=UA&SID=5F31tCvwnlgCPx6WZHU&field=SJ_ResearchArea_ResearchArea_en&yearSort=false.
73. Web of Science search: 11/27/20. http://apps.webofknowledge.com.ezp-prod1.hul.harvard.edu/summary.do?product=UA&doc=1&qid=8&SID=5F31tCvwnlgCPx6WZHU&search_mode=AdvancedSearch&update_back2search_link_param=yes.
74. PubMed search: 11/27/20. https://pubmed.ncbi.nlm.nih.gov/?term=%28global+warming%29&timeline=expanded&sort=date&sort_order=asc; PubMed search: 11/27/20. https://pubmed.ncbi.nlm.nih.gov/?term=(climate%20change)&sort=date&sort_order=asc&timeline=expanded.
75. PubMed search: 11/27/20. https://pubmed.ncbi.nlm.nih.gov/?term=(climate%20justice)%20AND%20(public%20health)&sort=date&sort_order=asc.
76. Friel S. *Climate Change and the People's Health*. New York: Oxford University Press, 2019.
77. Intergovernmental Panel on Climate Change (IPCC), United Nations. About. https://www.ipcc.ch/about/; accessed January 3, 2021.
78. Intergovernmental Panel on Climate Change (IPCC). https://www.ipcc.ch/site/assets/uploads/2018/03/ipcc_far_wg_III_full_report.pdf; accessed January 3, 2021.
79. Intergovernmental Panel on Climate Change (IPCC). https://www.ipcc.ch/site/assets/uploads/2018/03/ipcc_far_wg_III_full_report.pdf; accessed January 3, 2021.

80. Intergovernmental Panel on Climate Change (IPCC). https://www.ipcc.ch/site/assets/uploads/2018/05/2nd-assessment-en-1.pdf; pp. 35–36; accessed January 3, 2021.
81. Intergovernmental Panel on Climate Change (IPCC). https://www.ipcc.ch/site/assets/uploads/2018/05/2nd-assessment-en-1.pdf; pp. 35–36; accessed January 3, 2021.
82. Intergovernmental Panel on Climate Change (IPCC). Chapter 9: Human Health. In: TAR Climate Change 2001. United Nations, 2001; 452–485. https://www.ipcc.ch/site/assets/uploads/2018/03/wg2TARchap9.pdf; accessed January 3, 2021.
83. See https://www.ipcc.ch/report/ar4/wg2/ and for Chapter 8, see: https://www.ipcc.ch/site/assets/uploads/2018/02/ar4-wg2-chapter8-1.pdf. Confalonieri U, Menne B, Akhtar R, Ebi KL, Hauengue M, Kovats RS, Revich B, Woodward A. Human health. In: Parry ML, Canziani OF, Palutikof JP, van der Linden PJ, Hanson CE (eds). *Climate Change 2007: Impacts, Adaptation and Vulnerability. Contribution of Working Group II to the Fourth Assessment Report of the Intergovernmental Panel on Climate Change.* Cambridge: Cambridge University Press, 2007; 391–431 .
84. See https://www.ipcc.ch/report/ar5/wg2/ and for Chapter 11 see https://www.ipcc.ch/site/assets/uploads/2018/02/WGIIAR5-Chap11_FINAL.pdf;Smith KR, Woodward A, Campbell-Lendrum D, Chadee DD, Honda Y, Liu Q, Olwoch JM, Revich B, Sauerborn R. Human health: impacts, adaptation, and co-benefits. In: Field CB, Barros VR, Dokken DJ, Mach KJ, Mastrandrea MD, Bilir TE, Chatterjee M, Ebi KL, Estrada YO, Genova RC, Girma B, Kissel ES, Levy AN, MacCracken S, Mastrandrea PR, White LL (eds). *Climate Change 2014: Impacts, Adaptation, and Vulnerability. Part A: Global and Sectoral Aspects. Contribution of Working Group II to the Fifth Assessment Report of the Intergovernmental Panel on Climate Change.* Cambridge, New York: Cambridge University Press, 2014; 709–754.
85. Intergovernmental Panel on Climate Change (IPCC). https://www.ipcc.ch/site/assets/uploads/2018/11/AR6_WGII_outlines_P46.pdf; accessed January 3, 2021.
86. IPCC opens second draft of Working Group II Sixth Assessment Report for government and expert review. https://www.ipcc.ch/2020/11/27/ipcc-wgii-ar6-second-order-draft-review/; accessed January 3, 2021.
87. Watts N, Amann M, Arnell N, et al. The 2020 report of The *Lancet* Countdown on health and climate change: responding to converging crises [published online ahead of print December 2, 2020]. *Lancet* 2020;396. https://doi.org/10.1016/S0140-6736(20)32290-X.
88. Watts N, Amann M, Arnell N, et al. The 2020 report of The *Lancet* Countdown on health and climate change: responding to converging crises [published online ahead of print December 2, 2020]. *Lancet* 2020;396. https://doi.org/10.1016/S0140-6736(20)32290-X.

89. Watts N, Amann M, Arnell N, et al. The 2020 report of The *Lancet* Countdown on health and climate change: responding to converging crises [published online ahead of print December 2, 2020]. *Lancet* 2020;396. https://doi.org/10.1016/S0140-6736(20)32290-X.
90. Watts N, Amann M, Arnell N, et al. The 2020 report of The *Lancet* Countdown on health and climate change: responding to converging crises [published online ahead of print December 2, 2020]. *Lancet* 2020;396. https://doi.org/10.1016/S0140-6736(20)32290-X.
91. Watts N, Amann M, Arnell N, et al. The 2020 report of The *Lancet* Countdown on health and climate change: responding to converging crises [published online ahead of print December 2, 2020]. *Lancet* 2020;396. https://doi.org/10.1016/S0140-6736(20)32290-X.
92. Inside Climate News. A field guide to the U.S. environmental movement, April 8, 2015. https://insideclimatenews.org/content/field-guide-us-environmental-movement; accessed January 3, 2021.
93. National Wildlife Federation. Equity & Justice 2020 Strategic Plan. https://www.nwf.org/-/media/Documents/PDFs/Equity/NWF-Equity-and-Justice-Strategic-Plan.ashx?la=en&hash=F0CCD4173ED5F006EC4FC5208ECEDEE48143A77B; accessed January 3, 2021.
94. Sierra Club. Issues. https://www.sierraclub.org/explore-issues; accessed January 3, 2021.
95. NRDC. About. https://www.nrdc.org/about; accessed January 3, 2021.
96. NRDC. https://www.nrdc.org/about/healthy-people-thriving-communities; accessed January 3, 2021.
97. Sunrise Movement. About Sunrise. https://www.sunrisemovement.org/about/?ms=AboutTheSunriseMovement; accessed January 3, 2021.
98. Portier CJ, Thigpen Tart K, Carter SR, Dilworth CH, Grambsch AE, Gohlke J, Hess J, Howard SN, Luber G, Lutz JT, Maslak T, Prudent N, Radtke M, Rosenthal JP, Rowles T, Sandifer PA, Scheraga J, Schramm PJ, Strickman D, Trtanj JM, Whung P-Y. *A Human Health Perspective on Climate Change: A Report Outlining the Research Needs on the Human Health Effects of Climate Change.* Research Triangle Park, NC: Environmental Health Perspectives/National Institute of Environmental Health Sciences, April 22, 2010. doi:10.1289/ehp.1002272. www.niehs.nih.gov/climatereport; accessed January 3, 2021.
99. Crimmins A, Balbus J, Gamble JL, Beard CB, Bell JE, Dodgen D, Eisen RJ, Fann N, Hawkins MD, Herring SC, Jantarasami L, Mills DM, Saha S, Sarofim MC, Trtanj J, Ziska L (eds). *The Impacts of Climate Change on Human Health in the United States: A Scientific Assessment.* Washington, DC: US Global Change Research Program, 2016. http://dx.doi.org/10.7930/J0R49NQX; accessed January 3, 2021; quote from "home" webpage.

100. American Public Health Association. *Climate Change and Health Strategic Plan*. Washington, DC: APHA, August 2016. https://www.apha.org/-/media/files/pdf/topics/climate/apha_climate_change_strategic_plan.ashx?la=en&hash=03D148BBD2A45E2A2B98BC4C98D33F32118244E1; accessed January 3, 2021.
101. American Public Health Association. 2017 Annual Meeting, "Climate Changes Health," Atlanta, GA, November 4–8, 2017. https://apha.confex.com/apha/2017/meetingapp.cgi/Home/0; accessed January 3, 2021.
102. American Public Health Association. Center for Climate, Health and Equity. https://www.apha.org/topics-and-issues/climate-change/center; accessed January 3, 2021.
103. American Public Health Association. APHA's Center for Climate, Health and Equity announces members of its inaugural advisory board. June 3, 2020. http://publichealthnewswire.org/?p=cche-advisory-board; accessed January 3, 2021.
104. See, for example,National Resource Defense Council. Issue Brief: Climate Change and Health in Michigan. January 2019, IB: 18-09-A. https://www.nrdc.org/sites/default/files/climate-change-health-impacts-michigan-ib.pdf; accessed January 3, 2021; National Resource Defense Council. Issue Brief: Climate Change and Health in California. February 2019, IB: 18-10-A. https://www.nrdc.org/sites/default/files/climate-change-health-impacts-california-ib.pdf; accessed January 3, 2021; Limaye VS, Max W, Constible J, Knowlton, K. Estimating the health-related costs of 10 climate-sensitive U.S. events during 2012. *GeoHealth* 2019;3:245–265. https://doi.org/10.1029/2019GH000202.
105. See, for example,Russell D, Gawthorpe E, Penney V, Raj A, Hickey B. Climate change is killing Americans. Health Departments aren't equipped to respond. Center for Public Integrity, in collaboration with Columbia Journalism Investigations and *The Guardian*. June 16, 2020. https://publicintegrity.org/environment/hidden-epidemics/underfunded-unprepared-cdc-fight-against-climate-change-public-health-heat-death/; accessed January 3, 2021.
106. Climate Visuals. The evidence behind Climate Visuals. https://climatevisuals.org/evidence-behind-climate-visuals; accessed January 3, 2021; McLoughlin N, Corner A. The air that we breathe. *Climate Visuals*, February 18, 2020. https://climatevisuals.org/blogs/air-we-breathe-climate-and-health-imagery; accessed January 3, 2021.
107. See, among others,Mock B. Message from the EPA: It's about protecting people, not polar bears. *Grist*, September 23, 2013. https://grist.org/climate-energy/message-from-the-epa-its-about-protecting-people-not-polar-bears/; accessed January 3, 2021; Brownstone S. This weekend's "Shell No" rally focused on people, not polar bears. *The Stranger*, April 27, 2015. https://www.thestranger.com/blogs/slog/2015/04/27/22120225/this-weekends-shell-no-rally-focused-on-people-not-polar-bears; accessed January 3, 2021; Baugh AJ. *God and the Green*

Divide: Religious Environmentalism in Black and White. Berkeley: University of California Press, 2016 (see Chapter 1: "People, not polar bears: faith in place's first ten years," pp. 29–45);Shields FI. Why we're rethinking the images we use for our climate journalism. *The Guardian*, October 18, 2019. https://www.theguardian.com/environment/2019/oct/18/guardian-climate-pledge-2019-images-pictures-guidelines; accessed January 3, 2021; Ryan B. These days, it's not about the polar bears. *New York Times*, May 12, 2019. https://www.nytimes.com/2019/05/12/climate/climate-solutions-polar-bears.html; accessed January 3, 2021; Amstrup SC. Polar bears to people: momentum on climate change. *Huff Post*, April 20, 2015. https://www.huffpost.com/entry/polar-bears-to-people-momentum-on-climate-change_b_7100986; accessed January 3, 2021; Rabinowitz J. Climate risk and national security: people not polar bears. Initiative on Extreme Weather and Climate, Columbia University, New York City, September 23, 2016. http://extremeweather.columbia.edu/2016/09/23/climate-risk-and-national-security-people-not-polar-bears/; accessed January 3, 2021;Corner A. Climate visuals—catalyzing a new visual language for climate change. *Climate Visuals*, March 1, 2017. https://climatevisuals.org/blogs/climate-visuals-catalysing-new-visual-language-climate-change; accessed January 3, 2021;Chapman A, Corner A, Webster R, Markowitz E. Climate visuals: a mixed methods investigation of public perceptions of climate images in three countries. *Glob Environ Change* 2016;41(November):172–182. https://www.sciencedirect.com/science/article/abs/pii/S095937801630351X;Wang S, Corner A, Chapman A, Markowitz, E. Public engagement with climate imagery in a changing digital landscape. *Wires Climate Change* 2018;9(2):e509. https://onlinelibrary.wiley.com/doi/full/10.1002/wcc.509.

108. Inuit Circumpolar Council (ICC). Climate change in the Arctic: Human rights of Inuit interconnected with the world. December 10, 2003. https://www.inuitcircumpolar.com/press-releases/climate-change-in-the-arctic-human-rights-of-inuit-interconnected-with-the-world/; accessed January 3, 2021; Watt-Cloutier S. It's time to listen to the Inuit on climate change. *Canadian Geographic*, November 15, 2018. https://www.canadiangeographic.ca/article/its-time-listen-inuit-climate-change; accessed January 3, 2021; Mercer G. "Sea, ice, snow . . . it's all changing": Inuit struggle with a warming world. *The Guardian*, May 30, 2018. https://www.theguardian.com/world/2018/may/30/canada-inuits-climate-change-impact-global-warming-melting-ice; accessed January 3, 2021; Demuth B. *Floating Coast: An Environmental History of the Bering Strait*. New York: W. W. Norton, 2019; Nuttall M. Water, ice and climate change in northwest Greenland. *WIREs Water* 2020;7(3):e1433. https://doi-org.ezp-prod1.hul.harvard.edu/10.1002/wat2.1433;Watts P, Koutouki K, Booth S, Blum S. Inuit

food security in Canada: arctic marine ethnoecology. *Food Sec* 2017;9:421–440. https://doi-org.ezp-prod1.hul.harvard.edu/10.1007/s12571-017-0668-0.

109. Tyrrell M, Clark DA. What happened to climate change? CITES and the reconfiguration of polar bear conservation discourse. *Glob Environ Change* 2014;24:363–372. https://doi.org/10.1016/j.gloenvcha.2013.11.016.
110. Oreskes N, Conway EM. *Merchants of Doubt: How a Handful of Scientists Obscured the Truth on Issues from Tobacco Smoke to Global Warming.* New York: Bloomsbury Publishing USA, 2010; Michaels D. *Doubt Is Their Product: How Industry's Assault on Science Threatens Your Health.* New York: Oxford University Press, 2008; Michaels D. *The Triumph of Doubt: Dark Money and the Science of Deception.* New York: Oxford University Press, 2020; Markowitz G, Rosner, D. *Deceit and Denial: The Deadly Politics of Industrial Pollution.* Berkeley: University of California Press, 2007; Brandt A. *The Cigarette Century: The Rise, Fall, and Deadly Persistence of the Product That Defined America.* New York: Basic Books, 2007; Proctor R. *Golden Holocaust: Origins of the Cigarette Catastrophe and the Case for Abolition.* Berkeley: University of California Press, 2011; Proctor R, Schiebinger L. *Agnotology: The Making and Unmaking of Ignorance.* Stanford, CA: Stanford University Press, 2008.
111. Markowitz G, Rosner, D. *Deceit and Denial: The Deadly Politics of Industrial Pollution* (California/Milbank Books on Health and the Public; 6). Berkeley: University of California Press, 2007; Lartey J, Laughland O. Almost every household has someone that has died from cancer. *The Guardian*, May 6, 2019. https://www.theguardian.com/us-news/ng-interactive/2019/may/06/cancertown-louisana-reserve-special-report; accessed January 3, 2021; Laughland O, Lartey J. First slavery, then a chemical plant and cancer deaths: one town's brutal history. *The Guardian*, May 6, 2019. https://www.theguardian.com/us-news/2019/may/06/cancertown-louisiana-reserve-history-slavery; accessed January 3, 2021; Laughland O. Louisiana greenlights huge pollution-causing plastics facility in "Cancer Alley." *The Guardian*, January 8, 2020. https://www.theguardian.com/us-news/2020/jan/07/louisiana-formosa-plastics-facility-air-quality-permits-cancer-alley; accessed January 3, 2021; Laughland O, Holden E. In the most polluted part of America, residents now battle the US's biggest plastic plant. *The Guardian*, April 1, 2020. https://www.theguardian.com/us-news/2020/apr/01/cancer-town-chemical-plant-plastics-louisiana-toxic-pollution-greenhouse-gas; accessed January 3, 2021.
112. Drugmand D. Fossil fuel companies keep getting sued over climate impacts. Here's where cases stand. *Desmog*, October 7, 2020. https://www.desmogblog.com/2020/10/07/fossil-fuels-exxon-climate-lawsuits-update; accessed January 3, 2021; Ricker D. Lawyers are unleashing a flurry of lawsuits to step up the fight against climate change. *ABA Journal*, November 1, 2019. https://www.abajournal.com/

magazine/article/lawyers-are-unleashing-a-flurry-of-lawsuits-to-step-up-the-fight-against-climate-change; accessed January 3, 2021; Laville S. Governments and firms in 28 countries sued over climate crisis—report. *The Guardian*, July 4, 2019. https://www.theguardian.com/environment/2019/jul/04/governments-and-firms-28-countries-sued-climate-crisis-report; accessed January 3, 2021.

113. Our Children's Trust. Youth v. Gov, Juliana v. US. https://www.ourchildrenstrust.org/juliana-v-us; accessed January 3, 2021.
114. Our Children's Trust and Earth Guardians. National and global experts file briefs in support of Juliana v. United States, Youth-Led Climate Litigation. Press release, March 13, 2020. https://static1.squarespace.com/static/571d109b04426270152febe0/t/5e6c0ae9a53943154a6de234/1584138985726/2020.03.13.Juliana+Amicus.pdf; accessed January 3, 2021.
115. Our Children's Trust. Youth v. Gov, Oregon. Cherniak v. Brown. https://www.ourchildrenstrust.org/oregon; accessed January 3, 2021; Our Children's Trust. Youth v. Gov, Montana. Held v. State of Montana. https://www.ourchildrenstrust.org/montana; accessed January 3, 2021.
116. Our Children's Trust. Youth v. Gov, Colorado. Petition for rulemaking submitted to the Colorado Oil and Gas Conservation Commission (COGCC) in November 2019 and the COGCC's ongoing rulemaking to implement the mandates of Senate Bill 19-181. https://www.ourchildrenstrust.org/colorado; accessed January 3, 2021.
117. Union of Concerned Scientists. *The UCS Science Hub for Climate Litigation: Resources and Opportunities*. August 3, 2020. https://www.ucsusa.org/resources/science-hub-climate-litigation; accessed November 28, 2020.
118. A search on Web of Science on 12/5/20 using the search terms TS = (((fossil AND fuel) OR fracking OR (oil AND drilling)) AND (health AND hazards)) and spanning all databases yielded 1,935 entries primarily focused on environmental pollutants and climate change. See also: Ladd AE, Malin SA, Boudet H, Cable S, Gaustad B, Hall P. *Fractured Communities*. New Brunswick, NJ: Rutgers University Press, 2018; Nishime L, Williams HD (eds). *Racial Ecologies*. Seattle: University of Washington Press, 2018; Kerr J. *Introduction to Energy and Climate*. 1st ed. Milton: CRC Press, 2018; Selinus O, Alloway B, Centeno J, Finkelman R, Fuge R, Lindh U, Smedley P. *Essentials of Medical Geology*. Dordrecht: Springer Netherlands, 2013.
119. Ladd AE, Malin SA, Boudet H, Cable S, Gaustad B, Hall P. *Fractured Communities*. New Brunswick, NJ: Rutgers University Press, 2018; Nishime L, Williams HD (eds). *Racial Ecologies*. Seattle: University of Washington Press, 2018; Birn A, Pillay Y, Holtz T. *Textbook of Global Health*. 4th ed. New York: Oxford University Press, 2017 (see especially Chapters 9 and 10).

120. Ladd AE, Malin SA, Boudet H, Cable S, Gaustad B, Hall P. *Fractured Communities*. New Brunswick, NJ: Rutgers University Press, 2018; Nishime L, Williams HD (eds). *Racial Ecologies*. Seattle: University of Washington Press, 2018; Birn A, Pillay Y, Holtz T. *Textbook of Global Health*. 4th ed. New York: Oxford University Press, 2017 (see especially Chapters 9 and 10); Hirsch JK, Smalley KB, Selby-Nelson EM, Hamel-Lambert JM, Rosmann MR, Barnes TA, Abrahamson D, Meit SS, GreyWolf I, Beckmann S, LaFromboise T. Psychosocial impact of fracking: a review of the literature on the mental health consequences of hydraulic fracturing. *Int J Mental Health Addiction* 2018;16(1):1–5.
121. Huseth-Zosel AL, Secor-Turner M, Wen Q, Liu X, Jansen RJ. Associations between oil development and sexually transmitted infections: public health nurse perspectives [published online ahead of print November 20, 2020]. *Public Health Nurs*. doi:10.1111/phn.12836;Cunningham S, DeAngelo G, Smith B. Fracking and risky sexual activity [published online ahead of print May 15, 2020]. *J Health Econ* 2020;72:102322. doi:10.1016/j.jhealeco.2020.102322; Johnson NP, Warren JL, Elliott EG, Niccolai LM, Deziel NC. A multiregion analysis of shale drilling activity and rates of sexually transmitted infections in the United States. *Sex Transm Dis* 2020;47(4):254–260. doi:10.1097/OLQ.0000000000001127; Deziel NC, Brokovich E, Grotto I, Clark CJ, Barnett-Itzhaki Z, Broday D, Agay-Shay K. Unconventional oil and gas development and health outcomes: a scoping review of the epidemiological research [published online ahead of print January 8, 2020]. *Environ Res* 2020;182:109124. doi:10.1016/j.envres.2020.109124; Beleche T, Cintina I. Fracking and risky behaviors: evidence from Pennsylvania [published online ahead of print August 10, 2018]. *Econ Hum Biol* 2018;31:69–82. doi:10.1016/j.ehb.2018.08.001; Deziel NC, Humeau Z, Elliott EG, Warren JL, Niccolai LM. Shale gas activity and increased rates of sexually transmitted infections in Ohio, 2000–2016. *PLoS One* 2018;13(3):e0194203. doi:10.1371/journal.pone.0194203; Komarek T, Cseh A. Fracking and public health: evidence from gonorrhea incidence in the Marcellus Shale region. *J Public Health Policy* 2017;38(4):464–481. doi:10.1057/s41271-017-0089-5; Goldenberg S, Shoveller J, Koehoorn M, Ostry A. Barriers to STI testing among youth in a Canadian oil and gas community [published online ahead of print December 3, 2007]. *Health Place* 2008;14(4):718–729. doi:10.1016/j.healthplace.2007.11.005; Udoh IA, Stammen RM, Mantell JE. Corruption and oil exploration: expert agreement about the prevention of HIV/AIDS in the Niger Delta of Nigeria [published online ahead of print September 28, 2007]. *Health Educ Res* 2008;23(4):670–681. doi:10.1093/her/cym042.
122. A search on TS = ((fossil AND fuel) OR fracking OR (oil AND drilling)) AND ((sexually AND transmitted) OR syphilis OR gonorrhea OR HIV) yielded 37 entries on 12/5/20. Relevant examples include Johnson NP, Warren JL, Elliott

EG, Niccolai LM, Deziel NC. A multiregion analysis of shale drilling activity and rates of sexually transmitted infections in the United States. *Sex Transm Dis* 2020;47(4):254–260. doi:10.1097/OLQ.0000000000001127; Deziel NC, Brokovich E, Grotto I, Clark CJ, Barnett-Itzhaki Z, Broday D, Agay-Shay K. Unconventional oil and gas development and health outcomes: a scoping review of the epidemiological research [published online ahead of print January 8, 2020]. *Environ Res* 2020;182:109124. doi:10.1016/j.envres.2020.109124; Beleche T, Cintina I. Fracking and risky behaviors: evidence from Pennsylvania [published online ahead of print August 10, 2018]. *Econ Hum Biol* 2018;31:69–82. doi:10.1016/j.ehb.2018.08.001; Deziel NC, Humeau Z, Elliott EG, Warren JL, Niccolai LM. Shale gas activity and increased rates of sexually transmitted infections in Ohio, 2000–2016. *PLoS One* 2018;13(3):e0194203. doi:10.1371/journal.pone.0194203; Komarek T, Cseh A. Fracking and public health: evidence from gonorrhea incidence in the Marcellus Shale region. *J Public Health Policy* 2017;38(4):464–481. doi:10.1057/s41271-017-0089-5; Goldenberg S, Shoveller J, Koehoorn M, Ostry A. Barriers to STI testing among youth in a Canadian oil and gas community [published online ahead of print December 3, 2007]. *Health Place* 2008;14(4):718–729. doi:10.1016/j.healthplace.2007.11.005; Udoh IA, Stammen RM, Mantell JE. Corruption and oil exploration: expert agreement about the prevention of HIV/AIDS in the Niger Delta of Nigeria [published online ahead of print September 28, 2007]. *Health Educ Res* 2008;23(4):670–681. doi:10.1093/her/cym042.

123. Ruddell R, Britto S. A perfect storm: violence toward women in the Bakken oil patch. *Int J Rural Criminol* 2020;5(2):205–227; Deer S, Kronk Warner EA. Raping Indian Country. *Columbia J Gender Law* 2019;38(1):31–95; Grisafi L. Living in the blast zone: sexual violence piped onto native land by extractive industries. *Columbia J Law Soc Problems* 2019;53(4):509–539.
124. Komarek T, Cseh A. Fracking and public health: evidence from gonorrhea incidence in the Marcellus Shale region. *J Public Health Policy* 2017;38(4):464–481. doi:10.1057/s41271-017-0089-5.
125. Komarek T, Cseh A. Fracking and public health: evidence from gonorrhea incidence in the Marcellus Shale region. *J Public Health Policy* 2017;38(4):464–481. doi:10.1057/s41271-017-0089-5.
126. Cunningham S, DeAngelo G, Smith B. Fracking and risky sexual activity [published online ahead of print May 15, 2020]. *J Health Econ* 2020;72:102322. doi:10.1016/j.jhealeco.2020.102322.
127. Beleche T, Cintina I. Fracking and risky behaviors: evidence from Pennsylvania [published online ahead of print August 10, 2018]. *Econ Hum Biol* 2018;31:69–82. doi:10.1016/j.ehb.2018.08.001.

128. Deziel NC, Humeau Z, Elliott EG, Warren JL, Niccolai LM. Shale gas activity and increased rates of sexually transmitted infections in Ohio, 2000–2016. *PLoS One* 2018;13(3):e0194203. doi:10.1371/journal.pone.0194203.
129. Huseth-Zosel AL, Secor-Turner M, Wen Q, Liu X, Jansen RJ. Associations between oil development and sexually transmitted infections: public health nurse perspectives [published online ahead of print November 20, 2020]. *Public Health Nurs.* doi:10.1111/phn.12836.
130. Shandro JA, Veiga MM, Shoveller J, Scoble M, Koehoorn M. Perspectives on community health issues and the mining boom–bust cycle. *Resources Policy* 2011;36(2):178–186; Goldenberg SM, Shoveller JA, Ostry AC, Koehoorn M. Sexually transmitted infection (STI) testing among young oil and gas workers. *Can J Public Health* 2008;99(4):350–354; Udonwa NE, Ekpo M, Ekanem IA, Inem VA, Etokidem A. Oil doom and AIDS boom in the Niger Delta Region of Nigeria. *Rural Remote Health* 2004;4(2):273. www.rrh.org.au/journal/article/273.
131. Olaleye AO, Babah OA, Osuagwu CS, Ogunsola FT, Afolabi BB. Sexually transmitted infections in pregnancy—an update on Chlamydia trachomatis and Neisseria gonorrhoeae [published online ahead of print October 8, 2020]. *Eur J Obstet Gynecol Reprod Biol* 2020;255:1–12. doi:10.1016/j.ejogrb.2020.10.002.
132. Watts N, Amann M, Arnell N, et al. The 2020 report of The *Lancet* Countdown on health and climate change: responding to converging crises [published online ahead of print December 2, 2020]. *Lancet* 2020;396. https://doi.org/10.1016/S0140-6736(20)32290-X; Hogben M, Leichliter J, Aral SO. An overview of social and behavioral determinants of STI. In: Cristaudo A, Giuliani M. (eds). *Sexually Transmitted Infections: Advances in Understanding and Management.* Cham: Springer International Publishing, 2020; 25–45; Arcaya M, Raker EJ, Waters MC. The social consequences of disasters: individual and community change. *Annu Rev Sociol* 2020;46:671–691.
133. Cochrane reviews, search in "Title Abstract Keyword" for either "organic food" or "pesticide poisoning," at https://www.cochranelibrary.com/search; accessed December 23, 2020.
134. Web of Science (http://apps.webofknowledge.com.ezp-prod1.hul.harvard.edu/; accessed via Harvard Library system on December 23, 2020), "advanced search" for "all databases" for (1) TS = ((pesticide AND poisoning) AND (ecology OR ecosystem OR wildlife OR animals)); *N* = 104,092 articles, of which 5,482 were classified as "review" articles; (2) TS = ((organic AND food) AND (consumer AND health)); *N* = 5,288 articles, of which 577 were classified as "review" articles; (3) TS = ((organic AND food) AND (pesticide AND poisoning)); *N* = 3,103 articles, of which 252 were classified as "review articles; and (4) TS = ((organic AND food) AND (occupational AND health)); *N* = 1,536 articles, of which 245 were classified as "review articles."

135. Mostafalou S, Abdollahi M. Pesticides: an update of human exposure and toxicity. *Arch Toxicol* 2017;91:549–599; Mazlan N, Ahmed M, Muharam FM, Alam A. Status of persistent organic pesticide residues in water and food and their effects on environments and farmers: a comprehensive review in Nigeria. *Semina: Ciências Agrárias, Londrina* 2017;38(4):2221–2236. doi:10.5433/1679-0359.2017v38n4p2221.
136. Rana J, Paul J. Health motive and the purchase of organic food: a meta-analytic review. *Int J Consumer Studies* 2020;44:162–171. doi:10.1111/ijcs.12556;Reganold JP, Wachter JM. Organic agriculture in the twenty-first century. *Nature Plants* 2015;2(2):15221. https://doi.org/10.1038/nplants.2015.221;Rizzo G, Borrello M, Dara Guccione G, Schifani G, Cembalo L. Organic food consumption: the relevance of the health attribute. *Sustainability* 2020;12(2):595.
137. Brantsaeter AL, Ydersbond TA, Hoppin JA, Haugen M, Meltzer HM. Organic food in the diet: exposure and health implications. *Annu Rev Public Health* 2017;38:295–313.
138. Mie A, Anderson HR, Gunnarsson S, Kahl J, Kesse-Guyot E, Rembialkowska E, Quaglio G, Grandjean P. Human health implications of organic food and organic agriculture: a comprehensive review. *Env Health* 2017;16:11. doi:10.1186/s12940-017-0315-4.
139. Wyckhuys KAG, Aebi A, Bijleveld van Lexmond MFIJ, Bojaca CR, Bonmatin J-M, Furlan L, Guerro JA, Mai TV, Pham HV, Sanchez-Bayo F, Ikenaka Y. Resolving twin human and environmental health hazards of a plant-based diet. *Environ Intl* 2020;144:106081.
140. Office of the Alameda County District Attorney. Consumer, Environmental & Worker Protection Division (CEWPD). https://www.alcoda.org/cewpd/; accessed January 3, 2021.
141. NYC.gov. Consumer Affairs announces new, expanded mission that reflects the agency's work to protect and enhance the daily economic lives of New Yorkers. September 15, 2016. https://www1.nyc.gov/site/dca/media/pr091516.page; accessed January 3, 2021.
142. National Consumer's League. A look back at 100+ years of advocacy. https://nclnet.org/about-ncl/about-us/history/; accessed January 3, 2021; Klein JG, Smith NC, John A. Exploring motivations for participation in a consumer boycott. *Advances Consumer Re* 2002;29:363–369; Estefan K, Kuoni C, Raicovich L. *Assuming Boycott: Resistance, Agency and Cultural Production*. New York: OR Books, 2017; Tomlin KM. Assessing the efficacy of consumer boycotts of U.S. target firms: a shareholder wealth analysis. *South Econ J* 2019;86(2):503–529; Anderson M. Fair trade and consumer social responsibility. *Management Decision* 2018;56(3):634–651; Guthman J, Brown S. I will never eat another strawberry again: the biopolitics

of consumer-citizenship in the fight against methyl iodide in California. *Agriculture Human Values* 2015;33(3):575–585.

143. Equal Justice Initiative. *Reconstruction in America: Racial Violence After the Civil War, 1865–1876.* Montgomery, AL: Equal Justice Initiative, 2020. https://eji.org/report/reconstruction-in-america/; accessed January 3, 2021.

144. The literature is extensive; for introductions, among others, for diverse country contexts, see Starzmann M, Roby J (eds). *Excavating Memory: Sites of Remembering and Forgetting.* Gainesville: University Press of Florida, 2016; Tuhiwai Smith, L. *Decolonizing Methodologies.* London: Zed Books, 2012; Banivanua Mar T, Edmonds P (eds). *Making Settler Colonial Space.* London: Palgrave Macmillan UK, 2010; Dahlm A. *Empire of the People: Settler Colonialism and the Foundations of Modern Democratic Thought.* Lawrence: University Press of Kansas, 2018; Howe S. *Empire: A Very Short Introduction.* Oxford: Oxford University Press, 2002; for the United States, especially in relation to Indigenous peoples and enslaved Africans, see, for example,Dunbar-Ortiz R. *An Indigenous Peoples' History of the United States* (Revisioning American History). Boston: Beacon Press, 2014; Hoxie FE (ed). *The Oxford Handbook of American Indian History* (Oxford Handbooks). New York: Oxford University Press, 2016; Reséndez A. *The Other Slavery: The Uncovered Story of Indian Enslavement in America.* Boston: Houghton Mifflin Harcourt, 2016; Sleeper-Smith S, Barr J, O'Brien JM, Shoemaker N, Stevens SM (eds). *Why You Can't Teach United States History Without American Indians.* Chapel Hill: University of North Carolina Press, 2015; Du Bois WEB. *The Suppression of the African Slave Trade to the United States of America, 1638–1870.* New York: Oxford University Press, 2007 (originally published: New York: Longmans, Green and Co., 1896); Painter N. *Creating Black Americans: African-American History and Its Meanings, 1619 to the Present.* Oxford, New York: Oxford University Press, 2006; Williams H. *American Slavery: A Very Short Introduction* New York: Oxford University Press, 2014; Paquette R, Smith, M (eds). *The Oxford Handbook of Slavery in the Americas* (Oxford Handbooks). Oxford: Oxford University Press, 2010; New York Times. The 1619 Project. *New York Times,* launched August 2019. https://www.nytimes.com/interactive/2019/08/14/magazine/1619-america-slavery.html; accessed December 25, 2020; Hannah-Jones N. The idea of America. The New York Times 1619 Project. *New York Times,* August 14, 2019. https://www.nytimes.com/interactive/2019/08/14/magazine/black-history-american-democracy.html; accessed December 25, 2020; Spring J. *Deculturalization and the Struggle for Equality: A Brief History of the Education of Dominated Cultures in the United States.* 6th ed. Boston: McGraw-Hill Higher Education, 2010; Ortiz P. *An African American and Latinx History of the United States* (Revisioning American History). Boston: Beacon Press, 2018.

145. Hill A, Tiefenthaler A, Triebert C, Jordan D, Willis H, Stein R. How George Floyd was killed in police custody. Visual investigation. *New York Times*, May 31, 2020; updated November 5, 2020. https://www.nytimes.com/2020/05/31/us/george-floyd-investigation.html; accessed January 3, 2021; Buchanan L, Bui Q, Patel JK. Black Lives Matter may be the largest movement in U.S. history. *New York Times*, July 3, 2020. https://www.nytimes.com/interactive/2020/07/03/us/george-floyd-protests-crowd-size.html; accessed January 3, 2021; Safi M. George Floyd killing triggers wave of activism around the world. *The Guardian*, June 9, 2002. https://www.theguardian.com/us-news/2020/jun/09/george-floyd-killing-triggers-wave-of-activism-around-the-world; accessed January 3, 2021; Frazer-Carroll M. George Floyd was killed in America. His death has sparked a global movement. *Huff Post*, June 11, 2020. https://www.huffpost.com/entry/racism-protests-international-black-lives-matter_n_5ee0d5dcc5b6a457582a24d7; accessed January 3, 2021; CNN. Protests across the globe after George Floyd's death. *CNN*, June 13, 2020. https://www.cnn.com/2020/06/06/world/gallery/intl-george-floyd-protests/index.html; accessed January 3, 2021; Krieger N. ENOUGH: COVID-19, structural racism, police brutality, plutocracy, climate change—and time for health justice, democratic governance, and an equitable, sustainable future. *Am J Public Health* 2020;110(11):1620–1623. https://doi.org/10.2105/AJPH.2020.305886.
146. Krieger N. ENOUGH: COVID-19, structural racism, police brutality, plutocracy, climate change—and time for health justice, democratic governance, and an equitable, sustainable future. *Am J Public Health* 2020;110(11):1620–1623; quote: p. 1620. https://doi.org/10.2105/AJPH.2020.305886; Hill A, Tiefenthaler A, Triebert C, Jordan D, Willis H, Stein R. How George Floyd was killed in police custody. Visual investigation. *New York Times*, May 31, 2020; updated November 5, 2020. https://www.nytimes.com/2020/05/31/us/george-floyd-investigation.html; accessed January 3, 2021.
147. Equal Justice Initiative. *Reconstruction in America: Racial Violence After the Civil War, 1865–1876*. Montgomery, AL: Equal Justice Initiative, 2020. https://eji.org/report/reconstruction-in-america/; accessed January 3, 2021;Invisible Hate. https://invisiblehate.org/; accessed January 3, 2021.
148. Ortiz A, Diaz J. George Floyd protests reignite debate over Confederate statues. *New York Times*, June 3, 2020; updated September 12, 2020. https://www.nytimes.com/2020/06/03/us/confederate-statues-george-floyd.html; accessed January 3, 2021; Diaz J. Christopher Columbus statues in Boston, Minnesota, and Virginia are damaged. *New York Times*, June 10, 2020; updated July 24, 2020. https://www.nytimes.com/2020/06/10/us/christopher-columbus-statue-boston-richmond.html; accessed January 3, 2021; Miranda CA. At Los Angeles toppling of Junipero Serra statue, activists want the full story told. *Los Angeles*

Times, June 20, 2020. https://www.latimes.com/entertainment-arts/story/2020-06-20/statue-junipero-serra-monument-protest-activists-take-down-los-angeles; accessed January 3, 2021; Contreras R. Spanish colonial monuments fuel racial strife in U.S. Southwest. *PBS News Hour*, June 30, 2020. https://www.pbs.org/newshour/nation/spanish-colonial-monuments-fuel-racial-strife-in-u-s-southwest; accessed January 3, 2021; New York Times. How statues are falling around the world: statues and monuments that have long honored racist figures are being boxed up, spray-painted—or beheaded. *New York Times*, June 24, 2020; updated September 12, 2020. https://www.nytimes.com/2020/06/24/us/confederate-statues-photos.html; accessed January 3, 2021; Mervosh S, Romero S, Tompkins L. Reconsidering the past, one statue at a time. *New York Times*, June 16, 2020; updated June 25, 2020. https://www.nytimes.com/2020/06/16/us/protests-statues-reckoning.html; accessed January 3, 2021.

149. New York Times. How statues are falling around the world: statues and monuments that have long honored racist figures are being boxed up, spray-painted—or beheaded. *New York Times*, June 24, 2020; updated September 12, 2020. https://www.nytimes.com/2020/06/24/us/confederate-statues-photos.html; accessed January 3, 2021; Landler M. "Get rid of them": a statue falls as Britain confronts its racist history. *New York Times*, June 8, 2020; updated July 16, 2020. https://www.nytimes.com/2020/06/08/world/europe/edward-colston-statue-britain-racism.html; accessed January 3, 2021; Bhambra GK. A statue was toppled. Can we finally talk about the British Empire? *New York Times*, June 12, 2020. https://www.nytimes.com/2020/06/12/opinion/edward-colston-statue-racism.html; accessed January 3, 2021; Topple the Racists. A crowdsourced map of UK statues and monuments that celebrate slavery and racism. https://www.toppletheracists.org/; accessed January 3, 2021; Petrequin S. Belgium takes down statue, king expresses regret for colonial violence. *PBS News Hour*, June 30, 2020. https://www.pbs.org/newshour/world/belgium-takes-down-statue-king-expresses-regret-for-colonial-violence; accessed January 3, 2021; Corder M. Black Lives Matters spurs scrutiny of Dutch colonial past. *PBS News Hour*, June 19, 2020. https://www.pbs.org/newshour/world/black-lives-matter-spurs-scrutiny-of-dutch-colonial-past; accessed January 3, 2021; Brewis H. Memorials to colonial past vandalized in France as wave of anti-racist protests sees statues topple across the world. *Evening Standard*, June 22, 2020. https://www.standard.co.uk/news/world/statues-colonial-daubed-paint-france-a4476446.html; accessed January 3, 2021; Bond J. Bye Hamilton, hello Kirikiriroa? City mulls name change after statue's removal. *NewsHub RNZ*, June 13, 2020. https://www.newshub.co.nz/home/new-zealand/2020/06/bye-hamilton-hello-kirikiriroa-city-mulls-name-change-after-statue-s-removal.html; accessed January 3, 2021; BBC. Cecil Rhodes statue in Cape Town has head removed. *BBC News*, July 15, 2020. https://www.

bbc.com/news/world-africa-53420403; accessed January 3, 2021; Oduro K. Montreal city officials remove toppled statue of Sir John A. MacDonald. *Global News*, August 30, 2020. https://globalnews.ca/news/7306987/montreal-city-officials-remove-toppled-statue-of-sir-john-a-macdonald/; accessed January 3, 2021; BBC. Colombia: Indigenous protestors topple conquistador's statue. *BBC News*, September 17, 2020. https://www.bbc.com/news/world-latin-america-54186047; accessed January 3, 2021.

150. On June 6, 2020, the US Marine Corps issued a ban on Confederate battle flags, including on mugs, bumper stickers, and posters, stating that while the symbol might represent heritage to some, it was divisive and painful for others and had "all too often been co-opted by violent extremists and racist groups"; see Gross J. U.S. Marine Corps issues ban on Confederate battle flags. *New York Times*, June 6, 2020; updated June 10, 2020. https://www.nytimes.com/2020/06/06/us/marines-confederate-flag-ban.html; accessed January 3, 2021. On July 17, 2020, this ban was effectively extended to all US military bases, by virtue of a new policy that listed which flags and symbols could be permitted—with this list excluding the Confederate flag; seeLamonthe D. Defense secretary effectively bans Confederate flags from military bases while rejecting "divisive symbols." *Washington Post*, July 17, 2020. https://www.washingtonpost.com/national-security/confederate-flag-military-bases-ban/2020/07/17/301e9b48-c832-11ea-a9d3-74640f25b953_story.html; accessed January 3, 2021. However, this ban also unexpectedly extended to the LGBTQ Rainbow Pride flag and also sovereign Native Nation flags, with these bans now being contested by US members of Congress; see Johnson C. New military ban on Confederate flags also bans LGBTQ Pride flags. *Washington Blade*, July 17, 2020. https://www.washingtonblade.com/2020/07/17/new-military-ban-on-confederate-flags-also-bans-lgbtq-pride-flags/; accessed January 3, 2021; Modern Military Associations of America. Members of Congress demand reinstatement of Pride Flag on military installations. July 31, 2020. https://modernmilitary.org/2020/07/members-of-congress-demand-reinstatement-of-lgbtq-pride-and-sovereign-native-nations-flags-on-military-installations-explicit-ban-on-confederate-flag/; accessed January 3, 2021.

151. Edmonson C. House votes to remove Confederate statues from U.S. Capitol. *New York Times*, July 22, 2020. https://www.nytimes.com/2020/07/22/us/politics/confederate-statues-us-capitol.html; accessed January 3, 2021; Kurtz J. House votes to replace Taney bust from U.S. Capitol, replace it with Thurgood Marshall. *Maryland Matters*, July 23, 2020. https://www.marylandmatters.org/blog/house-votes-to-remove-taney-bust-from-u-s-capitol-replace-it-with-thurgood-marshall/; accessed January 3, 2021.

152. Kurtz J. House votes to replace Taney bust from U.S. Capitol, replace it with Thurgood Marshall. *Maryland Matters*, July 23, 2020. https://www.

marylandmatters.org/blog/house-votes-to-remove-taney-bust-from-u-s-capitol-replace-it-with-thurgood-marshall/; accessed January 3, 2021; Finkelman P. *Supreme Injustice: Slavery in the Nation's Highest Court.* Cambridge, MA: Harvard University Press, 2018; Taney R, Van Evrie J, Cartwright S. *The Dred Scott decision. Opinion of Chief Justice Taney, with an introduction by Dr. J.H. Van Evrie, and an appendix, containing an essay on the Natural history of the prognathous race of mankind, originally written for the New York day-book, by Dr. S.A. Cartwright.* New York: Van Evrie, Horton & Co, 1859.

153. Finkelman P. *Supreme Injustice: Slavery in the Nation's Highest Court.* Cambridge, MA: Harvard University Press, 2018; Taney R, Van Evrie J, Cartwright S. *The Dred Scott decision. Opinion of Chief Justice Taney, with an introduction by Dr. J.H. Van Evrie, and an appendix, containing an essay on the Natural history of the prognathous race of mankind, originally written for the New York day-book, by Dr. S.A. Cartwright.* New York: Van Evrie, Horton & Co, 1859.
154. Finkelman P. *Supreme Injustice: Slavery in the Nation's Highest Court.* Cambridge, MA: Harvard University Press, 2018; 172–173.
155. Edmonson C. House votes to remove Confederate statues from U.S. Capitol. *New York Times*, July 22, 2020. https://www.nytimes.com/2020/07/22/us/politics/confederate-statues-us-capitol.html; accessed January 3, 2021; Kurtz J. House votes to replace Taney bust from U.S. Capitol, replace it with Thurgood Marshall. *Maryland Matters*, July 23, 2020. https://www.marylandmatters.org/blog/house-votes-to-remove-taney-bust-from-u-s-capitol-replace-it-with-thurgood-marshall/; accessed January 3, 2021; Ball H. *A Defiant Life: Thurgood Marshall and the Persistence of Racism in America.* New York: Crown, 1998. Additionally, on December 21, the statue of Confederate general Robert E. Lee was removed, to be replaced by a statue of civil rights activist Barbara Johns, who, in 1951, at the age of 16, defied school segregation in Virginia; see Pietsch B. Robert E. Lee statue is removed from U.S. Capitol. *New York Times*, December 21, 2020. https://www.nytimes.com/2020/12/21/us/robert-e-lee-statue-us-capitol.html; accessed January 3, 2021; Guardian staff and agencies. Statue of civil rights pioneer Barbara Johns replaces Lee at US Capitol. *The Guardian*, December 21, 2020. https://www.theguardian.com/us-news/2020/dec/21/barbara-johns-robert-e-lee-capitol-washington; accessed January 3, 2021.
156. Stracqualursi V. Mississippi Ballot Measure 3: voters approve magnolia design as the new state flag. *CNN*, November 4, 2020. https://www.cnn.com/2020/11/03/politics/mississippi-new-state-flag-ballot-question/index.html; accessed January 3, 2021; Ramseth L. Mississippi voters approve new magnolia design for state flag. Here's what happens next. *Mississippi Clarion Ledger*, November 4, 2020. https://www.clarionledger.com/story/news/politics/elections/2020/11/04/new-mississippi-state-flag-election-results/6061248002/; accessed January

3, 2021; CNN. Mississippi votes 2-1 to keep the existing flag. *CNN*, April 17, 2001. https://www.cnn.com/2001/ALLPOLITICS/04/17/mississippi.flag.02/index.html; accessed January 3, 2021; Pittman A. "You white people don't get it": Mississippi's long, ugly road to changing its state flag. *Mississippi Free Press*, June 11, 2020. https://www.mississippifreepress.org/3710/you-white-people-dont-get-it-mississippis-long-ugly-road-to-changing-its-state-flag/; accessed January 3, 2021.

157. Stracqualursi V. Mississippi Ballot Measure 3: voters approve magnolia design as the new state flag. *CNN*, November 4, 2020. https://www.cnn.com/2020/11/03/politics/mississippi-new-state-flag-ballot-question/index.html; accessed January 3, 2021; Ramseth L. Mississippi voters approve new magnolia design for state flag. Here's what happens next. *Mississippi Clarion Ledger*, November 4, 2020. https://www.clarionledger.com/story/news/politics/elections/2020/11/04/new-mississippi-state-flag-election-results/6061248002/; accessed January 3, 2021.
158. Pietsch B. Princeton will remove Woodrow Wilson's name from school. *New York Times*, June 27, 2020. https://www.nytimes.com/2020/06/27/nyregion/princeton-university-woodrow-wilson.html; accessed January 3, 2021; Slotkin J. Princeton to remove Woodrow Wilson's name from Public Policy School. *NPR*, June 27, 2020. https://www.npr.org/sections/live-updates-protests-for-racial-justice/2020/06/27/884310403/princeton-to-remove-woodrow-wilsons-name-from-public-policy-school; accessed January 3, 2021.
159. Pietsch B. Princeton will remove Woodrow Wilson's name from school. *New York Times*, June 27, 2020. https://www.nytimes.com/2020/06/27/nyregion/princeton-university-woodrow-wilson.html; accessed January 3, 2021; Slotkin J. Princeton to remove Woodrow Wilson's name from Public Policy School. *NPR*, June 27, 2020. https://www.npr.org/sections/live-updates-protests-for-racial-justice/2020/06/27/884310403/princeton-to-remove-woodrow-wilsons-name-from-public-policy-school; accessed January 3, 2021; Yellin ES. *Racism in the Nation's Service: Government Workers and the Color Line in Woodrow Wilson's America*. Chapel Hill: University of North Carolina Press, 2013.
160. CBS SF Bay Area. Stanford President David Starr Jordan's name to be removed from campus spaces over role in eugenics movement. *CBS News*, October 8, 2020. https://sanfrancisco.cbslocal.com/2020/10/08/stanford-david-starr-jordan-eugenics-movement-campus-spaces-renamed/; accessed January 3, 2021; Espinosa M, Zaidel B. Stanford to rename spaces honoring David Starr Jordan, founding president and eugenicist. Statue of Jordan's mentor Louis Agassiz also to be relocated. *Stanford Daily*, October 7, 2020. https://www.stanforddaily.com/2020/10/07/stanford-to-rename-spaces-honoring-david-starr-jordan-founding-president-and-noted-eugenicist/; accessed January 3, 2021; Stanford Eugenics

History Project. David Starr Jordan. https://www.stanfordeugenics.com/david-starr-jordan; accessed January 3, 2021; Stanford Eugenics History Project. Quotes by David Starr Jordan. https://www.stanfordeugenics.com/jordan-quotes; accessed January 3, 2021; Maldonado B. Eugenics on the farm. *Stanford Daily*, October 7, 2019. https://www.stanforddaily.com/2019/10/07/eugenics-on-the-farm/; accessed January 3, 2021.

161. Caltech (California Institute of Technology). Caltech to remove the names of Robert A. Millikan and five other eugenics proponents from buildings, honors, and assets. *Caltech Weekly*, January 15, 2021. https://www.caltech.edu/about/news/caltech-to-remove-the-names-of-robert-a-millikan-and-five-other-eugenics-proponents; accessed February 14, 2021.
162. University of College London. UCL denames buildings named after eugenicists. *UCL News*, June 19, 2020. https://www.ucl.ac.uk/news/2020/jun/ucl-denames-buildings-named-after-eugenicists; accessed February 14, 2021; University College London. Bricks + mortals. A history of eugenics told through buildings. Hear the stories of UCL's pioneering eugenicists told through the landmark buildings and spaces named after them. https://www.ucl.ac.uk/culture/projects/bricks-mortals; accessed February 14, 2021.
163. University College London. UCL makes formal public apology for its history and legacy of eugenics. January 7, 2021. https://www.ucl.ac.uk/news/2021/jan/ucl-makes-formal-public-apology-its-history-and-legacy-eugenics; accessed February 14, 2021; Adams R. University College London apologises for role in promoting eugenics. *The Guardian*, January 7, 2021. https://www.theguardian.com/education/2021/jan/07/university-college-london-apologises-for-role-in-promoting-eugenics; accessed February 14, 2021.
164. University College London. UCL makes formal public apology for its history and legacy of eugenics. January 7, 2021. https://www.ucl.ac.uk/news/2021/jan/ucl-makes-formal-public-apology-its-history-and-legacy-eugenics; accessed February 14, 2021.
165. University College London. UCL makes formal public apology for its history and legacy of eugenics. January 7, 2021. https://www.ucl.ac.uk/news/2021/jan/ucl-makes-formal-public-apology-its-history-and-legacy-eugenics; accessed February 14, 2021.
166. I cannot divulge the name of the law firm who contacted me, given explicit instructions to keep the exchange confidential. With regard to timeline, on August 1, 2020, they wrote to me to ask if I could "speak to research on the connection between these two issues: segregation and symbols of white supremacy on the one hand, and adverse societal effects on the other." I did my literature review prior to speaking with them on August 3, 2020. While it was feasible to refer them to relevant literature on the health impacts of segregation, about which reams have

been published, I could find no public health studies with empirical data on the harms of being exposed to symbols of white supremacy. For my search, I employed PubMed, Web of Science (all databases), and Google Scholar, as well as searched the myriad books, articles, and journals located via the Harvard University HOLLIS system.

167. Pittman A. "You white people don't get it": Mississippi's long, ugly road to changing its state flag. *Mississippi Free Press*, June 11, 2020. https://www.mississippifreepress.org/3710/you-white-people-dont-get-it-mississippis-long-ugly-road-to-changing-its-state-flag/; accessed January 3, 2021; Miranda CA. At Los Angeles toppling of Junipero Serra statue, activists want the full story told. *Los Angeles Times*, June 20, 2020. https://www.latimes.com/entertainment-arts/story/2020-06-20/statue-junipero-serra-monument-protest-activists-take-down-los-angeles; accessed January 3, 2021; Bhambra GK. A statue was toppled. Can we finally talk about the British Empire? *New York Times*, June 12, 2020. https://www.nytimes.com/2020/06/12/opinion/edward-colston-statue-racism.html; accessed January 3, 2021.
168. Starzmann M, Roby J (eds). *Excavating Memory: Sites of Remembering and Forgetting.* Gainesville: University Press of Florida, 2016; Strauss J. Contested site or reclaimed space? Re-membering but not honoring the past on the empty pedestal. *History Memory* 2020;32(1):131–151; Kapp PH. Conservation, tradition and popular iconoclasm in North America. *Historic Env Policy Practice* 2020. doi:10.1080/17567505.2020.1810501;Lehrer E, Milton C, Patterson M (eds). *Curating Difficult Knowledge: Violent Pasts in Public Places.* London: Palgrave Macmillan, 2011; Lonetree A, Cobb A (eds). *The National Museum of the American Indian: Critical Conversations.* Lincoln: University of Nebraska Press, 2008; UNESCO. The Slave Route Project. https://en.unesco.org/themes/fostering-rights-inclusion/slave-route; accessed January 3, 2021; Witz L, Minkley G, Rassool, C. *Unsettled History: Making South African Public Pasts.* Ann Arbor: University of Michigan Press, 2017; Equal Justice Initiative. The National Memorial for Peace and Justice and Legacy Museum. https://museumandmemorial.eji.org/; accessed January 3, 2021; New York Times Style Magazine. America's monuments, reimagined for a more just future. *New York Times*, August 26, 2020. https://www.nytimes.com/2020/08/24/t-magazine/confederate-monuments-reimagined-racism.html; accessed January 3, 2021; Cheney-Rice Z. America is covered in Confederate statues. We can do better—and here's how. The Black Monuments Project. https://black-monuments.mic.com/; accessed January 3, 2021.
169. Invisible Hate. https://invisiblehate.org/; accessed January 3, 2021; Betha C. Toppling the statues, virtually. *New Yorker*, December 28, 2020 (online;

print: January 4 and 11, 2021 issue). https://www.newyorker.com/magazine/2021/01/04/toppling-the-statues-virtually; accessed January 3, 2021.

170. Wing C, Simon K, Bello-Gomez RA. Designing difference in difference studies: best practices for public health policy research. *Annu Rev Public Health* 2018;39(1):453–469; Lee M. *Matching, Regression Discontinuity, Difference in Differences, and Beyond*. New York: Oxford University Press, 2016.
171. Lai CK, Skinner AL, Cooley E, Murrar S, Brauer M, Devos T, et al. Reducing implicit racial preferences: II. Intervention effectiveness across time. *J Exp Psychol Gen* 2016;145(8):1001–1016.
172. Vuletich HA, Payne BK. Stability and change in implicit bias. *Psychol Sci* 2019;30(6):854–862.
173. Vuletich HA, Payne BK. Stability and change in implicit bias. *Psychol Sci* 2019;30(6):854–862.
174. Vuletich HA, Payne BK. Stability and change in implicit bias. *Psychol Sci* 2019;30(6):854–862; quote: p. 859.
175. Vuletich HA, Payne BK. Stability and change in implicit bias. *Psychol Sci* 2019;30(6):854–862; quote: p. 859.
176. Torpey J (ed). *Politics and the Past: On Repairing Historical Injustices*. Lanham, MD: Rowman & Littlefield Publishing Group, 2004; Darity WA, Mullen AK. *From Here to Equality: Reparations for Black Americans in the 21st Century*. Chapel Hill: University of North Carolina Press, 2020; Lenzerini F (ed). *Reparations for Indigenous Peoples: International and Comparative Perspectives*. Oxford: Oxford University Press, 2008; Tuck E, Yang KW. Decolonization is not a metaphor. *Decolonization Indigeneity Education Society* 2012;1(1):1–40; Shaheen-Hussain S, Blackstock C, Gabriel KE. *Fighting for a Hand to Hold*. Montreal: McGill-Queen's University Press, 2020; Hammonds EM, Reverby SM. Toward a historically informed analysis of racial health disparities since 1619. *Am J Public Health* 2019;109(10):1348–1349; Bassett MT, Galea S. Reparations as a public health priority—a strategy for ending black–white health disparities. *New Engl J Med* 2020;383(22):2101–2103; Williams DR, Collins C. Reparations: a viable strategy to address the enigma of African American health. *Am Behav Sci* 2004;47(7):977–1000.
177. Bjö LO (ed). *Photobiology*. New York: Springer New York, 2015; Cronin TW, Johnsen S, Marshall NJ, Warrant EJ. *Visual Ecology*. Princeton, NJ: Princeton University Press, 2014; Kumar V (ed). *Biological Timekeeping*. New Delhi: Springer, 2017.
178. Krieger N. History, biology, and health inequities: emergent embodied phenotypes and the illustrative case of the breast cancer estrogen receptor [published online ahead of print November 15, 2012]. *Am J Public Health* 2013;103(1):22–27. doi:10.2105/AJPH.2012.300967.

179. Bjö LO (ed). *Photobiology.* New York: Springer New York, 2015; Cronin TW, Johnsen S, Marshall NJ, Warrant EJ. *Visual Ecology.* Princeton, NJ: Princeton University Press, 2014; Foster RG, Kreitzman L. *Seasons of Life: The Biological Rhythms That Enable Living Things to Thrive and Survive.* New Haven, CT: Yale University Press, 2009; Gilbert SF, Epel D. *Ecological Developmental Biology: The Environmental Regulation of Development, Health, and Evolution.* Sunderland, MA: Sinaeur Associates, 2015; Angilletta MJ. *Thermal Adaptation: A Theoretical and Empirical Synthesis.* Oxford: Oxford University Press, 2009; Walker WH 2nd, Meléndez-Fernández OH, Nelson RJ, Reiter RJ. Global climate change and invariable photoperiods: a mismatch that jeopardizes animal fitness. *Ecol Evol* 2019;9(17):10044–10054. doi:10.1002/ece3.5537.
180. Scott A. *Fire: A Very Short Introduction.* Oxford: Oxford University Press, 2020; DiLaura D. A brief history of lighting. *Optics Photonics* 2008;September:21–28.
181. Jenkins N. *Energy Systems: A Very Short Introduction.* Oxford: Oxford University Press, 2019; DiLaura D. A brief history of lighting. *Optics Photonics* 2008;September:21–28.Nye DE. *American Illuminations: Urban Lighting, 1800–1920.* Cambridge, MA: MIT Press, 2019.
182. Sultan S. *Organism and Environment: Ecological Development, Niche Construction, and Adaptation.* Oxford: Oxford University Press, 2015 (see especially Chapter 6: "Community-level consequences of habitat construction and eco-devo responses," pp. 117–139).Walker WH 2nd, Meléndez-Fernández OH, Nelson RJ, Reiter RJ. Global climate change and invariable photoperiods: a mismatch that jeopardizes animal fitness. *Ecol Evol* 2019;9(17):10044–10054. doi:10.1002/ece3.5537; Renner SS, Zohner CM. Climate change and phenological mismatch in trophic interactions among plants, insects, and vertebrates. *Annu Rev Ecol Evol Systematics* 2018;49:165–182; Bogdziewicz M, Szymkowiak J, Bonal R, Hacket-Pain R, Espelta JM, Pesendorfer M, et al. What drives phenological synchrony? Warm springs advance and desynchronize flowering in oaks. *Agric For Meteorol* 2020;294:108140. https://doi.org/10.1016/j.agrformet.2020.108140;Flynn DFB, Wolkovich EM. Temperature and photoperiod drive spring phenology across all species in a temperate forest community. *New Phytol* 2018;219(4):1353–1362; Stevenson TJ, Visser ME, Arnold W, Barrett P, Biello S, Dawson A, Denlinger DL, Dominoni D, Ebling FJ, Elton S, Evans N. Disrupted seasonal biology impacts health, food security and ecosystems. *Proc R Soc B* 2015;282:20151453. http://dx.doi.org/10.1098/rspb.2015.1453.
183. Sultan S. *Organism and Environment: Ecological Development, Niche Construction, and Adaptation.* Oxford: Oxford University Press, 2015 (see especially Chapter 6: "Community-level consequences of habitat construction and eco-devo responses," pp. 117–139); Walker WH 2nd, Meléndez-Fernández OH, Nelson RJ, Reiter RJ. Global climate change and invariable photoperiods: a mismatch that

jeopardizes animal fitness. *Ecol Evol* 2019;9(17):10044–10054. doi:10.1002/ece3.5537; Renner SS, Zohner CM. Climate change and phenological mismatch in trophic interactions among plants, insects, and vertebrates. *Annu Rev Ecol Evol Systematics* 2018;49:165–182; Flynn DFB, Wolkovich EM. Temperature and photoperiod drive spring phenology across all species in a temperate forest community. *New Phytol* 2018;219(4):1353–1362;Friel S. *Climate Change and the People's Health*. New York: Oxford University Press, 2019; Schnitter R, Berry P. The climate change, food security and human health nexus in Canada: a framework to protect population health. *Int J Environ Res Public Health* 2019;16(14):2531. doi:10.3390/ijerph16142531;Myers SS, Smith MR, Guth S, Golden CD, Vaitla B, Mueller ND, Dangour AD, Huybers P. Climate change and global food systems: potential impacts on food security and undernutrition. *Annu Rev Public Health* 2017;38(1):259–277; Myers S. Food and nutrition on a rapidly changing planet (Chapter 5). In: Myers S, Frumkin H (eds). *Planetary Health*. Washington, DC: Island Press, 2020; 113–140; Stevenson TJ, Visser ME, Arnold W, Barrett P, Biello S, Dawson A, Denlinger DL, Dominoni D, Ebling FJ, Elton S, Evans N. Disrupted seasonal biology impacts health, food security and ecosystems. *Proc R Soc B* 2015;282:20151453. http://dx.doi.org/10.1098/rspb.2015.1453.

184. Jenkins N. *Energy Systems: A Very Short Introduction*. Oxford: Oxford University Press, 2019; Jelley N. *Renewable Energy: A Very Short Introduction*. Oxford: Oxford University Press, 2020; Jessel S, Sawyer S, Hernández D. Energy, poverty, and health in climate change: a comprehensive review of an emerging literature. *Front Public Health* 2019 Dec 12;7:357. doi:10.3389/fpubh.2019.00357; Friel S. *Climate Change and the People's Health*. New York: Oxford University Press, 2019.
185. Mizon B. *Light Pollution: Responses and Remedies*. 2nd ed. New York: Springer New York, 2012; Nadybal SM, Collins TW, Grineski SE. Light pollution inequities in the continental United States: a distributive environmental justice analysis [published online ahead of print October 1, 2020]. *Environ Res* 2020;189:109959.
186. Falcón J, Torriglia A, Attia D, Viénot F, Gronfier C, Behar-Cohen F, Martinsons C, Hicks D. Exposure to artificial light at night and the consequences for flora, fauna, and ecosystems. *Front Neurosci* 2020;14:602796. doi:10.3389/fnins.2020.602796;Borges RM. Dark matters: challenges of nocturnal communication between plants and animals in delivery of pollination services. *Yale J Biol Med* 2018;91(1):33–42; Macgregor CJ, Scott-Brown AS. Nocturnal pollination: an overlooked ecosystem service vulnerable to environmental change. *Emerg Top Life Sci* 2020;4(1):19–32. doi:10.1042/ETLS20190134;Sanders D, Frago E, Kehoe R, Patterson C, Gaston KJ. A meta-analysis of biological impacts of artificial light at night. *Nat Ecol Evol* 2020;5(1):74–81.
187. Zhang D, Jones RR, Powell-Wiley TM, Jia P, James P, Xiao Q. A large prospective investigation of outdoor light at night and obesity in the NIH-AARP Diet and

Health Study. *Environ Health* 2020;19(1):74. doi:10.1186/s12940-020-00628-4;Wu Y, Gui SY, Fang Y, Zhang M, Hu CY. Exposure to outdoor light at night and risk of breast cancer: a systematic review and meta-analysis of observational studies [published online ahead of print November 23, 2020]. *Environ Pollut.* doi:10.1016/j.envpol.2020.116114; Nadybal SM, Collins TW, Grineski SE. Light pollution inequities in the continental United States: a distributive environmental justice analysis [published online ahead of print October 1, 2020]. *Environ Res* 2020;189:109959;Hale L, Troxel W, Buysse DJ. Sleep health: an opportunity for public health to address health equity. *Annu Rev Public Health* 2020;41:81–99; Richmond RC, Anderson EL, Dashti HS, Jones SE, Lane JM, Strand LB, Brumpton B, Rutter MK, Wood AR, Straif K, Relton CL, Munafò M, Frayling TM, Martin RM, Saxena R, Weedon MN, Lawlor DA, Smith GD. Investigating causal relations between sleep traits and risk of breast cancer in women: mendelian randomisation study. *BMJ* 2019;365:l2327. doi:10.1136/bmj.l2327.

188. Zielinska-Dabkowska KM, Xavia K, Bobkowska K. Assessment of citizens' actions against light pollution with guidelines for future initiatives. *Sustainability* 2020;12(12):4997; Nadybal SM, Collins TW, Grineski SE. Light pollution inequities in the continental United States: a distributive environmental justice analysis [published online ahead of print October 1, 2020]. *Environ Res* 2020;189:109959;Hale L, Troxel W, Buysse DJ. Sleep health: an opportunity for public health to address health equity. *Annu Rev Public Health* 2020;41:81–99.
189. Bjö LO (ed). *Photobiology.* New York: Springer New York, 2015; Cronin TW, Johnsen S, Marshall NJ, Warrant EJ. *Visual Ecology.* Princeton, NJ: Princeton University Press, 2014; Gilbert SF, Epel D. *Ecological Developmental Biology: The Environmental Regulation of Development, Health, and Evolution.* Sunderland, MA: Sinaeur Associates, 2015; Land MF. *The Eye: A Very Short Introduction.* New York: Oxford University Press, 2014; Gibson JJ. *The Ecological Approach to Visual Perception* (Classic Edition; 1st ed.: 1979). New York: Psychology Press, Taylor and Francis, 2014; Reed E, Jones R (eds). *Reasons for Realism: Selected Essays of James Gibson.* New York, New York: Routledge, 2020; Blau JJC, Wagman JB (eds). *Perception as Information Detection: Reflections on Gibson's Ecological Approach to Visual Perception.* New York: Taylor and Francis, 2019.
190. Vitale S, Sperduto RD, Ferris FL III. Increased prevalence of myopia in the United States between 1971–1972 and 1999–2004. *Arch Opthalamol* 2009;127:1632–1639; National Center for Education Statistics. *The Nation's Report Card: Trends in Academic Progress 2012* (NCES 2013–456). Washington, DC: National Center for Education Statistics, Institute of Education Sciences, US Department of Education, 2013. https://nces.ed.gov/nationsreportcard/pubs/main2012/2013456.aspx; accessed December 28, 2020.

191. Vitale S, Sperduto RD, Ferris FL III. Increased prevalence of myopia in the United States between 1971–1972 and 1999–2004. *Arch Opthalamol* 2009;127:1632–1639; National Center for Education Statistics. *The Nation's Report Card: Trends in Academic Progress 2012* (NCES 2013–456). Washington, DC: National Center for Education Statistics, Institute of Education Sciences, US Department of Education, 2013. https://nces.ed.gov/nationsreportcard/pubs/main2012/2013456.aspx; accessed January 3, 2021.
192. Qiu M, Wang SY, Singh K, Lin SC. Racial disparities in uncorrected and undercorrected refractive error in the United States. *Invest Ophthalmol Vis Sci* 2014;55(10):6996–7005. doi:10.1167/iovs.13-12662.
193. Morgan IG, Ohno-Matsui K, Saw SM. Myopia. *Lancet* 2012;379:1739–1748;Holden BA, Fricke TR, Wilson DA, Jong M, Naidoo KS, Sankaridurg P, Wong TY, Naduvilath TJ, Resnikoff S. Global prevalence of myopia and high myopia and temporal trends from 2000 through 2050. *Ophthalmology* 2016;123(5):1036–1042. doi:10.1016/j.ophtha.2016.01.006.
194. Mountjoy E, Davies NM, Plotnikov D, Smith GD, Rodriguez S, Williams CE, Guggenheim JA, Atan D. Education and myopia: assessing the direction of causality by Mendelian randomisation. *BMJ* 2018 Jun 6;361:k2022. doi:10.1136/bmj.k2022. Erratum in: *BMJ* 2018 Jul 4;362:k2932; Cuellar-Partida G, Lu Y, Kho PF, Hewitt AW, Wichmann HE, Yazar S, Stambolian D, Bailey-Wilson JE, Wojciechowski R, Wang JJ, Mitchell P, Mackey DA, MacGregor S. Assessing the genetic predisposition of education on myopia: a Mendelian randomization study [published online ahead of print October 26, 2015]. *Genet Epidemiol* 2016;40(1):66–72. doi:10.1002/gepi.21936.
195. Pusti D, Benito A, Madrid-Valero JJ, Ordoñana JR, Artal P. Inheritance of refractive error in millennials. *Sci Rep* 2020 May 18;10(1):8173. doi:10.1038/s41598-020-65130-w.
196. Enthoven CA, Tideman JWL, Polling JR, Yang-Huang J, Raat H, Klaver CCW. The impact of computer use on myopia development in childhood: the Generation R study [published online ahead of print January 15, 2020]. *Prev Med* 2020 Mar;132:105988. doi:10.1016/j.ypmed.2020.105988; Cao K, Wan Y, Yusufu M, Wang N. Significance of outdoor time for myopia prevention: a systematic review and meta-analysis based on randomized controlled trials. *Ophthalmic Res* 2020;63(2):97–105; Eppenberger LS, Sturm V. The role of time exposed to outdoor light for myopia prevalence and progression: a literature review. *Clin Ophthalmol* 2020;14:1875–1890. doi:10.2147/OPTH.S245192; Deng L, Pang Y. Effect of outdoor activities in myopia control: meta-analysis of clinical studies. *Optom Vis Sci* 2019;96(4):276–282. doi:10.1097/OPX.0000000000001357.
197. Lingham G, Mackey DA, Lucas R, Yazar S. How does spending time outdoors protect against myopia? A review [published online ahead of print

November 13, 2019]. *Br J Ophthalmol* 2020;104(5):593–599. doi:10.1136/bjophthalmol-2019-314675.

198. Frumkin H, Bratman GN, Breslow SJ, Cochran B, Kahn PH Jr, Lawler JJ, Levin PS, Tandon PS, Varanasi U, Wolf KL, Wood SA. Nature contact and human health: a research agenda. *Environ Health Perspect* 2017;125(7):075001. doi:10.1289/EHP1663; Kondo MC, Fluehr JM, McKeon T, Branas CC. Urban green space and its impact on human health. *Int J Environ Res Public Health* 2018 Mar 3;15(3):445. doi:10.3390/ijerph15030445.
199. King M, Smith A, Gracey M. Indigenous health part 2: the underlying causes of the health gap. *Lancet* 2009;374(9683):76–85. doi:10.1016/S0140-6736(09)60827-8; Finn S, Herne M, Castille D. The value of traditional ecological knowledge for the environmental health sciences and biomedical research. *Environ Health Perspect* 2017 Aug 29;125(8):085006. doi:10.1289/EHP858; Greenwood M, de Leeuw S, Lindsay NM (eds). *Determinants of Indigenous Peoples' Health: Beyond the Social.* 2nd ed. Toronto: Canadian Scholars' Press, 2018;Commission of the Pan American Health Organization on Equity and Health Inequalities in the Americas. *Just Societies: Health Equity and Dignified Lives. Report of the Commission of the Pan American Health Organization on Equity and Health Inequalities in the Americas.* Washington, DC: PAHO, 2019. https://iris.paho.org/handle/10665.2/51571; accessed January 3, 2021.
200. "theory, n." *OED Online*, Oxford University Press. https://www.oed.com/view/Entry/200431; accessed January 3, 2021; Krieger N. *Epidemiology and the People's Health: Theory and Context.* New York: Oxford University Press, 2011; 17–28; Krieger N. Got theory? On the 21 st c. ce rise of explicit use of epidemiologic theories of disease distribution: a review and ecosocial analysis. *Curr Epidemiol Rep* 2014;1(1):45–56; Fleck L. *Genesis and Development of a Scientific Fact.* Chicago: University of Chicago Press, [1935] 1979; Löwy I. Fleck the public health expert: medical facts, thought collectives, and the scientist's responsibility. *Sci Tech Human Values* 2016;41(3):509–533.
201. Krieger N. *Epidemiology and the People's Health: Theory and Context.* New York: Oxford University Press, 2011; 17–28; Krieger N. Got theory? On the 21 st c. ce rise of explicit use of epidemiologic theories of disease distribution: a review and ecosocial analysis. *Curr Epidemiol Rep* 2014;1(1):45–56.
202. "theory, n." *OED Online*, Oxford University Press. https://www.oed.com/view/Entry/200431; accessed January 3, 2021.
203. Krieger N. *Epidemiology and the People's Health: Theory and Context.* New York: Oxford University Press, 2011; 17–28; Krieger N. Got theory? On the 21 st c. ce rise of explicit use of epidemiologic theories of disease distribution: a review and ecosocial analysis. *Curr Epidemiol Rep* 2014;1(1):45–56; Ziman J. *Real Science: What It Is, and What It Means.* Cambridge: Cambridge University

Press, 2000; Okasha S. *Philosophy of Science: A Very Short Introduction*. 2nd. ed. New York: Oxford University Press, 2016; Nagel ES. *Observation and Theory in Science*. Baltimore, MD: Johns Hopkins University Press, 2019 (1st printing: 1971); McMullin E. The virtues of a good theory. In: Curd M, Psillos S (eds). *The Routledge Companion to the Philosophy of Science*. 2nd ed. New York: Routledge, 2014; 561–571.

204. Gibson JJ. *The Ecological Approach to Visual Perception* (Classic Edition; 1st ed.: 1979). New York: Psychology Press, Taylor and Francis, 2014; Reed E, Jones R (eds). *Reasons for Realism: Selected Essays of James Gibson*. New York: Routledge, 2020.
205. Oreskes N. *Why Trust Science?* Princeton, NJ: Princeton University Press, 2019; Krieger N. *Epidemiology and the People's Health: Theory and Context*. New York: Oxford University Press, 2011; 17–28; Krieger N. Got theory? On the 21 st c. ce rise of explicit use of epidemiologic theories of disease distribution: a review and ecosocial analysis. *Curr Epidemiol Rep* 2014;1(1):45–56; Ziman J. *Real Science: What It Is, and What It Means*. Cambridge: Cambridge University Press, 2000.
206. Tollefson J. How Trump damaged science—and why it could take decades to recover. *Nature* 2020;586(7828):190–194. doi:10.1038/d41586-020-02800-9; Plumer B, Davenport C. Science under attack: how Trump is sidelining researchers and their work. *New York Times*, December 28, 2029. https://www.nytimes.com/2019/12/28/climate/trump-administration-war-on-science.html; accessed January 3, 2021; Specter M. Trump's unprecedented attacks on our public-health system. *New Yorker*, August 11, 2020. https://www.newyorker.com/news/daily-comment/trumps-unprecedented-attacks-on-our-public-health-system; accessed January 3, 2021; Krieger N. Climate crisis, health equity, and democratic governance: the need to act together. *J Public Health Policy* 2020;41(1):4–10. doi:10.1057/s41271-019-00209-x; Michaels D. *The Triumph of Doubt: Dark Money and the Science of Deception*. New York: Oxford University Press, 2020.
207. Benjamin R, Hinkson LR. What do we owe each other? Moral debts and racial distrust in experimental stem cell science. In: Ehlers N, Hinkson LR (eds). *Subprime Health: Debt and Race in US Medicine*. Minneapolis: University of Minnesota Press, 2017; 129–154; Jaiswal J, Halkitis PN. Towards a more inclusive and dynamic understanding of medical mistrust informed by science. *Behav Med* 2019;45(2):79–85. doi:10.1080/08964289.2019.1619511; Jaiswal J, LoSchiavo C, Perlman DC. Disinformation, misinformation and inequality-driven mistrust in the time of COVID-19: lessons unlearned from AIDS denialism. *AIDS Behav* 2020;24:2776–2780.
208. Huff C. Situating Donna Haraway in the life-narrative web. *a/b: Auto/Biography Stud* 2019;34(3):375–384. doi:10.1080/08989575.2019.1664167;Haraway

D. *Primate Visions: Gender, Race, and Nature in the World of Modern Science.* New York: Routledge, 1989; Haraway D. *Simians, Cyborgs, and Women: The Reinvention of Nature.* New York: Routledge, 1991; Haraway D. *When Species Meet.* Minneapolis: University of Minnesota Press, 2008; Haraway D. *Staying with the Trouble: Making Kin in the Chthulucene.* Durham, NC: Duke University Press, 2016.

209. Weigel M. Donna Haraway, interview: "a giant bumptious litter: Donna Haraway on truth, technology, and resisting extinction." *Logic* 2019 Dec 7;9. https://logicmag.io/nature/a-giant-bumptious-litter/; accessed January 3, 2021; see also Weigel M. Interview: feminist cyborg scholar Donna Haraway: "the disorder of our era isn't necessary." *The Guardian*, June 20, 2019. https://www.theguardian.com/world/2019/jun/20/donna-haraway-interview-cyborg-manifesto-post-truth; accessed January 3, 2021.

210. Yeager JM. Amazing Grace (how sweet the sound) (Chapter 46). In: *Early Evangelicalism: A Reader.* New York: Oxford University Press Online Scholarship, 2015. doi:10.1093/acprof:osobl/9780199916955.001.0001; accessed December 29, 2020.

211. Grady D. H. Jack Geiger, doctor who fought social ills, dies at 95. *New York Times*, December 28, 2020. https://www.nytimes.com/2020/12/28/health/h-jack-geiger-dead.html; accessed January 3, 2021; Hollands C. Leading the charge: how Dr. H. Jack Geiger and Tufts created the country's first community health centers—and launched a movement. *Tufts Magazine* 2018;Winter:16–21. https://tuftsmagazine.com/issues/magazine/leading-charge; accessed January 3, 2021; Geiger HJ. The first community health center in Mississippi: communities empowering themselves. *Am J Public Health* 1971;106(10:1738–1740; Ward T. *Out in the Rural: A Mississippi Health Center and Its War on Poverty*, with a foreword by H. Jack Geiger. New York: Oxford University Press, 2017.

212. Hollands C. Leading the charge: how Dr. H. Jack Geiger and Tufts created the country's first community health centers—and launched a movement. *Tufts Magazine* 2018;Winter:16–21. https://tuftsmagazine.com/issues/magazine/leading-charge; accessed January 3, 2021.

213. Krieger N. *Epidemiology and the People's Health: Theory and Context.* New York: Oxford University Press, 2011; 17–28; Krieger N. Got theory? On the 21 st c. ce rise of explicit use of epidemiologic theories of disease distribution: a review and ecosocial analysis. *Curr Epidemiol Rep* 2014;1(1):45–56; Krieger N, Davey Smith G. The tale wagged by the DAG: broadening the scope of causal inference and explanation for epidemiology. *Int J Epidemiol* 2016;45:1787–1808;Krieger N. Living and dying at the crossroads: racism, embodiment, and why theory is essential for a public health of consequence. *Am J Public Health* 2016;106:832–833; Krieger N. A critical research agenda for social justice and public health: an ecosocial proposal. In: Levy B (ed). *Social Injustice and Public*

Health. 3rd ed. New York: Oxford University Press, 2019; 531–552; Ziman J. *Real Science: What It Is, and What It Means*. Cambridge: Cambridge University Press, 2000; Nagel ES. *Observation and Theory in Science*. Baltimore, MD: Johns Hopkins University Press, 2019 (1st printing: 1971); McMullin E. The virtues of a good theory. In: Curd M, Psillos S (eds). *The Routledge Companion to the Philosophy of Science*. 2nd ed. New York: Routledge, 2014; 561–571; Wemrell M, Merlo J, Mulinari S, Homborg A-C. Contemporary epidemiology: a review of critical discussions within the discipline and a call for further dialogue with social theory. *Sociol Compass* 2016;10:153–171; Solar O, Irwin A. *A Conceptual Framework for Action on the Social Determinants of Health*. Social Determinants of Health Discussion Paper 2 (Policy and Practice). Geneva: World Health Organization, 2010. http://www.who.int/social_determinants/corner/SDHDP2.pdf; accessed January 3, 2021.

214. Krieger N. *Epidemiology and the People's Health: Theory and Context*. New York: Oxford University Press, 2011; 17–28; Krieger N. Got theory? On the 21 st c. ce rise of explicit use of epidemiologic theories of disease distribution: a review and ecosocial analysis. *Curr Epidemiol Rep* 2014;1(1):45–56; Krieger N, Davey Smith G. The tale wagged by the DAG: broadening the scope of causal inference and explanation for epidemiology. *Int J Epidemiol* 2016;45:1787–1808;Krieger N. Living and dying at the crossroads: racism, embodiment, and why theory is essential for a public health of consequence. *Am J Public Health* 2016;106:832–833; Krieger N. A critical research agenda for social justice and public health: an ecosocial proposal. In: Levy B (ed). *Social Injustice and Public Health*. 3rd ed. New York: Oxford University Press, 2019; 531–552; Ziman J. *Real Science: What It Is, and What It Means*. Cambridge: Cambridge University Press, 2000; Nagel ES. *Observation and Theory in Science*. Baltimore, MD: Johns Hopkins University Press, 2019 (1st printing: 1971); McMullin E. The virtues of a good theory. In: Curd M, Psillos S (eds). *The Routledge Companion to the Philosophy of Science*. 2nd ed. New York: Routledge, 2014; 561–571; Fleck L. *Genesis and Development of a Scientific Fact*. (1935). Chicago: University of Chicago Press, 1979; Löwy I. Fleck the public health expert: medical facts, thought collectives, and the scientist's responsibility. *Sci Tech Human Values* 2016;41(3):509–533.
215. Krieger N. *Epidemiology and the People's Health: Theory and Context*. New York: Oxford University Press, 2011; 31; Tremblay MC, Parent AA. Reflexivity in PHIR: let's have a reflexive talk! *Can J Public Health* 2014; 105(3):e221–223.
216. Krieger N. *Epidemiology and the People's Health: Theory and Context*. New York: Oxford University Press, 2011; 124.
217. Sydenstricker E. *Health and Environment*. New York: McGraw-Hill, 1933; 206.

218. Sydenstricker E. *Health and Environment*. New York: McGraw-Hill, 1933; 209–210.
219. World Social Forum Charter of Principles. April 9, 2001. http://ciranda.net/World-Social-Forum-Charter-of?lang=pt_br.; accessed January 3, 2021; Krieger N. ENOUGH: COVID-19, structural racism, police brutality, plutocracy, climate change—and time for health justice, democratic governance, and an equitable, sustainable future. *Am J Public Health* 2020; 110(11):1620–1623. https://doi.org/10.2105/AJPH.2020.305886.

INDEX

For the benefit of digital users, indexed terms that span two pages (e.g., 52–53) may, on occasion, appear on only one of those pages.

Tables, figures and boxes are indicated by *t*, *f* and *b* following the page number